Scripture and the Sword:

Tactical, Practical and Spiritual
Lessons from the Bible's
Greatest Warriors

Scripture quotations taken from the New American Standard Bible® (NASB®), Copyright © 1960, 1971, 1977, 1995, 2020 by The Lockman Foundation. Used by permission. All rights reserved. lockman.org.

ISBN: 979-8-9998282-0-0 (paperback)

ISBN: 979-8-9998282-1-7 (hardback)

ISBN: 979-8-9998282-2-4 (eBook)

Library of Congress Control Number: 2025917477

For professional inquiries: scriptureandthesword@gmail.com

TABLE OF CONTENTS

PART ONE:
JOSHUA—CARRYING THE TORCH OF MOSES

JOSHUA INTRODUCTION

During Israel's time in the wilderness, one leader consistently stood apart. Joshua was more than a faithful servant; he was a warrior forged through hardship, discipline, and unwavering obedience to the Lord. His rise began in the crucible of battle. When the Amalekites attacked the Israelites at Rephidim, Moses turned to Joshua. The harsh desert conditions had taken their toll, and the people's desperation bordered on rebellion (Exodus 17). Under the guidance of Moses and by the power of the Holy Spirit, Joshua led the charge and defeated these ferocious foes. It was the first of many victories to come.

As time passed, Moses appointed Joshua as his assistant (Exodus 24:13). Joshua accompanied him partway up Mount Sinai, where Moses received the tablets of the law from God. This moment changed the destiny of Israel. When the people fell into idolatry and worshiped the golden calf, Joshua remained loyal. He did not follow the crowd into rebellion. Later, when Moses established the Tent of Meeting to speak with the Lord, Joshua would remain behind even after Moses returned to the camp; presumably to guard the sacred place (Exodus 33).

The book of Numbers reveals even more about Joshua's emerging leadership. The Lord told Moses to send scouts into the new frontier; one from each tribe (Numbers 13:1–2). Moses selected Joshua, who at that time was still called Hoshea, meaning "Salvation." Moses gave him a new name: Joshua, meaning "the Lord Saves" (Numbers 13:16). It was a prophetic renaming. Representing the tribe of Ephraim, Joshua was one of twelve scouts selected for a daring forty-day mission. They crossed into enemy territory to survey the land and gather vital intelligence. What they discovered stunned the camp. The land was rich and bountiful, just as God had promised. "So they reported to him and said, 'We came into the land where you sent us, and it certainly does flow with milk and honey, and this is its fruit'" (Numbers 13:27).

But abundant food was not the only thing they found. The land was also filled with massive, fortified cities—and giants. Despite the Lord's clear command, most of the scouts allowed fear to overtake them. They spread panic among the people. "The men who had gone up with him said, 'We are not able to go up against the people, because they are too strong for us'" (Numbers 13:31). They added, "We also saw the Nephilim there (the sons of Anak are part of the Nephilim); and we were like grasshoppers in our own sight, and so we were in their sight" (Numbers 13:33). The Lord had promised victory, but the majority chose fear over faith.

Only two remained faithful: Caleb and Joshua.

Despite all the miracles the Lord had performed, many in the tribe still lacked faith. Overwhelmed by fear, they saw themselves as small and powerless before the giants of Canaan. Panic spread like wildfire. Screams and weeping erupted throughout the camp. The people turned on Moses: "And all the sons of Israel grumbled against Moses and Aaron; and the entire congregation said to them, 'If only we had died in the land of Egypt! Or even if we had died in this wilderness! So why is the Lord bringing us into this land to fall by the sword? Our wives and our little ones will become plunder! Would it not be better for us to return to Egypt?'" (Numbers 14:2–3).

Fear is one of the enemy's greatest weapons. *De Re Militari* affirms this timeless truth: "A prudent general will also try to sow dissention among his adversaries, for no nation ... can be completely ruined by its enemies unless its fall be facilitated by its own distraction" (Vegetius 66). That distraction had taken root. After decades of struggling in the desert, the Israelites were teetering on the edge of self-destruction.

A full mutiny began to stir. Some even wanted to replace Moses and return to Egypt (Numbers 14:4). Their fear had a stranglehold on them; many preferred the predictability of slavery to the uncertainty of freedom. It was then that Joshua and Caleb stepped forward. While the mob unraveled, these two men stood firm. They remained faithful to the Lord and rose in defense of Moses. Faith formed the bedrock of their courage, and that courage empowered them to speak the truth, even in the face of death. The crowd considered stoning them (Numbers 14:10).

Though still early in his leadership, Joshua displayed rare qualities. According to *Principles of War*, "no military leader can ever become great without a dash of boldness in him" (Clausewitz 3). Joshua met that standard with conviction. He and Caleb addressed the mob head on. "If the Lord is pleased with us, then He will bring us into this land and give it to us—a land which flows with milk and honey. Only do not rebel against the Lord; and do not fear the people of the land, for they will be our prey. Their protection is gone from them, and the Lord is with us; do not fear them" (Numbers 14:8–9).

There were a few exceptions, however. Joshua, Caleb, and all those under twenty years old received God's blessing (Numbers 14:29–31). They were the future of Israel. As a sign of judgment, the ten scouts who had spread fear throughout the camp were struck with a plague and died (Numbers 14:37). With the agitators gone, the mission carried on.

But the rebellion had not yet run its course. The very next day, some Israelites attempted to enter the Promised Land in defiance of the Lord's command. Moses warned them, to no avail. This mission would be their last. Arrogantly, they pressed forward anyway. They marched uphill toward Canaan, but the hostile locals descended from the high ground and slaughtered them (Numbers 14:45).

Their pride blinded them to one of the oldest and most basic military axioms: "not to advance uphill against the enemy, nor to oppose him when he comes downhill" (Tzu 15). *De Re Militari* elaborates, "The highest ground is reckoned the best. Weapons thrown from a height strike with greater force; and the party above their antagonists can repulse and bear them down with greater impetuosity, while they who struggle with the ascent have both the ground and the enemy to contend with" (Vegetius 70). Their arrogance and disobedience led them to ruin. They had been warned. They had no one to blame but themselves.

After Joshua's bold stand, the Lord confirmed his future leadership. God instructed Moses to appoint him as the next shepherd of Israel, a man who would lead and protect the people. This was because Joshua possessed a spirit of leadership (Numbers 27:18). Later, God reiterated this appointment: "But charge Joshua and encourage him and strengthen him, for he will cross the Jordan ahead of this people, and he will enable them to inherit the land which you will see" (Deuteronomy 3:28). Moses obeyed. He appointed Joshua in front of the entire nation as he said: "And the Lord is the one who goes before you; He will be with you—He will not fail you or forsake you. Do not fear or be dismayed" (Deuteronomy 31:8).

The Art of War warns, "Unhappy is the fate of one who tries to win his battles and succeed in his attacks without cultivating the spirit of enterprise; for the result is waste of time and general stagnation" (Tzu 27). Through hardship, testing, and trust in God, Joshua had cultivated that spirit well. He was not merely a military leader. He was also a man of faith, discipline, and divine favor.

After decades of wandering, the time had finally come. Joshua was ready. With faith in their hearts and the Lord at their side, the Israelites stood at the threshold of the Promised Land.

JOSHUA

Joshua's rise began during Israel's darkest days in Egypt. In his early life, he and the other Israelites suffered as slaves under the Pharaoh's brutal regime. He witnessed the plagues and the miracles of the Lord firsthand. These events greatly influenced the young Joshua. During the time of Moses, he remained steadfast in his faith, while many of the Israelites fell by the wayside. After emancipation, Joshua lived a life shaped by trial, conquest, and obedience.

These early events shaped Joshua and forged him into an excellent leader. After the death of Moses, it was his time to lead the charge into the Promised Land. The Israelites had endured forty years of wandering in the desert, yet there were new challenges awaiting them in the Promised Land. According to *De Re Militari*, "Achieving victory in war does not depend entirely upon numbers or simple courage; only skill and discipline will ensure it" (Vegetius 3). Fortunately for the Israelites, Joshua had cultivated all these virtues during his tenure serving Moses.

Shortly after the death of Moses, the Lord called on Joshua: "Moses My servant is dead; now then, arise, cross this Jordan, you and all this people, to the land which I am giving to them, to the sons of Israel. Every place on which the sole of your foot steps, I have given it to you, just as I spoke to Moses" (Joshua 1:2–3). After giving more details of the territory, God elaborated further. "No one will be able to oppose you all the days of your life. Just as I have been with Moses, I will be with you; I will not desert you nor abandon you" (Joshua 1:5).

The Lord patiently encouraged Joshua to lead boldly. He also gave Joshua the keys to success while establishing the new territory. He said,

"This Book of the Law shall not depart from your mouth, but you shall meditate on it day and night, so that you may be careful to do according to all that is written in it; for then you will make your way prosperous, and then you will have good success" (Joshua 1:8). During His brief speech to Joshua, He told him to be strong and courageous three times (Joshua 1:6, 7, 9). God makes it evidently clear that He appreciates these virtues. The Lord does not call us to pray for easier circumstances. Rather, we are to pray for the strength and courage to carry out the Lord's work.

Invigorated by the Holy Spirit, Joshua assumed command and rallied the troops. He commanded his officers to prepare to enter the Promised Land (Joshua 1:11). While two and a half tribes had territory east of the Jordan (Numbers 32:18–19), they were still required to provide troops. Joshua reminded them of the agreement they made with Moses (Joshua 1:14–15).

After Joshua issued his orders, the troops acknowledged their leader. "They answered Joshua, saying, 'All that you have commanded us we will do, and wherever you send us we will go. Just as we obeyed Moses in all things, so we will obey you; only may the Lord your God be with you as He was with Moses'" (Joshua 1:16–17). At the same time, the troops acknowledged the death penalty for treason. "Anyone who rebels against your command and does not obey your words in all that you command him, shall be put to death; only be strong and courageous" (Joshua 1:18). Joshua had led by example all his career. He created an environment of obedience, readiness, and discipline. He had successfully cultivated a spirit of enterprise in his fighting men.

The energized troops were ready for battle. As the army was preparing its provisions, it was also time to gather intelligence on the enemy. Claiming the Promised Land was no small venture. Gathering intel was crucial before all the troops crossed the Jordan river. To aid this, Joshua secretly sent out two spies to their first target, the city of Jericho (Joshua 2:1). According to *The Art of War*, "Spies are a most important element in war, because on them depends an army's ability to move" (Tzu 29). Though Scripture never records their names, their mission carried life-or-death significance. Joshua must have thought highly of these two brave men.

In Jericho, the two spies met a local woman named Rahab, who was a prostitute. As we will soon see, God often uses unlikely heroes. Word had spread in Jericho that spies were around (Joshua 2:2). The king of Jericho sent a message to Rahab, and ordered her to hand over the spies. The king said to her, "Bring out the men who have come to you, who have entered your house, for they have come to spy out all the land" (Joshua 2:3).

Amazingly, Rahab defied the king's order. She misdirected the king's forces away from the Israelites. She told them she had seen the men but claimed to know nothing about them. Wisely, she misdirected the pursuers away: "when it was time to shut the gate at dark ... the men went out; I do not know where the men went. Pursue them quickly, for you will overtake them" (Joshua 2:5).

However, she had actually hidden the spies in her house. She had concealed them under stalks of flax on her rooftop (Joshua 2:6). Why would she risk her life to protect two foreign spies? The king's men pursued this false lead outside of the city, and the gates shut behind them. With the immediate danger gone, Rahab went to the roof and explained herself to the two Israelite spies.

She addressed the two Israelites. "I know that the Lord has given you the land, and that the terror of you has fallen on us, and that all the inhabitants of the land have despaired because of you" (Joshua 2:9). Rahab continued. She recounted the Lord's miracles during Israel's escape from Egypt. She described the battles the Israelites had fought and won by the Lord's hand (Joshua 2:10). She revealed how these events had shattered the morale of her people. "When we heard these reports, our hearts melted and no courage remained in anyone any longer because of you; for the Lord your God is God in heaven above and on earth below" (Joshua 2:11). Her heart turned to the Lord. She now honored Him.

While the residents of Jericho may have melted in fear, Rahab stood out. Indeed, she had embraced a strong and courageous spirit that God had encouraged in Moses and Joshua. In fact, several events foreshadowed this during Moses' leadership. In Exodus 23:27, the Lord told Moses: "I will send My terror ahead of you, and throw into confusion all the people among whom you come, and I will make all your enemies turn their backs to you." Additionally, Deuteronomy 2:25

states: "This day I will begin to put the dread and fear of you upon the faces of people everywhere, who, when they hear the news of you, will tremble and be in anguish because of you."

At first, Rahab seemed unlikely to be an ally of the Lord. Now we can see some of His divine wisdom in choosing her to aid the Israelites. Her profession offered a believable pretext for sheltering the spies. Recall, she didn't deny that the two men had visited her. Rather she said she hadn't gathered many personal details about who they were or where they had come from. Given her profession as a prostitute, this made sense. She now had plausible deniability about their intentions and identities. She used this plausible deniability to misdirect the king's soldiers.

Rahab's passive defiance echoes that of some of the early Israelites; for example, the midwives who defied the Pharaoh's orders to kill the newborn Hebrew boys (Exodus 1:16–22). They had a dilemma: to obey unjust earthly authorities or to obey the word of God. The midwives chose wisely, and God repaid them bountifully (Exodus 1:20–21). Moses' own parents followed their brave example, which ended up saving his life (Exodus 2:1–10).

Other, much later Biblical figures noted Rahab's courage and faithfulness. The early church leader James praised her highly. "You see that a person is justified by works and not by faith alone. In the same way, was Rahab the prostitute not justified by works also when she received the messengers and sent them out by another way? For just as the body without the spirit is dead, so also faith without works is dead" (James 2:24–26).

Throughout the Bible, a perilous dilemma reoccurs. Obey unjust orders to spare your life or defy the unjust orders and risk your life to follow God's word. The apostles provide clarity on this matter in the book of Acts. Preaching the Gospel of Christ led to friction and quickly drew the attention of the authorities. They were imprisoned for their faith. An angel of the Lord liberated them shortly after (Acts 5:18–19). Yet once again, the authorities rounded them up. This time, the locals (the Sadducees) were even angrier.

The Sadducees considered the death penalty for teaching the works of Jesus Christ. Things were at a crossroads. During an interrogation, the apostles had enough. With great courage and faith, the apostles boldly declared their allegiance: "We must obey God rather than men. The God of our fathers raised up Jesus, whom you put to death by hanging Him on a cross. He is the One whom God exalted to His right hand as a Prince and a Savior, to grant repentance to Israel, and forgiveness of sins. And we are witnesses of these things; and so is the Holy Spirit, whom God has given to those who obey Him" (Acts 5:29–32).

After Rahab explained herself to the Israelite spies, she asked for their protection for her and her family. She pleaded for their help. The two spies happily agreed. "And the men said to her, 'Our life for yours even to death! If you do not tell this business of ours, then when the Lord gives us the land we will deal kindly and faithfully with you'" (Joshua 2:14).

The spies had now made a sworn agreement with Rahab. The Israelites treated oaths with utmost seriousness. Previously, Moses had said to the heads of the tribes of Israel: "If a man makes a vow to the Lord, or takes an oath to put himself under a binding obligation, he shall not break his word; he shall act in accordance with everything that comes out of his mouth" (Numbers 30:2). Moses elaborated on this further. "When you make a vow to the Lord your God, you shall not delay to pay it, for the Lord your God will certainly require it of you, and it will be a sin for you. However, if you refrain from making vows, it will not be a sin for you. You must be careful to carry out what has gone from your lips, just as you have voluntarily vowed to the Lord your God, and you must fulfill it" (Deuteronomy 23:21–23). Later, we will examine the consequences of hasty oaths in great detail.

The two spies indicated their good faith by swearing to the Lord. This also correlates with *The Art of War*, "Peace proposals unaccompanied by a sworn covenant indicate a plot" (Tzu 18). As a sign of their agreement, Rahab tied a scarlet cord around her window (Joshua 2:17–18). Rahab happily agreed to these terms and sent the men on their way. She let down a rope from her window, so that the men could escape undetected. She advised them to hide in the hill country for three days to evade the king's guard.

The Israelite spies agreed and executed their escape with tactical precision. "So they departed and came to the hill country, and remained there for three days, until the pursuers returned. Now the pursuers had searched for them all along the road, but had not found them" (Joshua 2:22). This correlates with Sun Tzu's advice about finding spies: "If in the neighborhood of your camp there should be any hilly country, ponds surrounded by aquatic grass, hollow basins filled with reeds, or woods with thick undergrowth, they must be carefully routed out and searched; for these are places where men in ambush or insidious spies are likely to be lurking" (Tzu 17).

After evading their pursuers, the spies returned to camp. With the scarlet cord in place, Rahab and her family were safe. The spies briefed their leader, Joshua, on the situation. They told how the people quaked in fear (Joshua 2:24). According to *De Re Militari*, "in every kind of warfare it is deemed of the highest importance to spy out and get to know thoroughly the habits of the enemy" (Vegetius 107). Armed with intel on the enemy forces, now it was time for the Israelites to cross the Jordan river and begin the fight of their lives.

ECHOES OF EGYPT: THE MIRACULOUS CROSSING OF THE JORDAN RIVER

Early the next morning, Joshua and his troops made camp on the banks of the Jordan river. The officers moved through the camp to give instructions for the crossing. "Then Joshua said to the people, 'Consecrate yourselves, for tomorrow the Lord will do miracles among you'" (Joshua 3:5). This moment echoed the words of Moses before the Lord gave him the Ten Commandments at Mount Sinai (Exodus 19:10).

Shortly before the crossing, the Lord spoke to Joshua. "Now the Lord said to Joshua, 'This day I will begin to exalt you in the sight of all Israel, so that they will know that just as I have been with Moses, I will be with you. So you shall command the priests who are carrying the ark of the covenant, saying, "When you come to the edge of the waters of the Jordan, you shall stand still in the Jordan"'" (Joshua 3:7–8). The priests obeyed. As soon as their feet touched the water, the river stopped flowing. The Holy Spirit intervened. The water piled up in a great heap a long distance away (Joshua 3:15–16). Just as He had parted the Red Sea for Moses, the Lord made a way through the Jordan for Joshua. The priests stood on dry ground in the riverbed as the people crossed. "About forty thousand equipped for war crossed over before the Lord to the desert plains of Jericho" (Joshua 4:13).

After the nation had finished crossing, the Lord instructed Joshua to choose one man from each of the twelve tribes to gather stones from the riverbed. These twelve stones would serve as a lasting memorial

11

(Joshua 4:2–3). Joshua obeyed and relayed the instructions. Once the stones were gathered, the priests carrying the ark stepped out of the river. "It came about when the priests who carried the ark of the covenant of the Lord had come up from the middle of the Jordan, and the soles of the priests' feet were lifted up to the dry ground, that the waters of the Jordan returned to their place, and went over all its banks as before" (Joshua 4:18).

The crossing of the river symbolized a point of no return. A new era had begun. "On that day the Lord exalted Joshua in the sight of all Israel, so that they revered him, just as they had revered Moses all the days of his life" (Joshua 4:14).

As Passover approached, Joshua used the stones to build a monument to the Lord. When he finished, he addressed the people: "And he said to the sons of Israel, 'When your children ask their fathers in time to come, saying, "What are these stones?" then you shall inform your children, saying, 'Israel crossed this Jordan on dry ground'" (Joshua 4:21–22). Joshua continued, connecting this miracle to the Red Sea crossing, and concluded by giving glory to God. "For the Lord your God dried up the waters of the Jordan before you until you had crossed, just as the Lord your God had done to the Red Sea, which He dried up before us until we had crossed; so that all the peoples of the earth may know that the hand of the Lord is mighty, so that you may fear the Lord your God forever" (Joshua 4:23–24). By involving one man from each tribe, he cultivated unity and shared responsibility. The monument was not built for glory or conquest, but so that future generations would remember the Lord's grace.

This stands in contrast to King Saul, who later built a monument in his own honor. We will explore this in great detail during the *Saul* chapter. As *The Art of War* says, "The general who advances without coveting fame and retreats without fearing disgrace, whose only thought is to protect his country and do good service for his sovereign, is the jewel of the kingdom" (Tzu 21). Joshua embodied this ideal.

Rahab had earlier told the spies how God's miracles had struck fear into the people of Canaan. The Jordan crossing intensified that fear. "Now it came about when all the kings of the Amorites who were beyond the Jordan to the west, and all the kings of the Canaanites who were by the sea, heard how the Lord had dried up the waters of the

Jordan before the sons of Israel until they had crossed, that their hearts melted and there was no spirit in them any longer because of the sons of Israel" (Joshua 5:1).

Soon after, they celebrated Passover. Joshua made sure the people gave thanks to God for delivering them from Egypt. This marked a milestone. The people ate from the land of Canaan, and the manna from heaven ceased. It was a new beginning. While their food supply had stabilized, the battle for the Promised Land was just beginning.

THE FALL OF JERICHO AND SPIRITUAL SYMBOLISM

Chapters 6–8 of Joshua are profoundly symbolic. They beautifully illustrate a journey towards the glory of God. This journey begins to unfold at the end of Chapter 5. While near the city of Jericho, Joshua encountered a figure with a drawn sword in his hand. Joshua approached and asked, "Are you for us or for our enemies?" (Joshua 5:13). The mysterious figure replied: "'No; rather I have come now as captain of the army of the Lord.' And Joshua fell on his face to the ground, and bowed down, and said to him, "What has my lord to say to his servant?"'" (Joshua 5:14). This wasn't a man, but a divine messenger sent from the Lord! In awe, he fell to the ground in reverence of this warrior angel. "And the captain of the Lord's army said to Joshua, 'Remove your sandals from your feet, for the place where you are standing is holy.' And Joshua did so" (Joshua 5:15). God said the same thing to Moses when he encountered the burning bush (Exodus 3:5). Joshua continued down this holy path.

This encounter reveals where the Israelites stood with the Lord. The lesson is not that God was on Israel's side; but that Israel was called to be on God's side, fighting His battles under His command. Joshua had mistakenly placed his own perspective first by asking who the angel fought for. The reply of "neither" shows that we must either accept or reject the ways of the Lord. When we prioritize ourselves over God, we invite disaster.

Chapters 6–8 portray this truth with clarity. Chapter 6 shows that if we stay true to the word of the Lord, He is willing to work with us in miraculous ways. Chapter 7 reveals the consequences of disobedience and sin. It also provides a path to repentance. Finally, Chapter 8 shows that redemption is possible. Through God's grace, we can rise from failure and move closer to salvation. Let's examine further.

Armed with this new understanding, Joshua and his troops set forth toward Jericho. The city was heavily fortified. Its occupants cowered in fear of the Israelites. They had heard of the Lord's mighty power and lacked the courage to face Joshua and his army. The city gates were securely shut, and no one could enter or leave Jericho (Joshua 6:1).

A properly fortified base is one of the most difficult challenges a military commander can face. One potential solution is to starve the enemy out. *De Re Militari* states, "An army unsupplied with grain and other necessary provisions will be vanquished without striking a blow" (Vegetius 88). Additionally, "To distress the enemy more by famine than the sword is a mark of consummate skill" (Vegetius 90). Yet this strategy has drawbacks. As *The Art of War* states, "There is no instance of a country having benefited from prolonged warfare" (Tzu 5). A siege demands time; the occupants of Jericho had clearly prepared for one. It is no easy feat to keep your own troops provisioned in enemy territory. A prolonged campaign can also leave your forces overextended and vulnerable to outside attack.

The Art of War makes it abundantly clear that besieging a walled city should be a last resort. "Thus the highest form of generalship is to balk the enemy's plans; the next best is to prevent the junction of the enemy's forces; the next in order is to attack the enemy's army in the field; and the worst policy of all is to besiege walled cities" (Tzu 6). Moreover, "The rule is, not to besiege walled cities if it can possibly be avoided" (Tzu 6).

What was Joshua to do? The Lord responded with a divine command. He gave Joshua specific instructions that defied conventional military wisdom. Though the orders seemed unusual, Joshua remained faithful. He gathered his army and relayed the unconventional plan. "So Joshua the son of Nun called the priests and said to them, 'Take up the ark of the covenant, and have seven priests carry seven trumpets of rams' horns in front of the ark of the Lord.' Then he said to the people, 'Go forward and march around the city, and let the armed men go on ahead of the ark of the Lord'" (Joshua 6:6–7). He further instructed, "You shall not shout nor let your voice be heard, nor let a word proceed from your mouth, until the day I tell you, 'Shout!' Then you shall shout!'" (Joshua 6:10). With discipline and reverence, the people obeyed. Joshua also issued a solemn warning: disobedience would bring destruction upon the entire camp (Joshua 6:18–19).

For six days, they marched around the city once per day. On the seventh day, the moment of divine deliverance arrived. They marched around Jericho seven times. On the final circuit, the priests blew their trumpets, and Joshua addressed the army: "Shout! For the Lord has given you the city!" (Joshua 6:16).

A miracle followed. "So the people shouted when the priests blew the trumpets; and it came about when the people heard the sound of the trumpet, that the people shouted with a great shout and the wall fell down flat. Then the people went up into the city, every man straight ahead, and took the city" (Joshua 6:20). With the walls destroyed, the Israelites captured the city with ease. Joshua ordered the two spies to rescue Rahab and her family, who had remained faithful (Joshua 6:22–25). Once they were safe, Joshua's army razed the city. "Then they burned the city with fire, and all that was in it. Only the silver and gold, and the articles of bronze and iron, they put into the treasury of the house of the Lord" (Joshua 6:24).

Joshua concluded the victory with a prophetic oath: "Then Joshua made them take an oath at that time, saying, 'Cursed before the Lord is the man who rises up and builds this city Jericho; with the loss of his firstborn he will lay its foundation, and with the loss of his youngest son he will set up its gates'" (Joshua 6:26). Generations later, this curse came to pass during King Ahab's reign (1 Kings 16:34).

Following this miraculous triumph, Joshua's fame spread throughout Canaan.

One of the central themes in *Joshua* is the strength to remain faithful in the face of danger. Rahab embodied this courage. She sheltered the spies, trusted the Lord, and followed through on her oath. As a result, she and her family were spared.

Her legacy is profound. After joining the Israelites, Rahab gave birth to Boaz, a central figure in the Book of Ruth and the great-grandfather of King David (Matthew 1:5–6; Ruth 4:21–22). Generations later, she became part of the royal lineage that led to Jesus Christ Himself

(Matthew 1:1–16). Rahab's faith not only saved lives; it altered the course of history.

Her story reminds us that God does not require perfection. He requires repentance. In His wisdom, the Lord often works through those the world overlooks. Rahab's past did not disqualify her from being a vessel of grace. As C. S. Lewis wrote in *Mere Christianity*: "The sins of the flesh are bad, but they are the least bad of all sins. All the worst pleasures are purely spiritual: the pleasure of putting other people in the wrong, of bossing and patronising and spoiling sport, and back-biting, the pleasures of power, of hatred.... That is why a cold, self-righteous prig who goes regularly to church may be far nearer to hell than a prostitute. But, of course, it is better to be neither" (Lewis 102).

Rahab's redemption reminds us never to underestimate repentance or overlook the unexpected. God's plan often advances through the most unlikely of heroes. Thus, treat people well. You never know what role they might be playing in God's plan.

Joshua, too, exemplified obedience under pressure. Though the Lord's commands defied conventional wisdom, Joshua did not waver. He led with discipline, set the standard for his troops, and stayed true to God's word. Through this faith-driven obedience, God worked in miraculous ways, tearing down walls no human force could breach.

THE CONSEQUENCES OF SIN

The Israelites were riding high after their miraculous victory. They had crossed the Jordan and entered the Promised Land. Morale was high, and God's presence was clear. Yet, beneath this triumph, trouble was brewing. Just as the previous generation had turned its backs on the Lord by worshiping the golden calf, some from the new generation faltered as well (Exodus 32:1–6).

Achan, son of Karmi, violated the Lord's command. Against Joshua's direct orders, he stole plunder from Jericho and buried it beneath his tent. The directive had been clear: Rahab was to be protected, and the devoted things of the city were to be left untouched (Joshua 6:18–19). Achan ignored the latter. Despite the warnings, the allure of gold and silver proved too great. He took what belonged to the Lord and hid it, hoping no one would know.

But the Lord knew. Achan's sin wasn't just theft; it was a breach of covenant. His disobedience angered the Lord (Joshua 7:1). At the time, Joshua had no idea. He prepared for the next phase of the campaign, unaware that one man's sin had tainted the whole camp. With Jericho behind him, Joshua sent men to scout the hill country around Ai. The spies returned with confidence. "Then they returned to Joshua and said to him, 'Do not have all the people go up; have only about two or three thousand men go up and attack Ai; do not trouble all the people there, for they are few'" (Joshua 7:3).

But the mission went horribly wrong. The soldiers of Ai offered stiff resistance and drove back the Israelites. "And the men of Ai struck and killed about thirty-six of their men, and pursued them from the gate as far as Shebarim and struck them on the mountainside; and the hearts of the people melted and became like water" (Joshua 7:5). Thirty-six men fell that day. The abrupt defeat stunned the nation. Two hard lessons emerged. First, never underestimate the enemy. Second, intel is important, but it's not the end all be all. Expect the unexpected.

For Joshua and his people, the loss was crushing. After Jericho, they thought victory was assured. Instead, they were humiliated. "Then Joshua tore his clothes and fell to the ground on his face before the ark of the Lord until the evening, both he and the elders of Israel; and they put dust on their heads" (Joshua 7:6). The Israelite forces were greatly outnumbered in this new land. Joshua worried that the tides were turning against them.

Joshua cried out in anguish. How could the same God who crumbled Jericho's walls now allow such a defeat? "O Lord, what can I say since Israel has turned their back before their enemies? For the Canaanites and all the inhabitants of the land will hear about it, and they will surround us and eliminate our name from the earth. And what will You do for Your great name?" (Joshua 7:8–9).

The Lord responded to Joshua's prayer with resolute clarity: "So the Lord said to Joshua, 'Stand up! Why is it that you have fallen on your face?'" (Joshua 7:10). God revealed that Israel had sinned. Some among them had violated His covenant. This moral corruption had compromised the entire nation. "Israel has sinned, and they have also violated My covenant which I commanded them. And they have even taken some of the things designated for destruction, and have both stolen and kept it a secret. Furthermore, they have also put them among their own things" (Joshua 7:11). Because of this, the Israelites had become liable to destruction. The Lord was not interested in ritual lamentation. HE didn't want a groveling apology. HE demanded decisive action. The Lord warned Joshua, "Therefore the sons of Israel cannot stand before their enemies; they turn their backs before their enemies, because they have become designated for destruction. I will not be with you anymore unless you eliminate from your midst the things designated for destruction" (Joshua 7:12).

Upon receiving this warning, Joshua immediately addressed the people. He called for them to consecrate themselves in preparation for the next day. The guilty party would soon be revealed. The Lord gave Joshua marching orders: "Stand up! Consecrate the people and say,

'Consecrate yourselves for tomorrow, because the Lord, the God of Israel, has said this: "There are things designated for destruction in your midst, Israel. You cannot stand against your enemies until you have removed the designated things from your midst"'" (Joshua 7:13). Joshua dutifully obeyed. After a lengthy investigation, all signs pointed toward Achan. Joshua confronted him directly. "Then Joshua said to Achan, 'My son, I implore you, give glory to the Lord, the God of Israel, and give praise to Him; and tell me now what you have done. Do not hide it from me'" (Joshua 7:19).

Achan did not resist. He confessed immediately. "So Achan answered Joshua and said, 'Truly, I have sinned against the Lord, the God of Israel, and this is what I did: when I saw among the spoils a beautiful robe from Shinar, two hundred shekels of silver, and a bar of gold fifty shekels in weight, then I wanted them and took them; and behold, they are hidden in the ground inside my tent, with the silver underneath'" (Joshua 7:20–21). Two hundred shekels of silver amounts to roughly five pounds, and the gold bar about one and a quarter. Joshua sent messengers to retrieve the stolen goods. They uncovered the plunder and laid it out before the entire assembly of Israel. The evidence was undeniable.

The consequence was swift and severe. The people of Israel stoned Achan (Joshua 7:25). Perhaps Solomon had this very incident in mind when he later wrote, "Ill-gotten gains do not benefit, But righteousness rescues from death" (Proverbs 10:2). There are several lessons to draw from this episode. When we act selfishly, the damage often extends beyond ourselves. Our sins ripple outward. Furthermore, true repentance requires action. When Joshua collapsed in sorrow, the Lord didn't coddle him; He told him to *stand up* and purge the sin from their ranks. *Words may confess, but only action redeems*. Last, temptation often begins with what we allow ourselves to see. Achan only desired the silver and gold after laying eyes on it. The ancient phrase rings true: *out of sight, out of mind*. This is a practical warning for us today, guard what you expose yourself to; covetousness is downstream of proximity.

THE BATTLE OF AI

The atmosphere in the camp was heavy. Justice had been served, but the cost weighed on every heart. Many lives were lost due to one man's selfish actions. First, three dozen men fell in battle. Achan was executed shortly after. With judgment complete, it was time to prepare for the next battle. The Lord made His intentions clear. "Now the Lord said to Joshua, 'Do not fear or be dismayed. Take all the people of war with you. Arise, go up to Ai; see, I have handed over to you the king of Ai, his people, his city, and his land'" (Joshua 8:1).

The Lord encouraged Joshua further. He said the king of Ai would fall just like the king of Jericho. This time, however, the Israelites were allowed to carry off livestock and plunder for themselves (Joshua 8:2). The irony stings: had Achan remained faithful, he would have shared in the spoils lawfully and honorably. As Proverbs states, "The wages of the righteous is life, The income of the wicked, punishment" (Proverbs 10:16).

Prior to the battle of Jericho, the Lord gave highly specific instructions to Joshua. Those detailed commands were part of His miraculous plan. This time, however, the Lord issued a simple directive: set an ambush behind the city. Joshua received his marching orders and acted swiftly. He selected thirty thousand of his best fighting men and sent them out at night with the plan. They were to position themselves behind the city and remain close, alert, and ready (Joshua 8:2–5). While the ambush party waited, Joshua advanced with the main force.

He addressed his soldiers plainly: "Then I and all the people who are with me will approach the city. And when they come out to meet us as they did the first time, we will flee before them. They will come out after us until we have lured them away from the city, for they will say, 'They are fleeing before us just as they did the first time.' So we will flee before them" (Joshua 8:5–6). Joshua put himself at risk to carry out the deception. He didn't ask his men to do anything he wouldn't do himself. Seeing as the Canaanites had already driven out the Israelites from this territory once, they were likely confident they could do it again. But once they were drawn out from the safety of the city, the real strike would begin.

This was a wise plan. The lure-to-ambush strategy echoes *De Re Militari*: "draw the enemy into an ambuscade or more advantageous position where you may easier defeat them in case they follow you" (Vegetius 82). Joshua ordered the ambush party to strike Ai like a viper—silent, swift, and deadly. As Vegetius also notes, "Good officers decline general engagements where the danger is common and prefer the employment of stratagem and finesse to destroy the enemy as much as possible in detail and intimidate them without exposing our own forces" (Vegetius 61). Joshua was a good officer, indeed.

To conclude the briefing, Joshua reminded his troops of the good graces of the Lord. "Then you shall rise from your ambush and take possession of the city, for the Lord your God will hand it over to you. Then it will be when you have seized the city, that you shall set the city on fire. You shall do it in accordance with the word of the Lord. See, I have commanded you" (Joshua 8:7–8).

With the strategy in place and hearts steeled, Joshua dispatched the ambush party under the cover of night.

The following day, Joshua mustered his army in plain view of Ai. His main force took position just north of the city, across a valley. Meanwhile, he ordered an additional five thousand troops to reinforce the ambush team, positioning them west of the city. With all pieces in place, the trap was set.

When the king of Ai spotted Joshua's forces, he took the bait. Confident from his prior victory, he assumed another easy rout was at hand. Arrogantly, the king and his men charged out to meet Israel in open battle (Joshua 8:14). This move was both rash and unnecessary. As the defending force, they had every advantage, yet they abandoned it. Unbeknownst to them, they were marching into a carefully laid snare.

The king of Ai made the fatal mistake of underestimating his opponent. Joshua followed a classic strategy found in *The Art of War*: "If your opponent is of choleric temper, seek to irritate him. Pretend to be weak, that he may grow arrogant" (Tzu 4). Joshua and his men executed the plan precisely. They allowed themselves to be pushed back and feigned retreat toward the wilderness (Joshua 8:15). The trap was sprung. "And all the people who were in the city were called together to pursue them, and they pursued Joshua and were lured away from the city. So not a man was left in Ai or Bethel, but they had all gone out after Israel, and they left the city unguarded and pursued Israel" (Joshua 8:16–17).

In grappling martial arts, this principle is known as *kuzushi*. The term, originating in Japan, means "to unbalance." At certain times in combat, you must meet force with force, but kuzushi is different. It involves redirecting an opponent's momentum to your advantage. The more aggressively they come at you, the easier they are to topple. A small man can throw a larger man decisively by letting his opponent overcommit.

Beginners often use this as a counter when their opponent charges recklessly. But seasoned practitioners elevate the principle. Rather than waiting for a mistake, they *invite* it; laying traps that lure their opponent into overextending. Once unbalanced, the opponent falls with little resistance. While kuzushi typically refers to one-on-one fighting, the principle applies to large-scale combat as well.

This is exactly what Joshua did to the king of Ai. Emboldened by overconfidence, the king believed Israel to be weak. His arrogance led him to abandon his stronghold and commit all his forces. In doing so, he handed Joshua the keys to the kingdom.

Composure is vital in warfare. Letting pride dictate decisions is dangerous. As *The Art of War* reminds us: "To secure ourselves against defeat lies in our own hands, but the opportunity of defeating the enemy is provided by the enemy himself" (Tzu 7). Therefore, *never give the enemy leverage against you.*

The city was empty. Now was the time for the killing blow. To execute this final strike, Israel embodied the teaching of legendary swordsman Miyamoto Musashi: "When the enemy attacks, remain undisturbed but feign weakness. As the enemy reaches you, suddenly move away indicating that you intend to jump aside, then dash in attacking strongly as soon as you see the enemy relax" (Musashi 22).

Though Musashi wrote this for the lone swordsman, the principle scales to armies as well. Joshua gave the signal for the ambush party to strike. "Then the Lord said to Joshua, 'Reach out with the sword that is in your hand toward Ai, for I will hand it over to you.' So Joshua reached out with the sword that was in his hand toward the city" (Joshua 8:18). Without hesitation, Joshua raised the javelin high. This moment echoed the early days of his military career. Recall the battle against Amalek, when Moses stood on the hilltop with the staff of God in hand. Whenever his hands were raised in worship, Israel prevailed (Exodus 17:8–13). As with Moses' staff, so too with Joshua's javelin. The Spirit of the Lord remained with His people.

The moment Joshua lifted his weapon, the ambush force surged forward. They embodied another classic maxim: "Attack him where he is unprepared, appear where you are not expected" (Tzu 4). With no defenders left behind, the Israelites entered Ai unopposed and set it ablaze. Black smoke rose quickly into the sky, thick and suffocating. Flames consumed the city as its warriors remained distracted, still chasing after Joshua's decoy force. They operated on pure impulse rather than defensive strategy. Indeed, "When some are seen advancing and some retreating, it is a lure" (Tzu 18). The soldiers of Ai ignored this timeless rule and were now surrounded. Their only option was to fight to the death.

This marked the turning point of the battle. Shocked and disoriented, the enemy had lost cohesion. As Vegetius wrote, "For part of the victory consists in throwing the enemy into disorder before you engage them" (Vegetius 75). Joshua's men capitalized immediately. The Israelite force that had fled into the wilderness turned and pressed back toward Ai. At the same time, the ambush party struck from the rear. The Canaanites were trapped between two armies, boxed in by the very overconfidence that led them into the field.

This maneuver echoes the fate of Pharaoh's army. As Moses fled Egypt, the Egyptian chariots gave chase. At God's command, Moses stretched out his hand and the parted waters of the Red Sea collapsed inward (Exodus 14:26–28). The sea swallowed the horsemen and chariots alike, drowning them beneath the waves. Likewise, Joshua's trap embodied this vengeful sea. His forces fell upon the enemy like a crashing tide. His soldiers fell upon their enemies like a tsunami. Just as the waves swept the Egyptians to sea, the Canaanites collapsed under this double-sided assault.

With this two-sided attack, the Israelite forces epitomized a key principle from *Principles of War*: "With overwhelming energy, pursue our successes. Only the pursuit of a defeated enemy will give us the fruits of victory" (Clausewitz 13). Surrounding an enemy is effective, but dangerous. It must be executed with precision and absolute resolve. There is nothing more lethal than a cornered, wounded animal. Fighting men are no different. Indeed, "When men find they must inevitably perish, they willingly resolve to die with their comrades and with their arms in their hands" (Vegetius 80).

To neutralize this threat, Israel adopted a ruthless clarity of purpose. It embodied the words of Musashi: "Be intent solely on killing the enemy. Do not try to cut strongly and, of course, do not think of cutting weakly. You should only be concerned with killing the enemy" (Musashi 33). And so they advanced. With the trap fully sprung, the Canaanites had nowhere to run. Just as the python coils around its prey, tightening with every breath, so too did Israel encircle Ai. Each fallen Canaanite marked another tightening loop.

The Israelites stayed true to their blades. Thousands of sword points sliced and stabbed with unrelenting force, gnawing like the jagged teeth of ravenous sharks. Just like a feeding frenzy, they struck until nothing remained but blood and silence. They had fully absorbed Musashi's mindset: "In large-scale strategy, when the enemy starts to collapse you must pursue him without letting the chance go. If you fail to take advantage of your enemies' collapse, they may recover" (Musashi 24).

Joshua did not waver. He held his javelin high until the final enemy fell. "The others came out from the city to confront them, so that they were trapped in the midst of Israel, some on this side and some on that side; and they killed them until there was not one left who escaped or survived. But they captured the king of Ai alive and brought him to Joshua" (Joshua 8:22–23).

Imagine the scene.

Ai, once fortified and arrogant, now lay in ruins. Flames crackled over charred buildings. A smoky haze blanketed the city. The fog of war lifted quickly, but the bloodstained ground would remain. The ruins stood as a blackened monument, a brutal reminder of Israel's power, resolve, and obedience.

Their redemption was complete.

Where Achan's sin had led to humiliation, Israel's repentance led to restoration. They had purged the corruption from within. Not a single Israelite died in the destruction of Ai. Approximately twelve thousand Canaanites fell that day. The king of Ai alone survived—briefly. Joshua executed the king and impaled his body on a pole. After the soldiers and people beheld their fallen enemy, Joshua ordered the corpse buried under a heap of stones, just as the Law prescribed (Joshua 8:29; Deuteronomy 21:22–23). Fittingly, the name *Ai* roughly translates to "The Ruin."

This battle revealed the enduring strength of courage, discipline, and faith. In the face of sin and failure, Israel repented, regrouped, and struck back. Chapter 8 of *Joshua* reminds us that when we purge corruption and remain steadfast, we too can rise from failure and reclaim the battlefield.

GETTING BACK ON TRACK

After the bloodshed at Ai, Joshua didn't pause to celebrate. He turned instead to the Lord. Before moving onward, he built an altar on Mount Ebal according to Mosaic Law. Made from uncut stones, untouched by iron tools, the altar was used to offer sacrifices to the Lord (Joshua 8:30–31). There, in the presence of all Israel, Joshua engraved a fresh copy of the Law of Moses into stone, a powerful symbol of its permanence (Joshua 8:32). Joshua was a true warrior of the Lord. He never sought his own glory, and nor did he grow arrogant in victory. All he did was in service to God.

The people of Israel assembled to witness Joshua's act of devotion. As instructed by Moses, they arranged themselves on two opposing hills: half towards Mount Gerizim and half in front of Mount Ebal (Joshua 8:33). Between them stood the Levitical priests with the Ark of the Covenant. The symbolism was potent. One mountain represented the blessings for obedience; the other, the curses for rebellion. These twin peaks stood as a fork in the road. The Law of Moses was the map to navigate that terrain.

Mount Ebal was the mountain of warning, a symbol of the devastation that awaited disobedience—including disease, disaster, and destruction (Deuteronomy 28:59–61). Mount Gerizim represented the rewards of faithfulness: protection from enemies, fruitful harvests, and lasting prosperity (Deuteronomy 28:7, 28:11).

Joshua followed the example of Moses and renewed the covenant publicly. This ceremony was a national recommitment to the Lord. It underscored both the consequences of sin and the path to redemption. Israel had already experienced both. It saw the miracle at Jericho, the loss caused by Achan's sin, and the redemption at Ai. Joshua upheld the spiritual discipline of his people. "There was not a word of all that Moses had commanded which Joshua did not read before all the assembly of Israel with the women, the little ones, and the strangers who were living among them" (Joshua 8:35).

THE GIBEONITE TREATY AND MORAL DILEMMAS

Before the dust could settle on the ruins of Ai, conflict was already stirring again. The kings west of the Jordan heard of Israel's victories at Jericho and Ai. In response, they formed a coalition to wage war against Joshua and Israel (Joshua 9:1–2). But one group stood apart: the Gibeonites. Rather than fight, they sought mercy. Having heard of the Lord's power and the destruction of other cities, they feared Israel's God.

In a desperate bid for survival, they crafted a clever deception. Disguised in worn clothes and carrying moldy bread and cracked wineskins, they approached Joshua, claiming to be envoys from a distant land (Joshua 9:4–6). Their act was convincing. Joshua questioned them, but the Gibeonites held firm, praising the Lord's victories in Egypt and the defeat of the Amorite kings (Joshua 9:9–10). Without consulting the Lord, Joshua and the Israelite leaders made a peace treaty with them, ratified by oath (Joshua 9:15).

This was no small matter. Israel took oaths with deadly seriousness. Only days later, the truth emerged; the Gibeonites were neighbors. The people grumbled against the leadership. But the leaders stood firm: after rendering the holy oath, they could not harm them. To break their word would mean dishonoring God. Instead, the leaders chose mercy with mutual benefit. Instead of harming them, they were utilized as woodcutters and water carriers in service of the tabernacle (Joshua 9:19–21).

This episode presents a crucial distinction between being peaceful and being harmless. Peacefulness is a *choice*, an act of will made from a position of strength. Harmlessness, on the other hand, is a condition of weakness. To be able to protect truly, one must first understand how to attack effectively. How can one properly bear a shield if one is ignorant of swordsmanship? Joshua was no tyrant. He had the means to destroy the Gibeonites, but instead, he honored his word. He took responsibility for the hasty treaty and chose de-escalation, not because he had to, but because that decision honored the Lord.

After the treaty was confirmed, Joshua confronted the Gibeonites. They confessed the truth without protest. "So they answered Joshua and said, 'Since your servants were fully informed that the Lord your God had commanded His servant Moses to give you all the land, and to destroy all the inhabitants of the land before you, we feared greatly for our lives because of you, and did this thing'" (Joshua 9:24). Then they reaffirmed their loyalty: "And now behold, we are in your hands; do to us as it seems good and right in your sight to do" (Joshua 9:25).

The arrangement proved beneficial. The Gibeonites served the tabernacle and contributed labor to the nation. More importantly, they received Israel's protection. Gibeon was no small village; it was a large, fortified city of skilled warriors (Joshua 10:2). While the Gibeonites were fighters, Israel's numbers and divine backing offered far greater security. With this new alliance, Israel's strength grew even greater.

But the peace would not last long. News of this treaty alarmed King Adoni-Zedek of Jerusalem. With Jericho, Ai, and now Gibeon aligned with Israel, the tides of power were shifting fast.

THE SOUTHERN CAMPAIGN

Fueled by concern, the king of Jerusalem sought help from nearby Amorite kings. "Come up to me and help me, and let's attack Gibeon, for it has made peace with Joshua and with the sons of Israel" (Joshua 10:4). Individually, they stood little chance. But together, perhaps they could challenge the new Israel–Gibeon alliance. They forged a coalition of five kings. Unlike the Gibeonites, who sought peace, this group rushed to war. It was a do-or-die gambit. "So the five kings of the Amorites, the king of Jerusalem, the king of Hebron, the king of Jarmuth, the king of Lachish, and the king of Eglon, gathered together and went up, they with all their armies, and camped by Gibeon and fought against it" (Joshua 10:5). This direct assault on a fortified city left them dangerously exposed.

In desperation, the Gibeonites sent word to Joshua, begging for help (Joshua 10:6). Without hesitation, Joshua honored the treaty. He and all his warriors marched to defend their new allies (Joshua 10:7). Just as before Jericho and Ai, the Lord encouraged him: "And the Lord said to Joshua, 'Do not fear them, for I have handed them over to you; not one of them will stand against you'" (Joshua 10:8). Armed with faith and fighting men, Joshua advanced toward Gibeon.

His army marched through the night. Their discipline and experience paid off. The Amorites, unprepared for such a swift response, were thrown into chaos (Joshua 10:9–10). Israel seized the initiative—and never gave it back. Its recent campaigns had forged it into a battle-hardened force. Now, it was executing strategy with precision. Joshua's tactics echoed *De Re Militari*: "Those designs are best which the enemy are entirely ignorant of till the moment of

31

execution. Opportunity in war is often more to be depended on than courage" (Vegetius 87). Israel possessed both.

It also had divine support. Having purged corruption from its ranks, Israel once again fought on holy ground. And once again, the Lord intervened miraculously. Joshua prayed boldly, "Then Joshua spoke to the Lord on the day when the Lord turned the Amorites over to the sons of Israel, and he said in the sight of Israel, 'Sun, stand still at Gibeon, And moon, at the Valley of Aijalon!'" (Joshua 10:12). God answered. "So the sun stood still, and the moon stopped, Until the nation avenged themselves of their enemies" (Joshua 10:13). Just as at Jericho, the Lord fought alongside them.

With time extended, Israel pressed forward. It pursued the Amorites with the sword. *De Re Militari* advises, "A golden bridge should be made for a flying enemy" (Vegetius 81). Joshua followed suit accordingly and the pursuit intensified. The fleeing Amorites *thought* they could reach safety, yet they didn't factor in divine intervention. The Amorites ran, but they could not escape. The Lord hurled massive hailstones with lethality. The skies became a second battlefield.

Once again, God used the weather as a weapon. He hurled hailstones that killed more Amorites than the swords of Israel (Joshua 10:11). He had done the same in Egypt, and He would again during the time of the judges (Exodus 9:23–24, Judges 5:20–21). With this fierce combination of blade and storm, the skirmish ended decisively. After the miraculous victory, Joshua's army returned to its camp.

But the broader conflict wasn't over. The five Amorite kings had fled the battlefield and sought shelter in a cave at Makkedah. Joshua acted swiftly. He ordered his men to trap them inside by rolling large rocks to cover the mouth of the cave. Additionally, guards were posted (Joshua 10:18). Then, without delay, he turned his focus to the remaining Amorite troops. "But do not stay there yourselves; pursue your enemies and attack them from behind. Do not allow them to enter their cities, for the Lord your God has handed them over to you" (Joshua 10:19). Joshua personified a principle from *Principles of War*: "With overwhelming energy, pursue our successes. Only the pursuit of a defeated enemy will give us the fruits of victory" (Clausewitz 13).

That day, Israel was fruitful indeed as it made short work of its foes. Though a few enemy stragglers escaped to fortified cities, Joshua's army returned to Makkedah untouched. No one dared to challenge Israel after such a one-sided defeat (Joshua 10:20–21). The battle in the field was over—but the five kings still had to face justice.

Joshua gave the order to retrieve the five kings (Joshua 10:22). His men complied. In front of the assembled troops, Joshua had his commanders place their feet on the necks of the kings. "When they brought these kings out to Joshua, Joshua called for all the men of Israel, and said to the leaders of the men of war who had gone with him, 'Come forward, put your feet on the necks of these kings.' So they came forward and put their feet on their necks. Joshua then said to them, 'Do not fear or be dismayed! Be strong and courageous, for the Lord will do this to all your enemies with whom you fight'" (Joshua 10:24–25). After the execution, the bodies were impaled and displayed until sunset, in accordance with the Law. Then the corpses were cast back into the cave and sealed with stones, an unspoken warning to all who might oppose Israel (Joshua 10:26–27).

Joshua had all his men witness the execution. Why? One reason may have been to profess the Lord's strength. The point was clear: the Israelites were fighting on the Lord's side, not the other way around. After cleansing the camp of corruption, all of their victories had been swift and total. Joshua reinforced the importance of staying submitted to God's word.

Another reason may have been to steel the troops for the brutal campaign ahead. *De Re Militari* gives insight: "For troops that have never been in action or have not for some time been used to such spectacles, are greatly shocked at the sight of the wounded and dying; and the impressions of fear they receive dispose them rather to fly than fight" (Vegetius 66). Joshua made sure his men were ready.

With the five enemy kings eliminated, Joshua continued his southern campaign without pause. The first to fall was Makkedah. "Now Joshua captured Makkedah on that day, and struck it and its king with the edge of the sword; he utterly destroyed it and every person who was in it. He left no survivor. So he did to the king of Makkedah just as he had done to the king of Jericho" (Joshua 10:28). He struck city after city, leaving no enemy behind. Altogether, he overtook seven kingdoms (Joshua 10:28–43).

Joshua did exactly as Moses had commanded: "Only in the cities of these peoples that the Lord your God is giving you as an inheritance, you shall not leave anything that breathes alive. Instead, you shall utterly destroy them, the Hittite and the Amorite, the Canaanite and the Perizzite, the Hivite and the Jebusite, just as the Lord your God has commanded you" (Deuteronomy 20:16–17). With the southern region of Canaan fully subdued, Joshua led the army back to Gilgal (Joshua 10:43).

THE NORTHERN CAMPAIGN

T he Southern Campaign had been highly successful, but the conquest of Canaan was far from over. The fighting continued over a long period (Joshua 11:18). Trouble was again brewing. Jabin, king of Hazor, heard of Joshua's victories and took action. *(Note: This is not the same Jabin mentioned in Judges 4:2—this likely refers to a dynastic title rather than a specific individual.)*

Jabin sent word to kings across the eastern, western, and northern regions of Canaan. Together they formed a coalition that was even greater than the alliance of the five Amorite kings. This new force was the largest the Israelites had yet encountered. "Then they came out, they and all their armies with them, as many people as the sand that is on the seashore, with very many horses and chariots" (Joshua 11:4). The Canaanite army assembled at the Waters of Merom, as it prepared to crush Israel.

But Israel had been forged through fire. Indeed, "Achieving victory in war does not depend entirely upon numbers or simple courage; only skill and discipline will ensure it" (Vegetius 3). Joshua had been cultivating discipline throughout the campaign. His soldiers were now battle-hardened and spiritually refined. Most importantly, they were aligned with God's purpose.

The Lord gave Joshua clear orders: "Do not be afraid of them, because by this time tomorrow I will hand all of them, slain, over to Israel. You are to hamstring their horses and burn their chariots" (Joshua 11:6). With divine assurance, Joshua acted without delay. Swiftness is essential in warfare, and Joshua embodied this principle. The very next day, he launched a surprise assault. "So Joshua and all the people of war with him came upon them suddenly at the waters of Merom, and attacked them" (Joshua 11:7). The Lord guided them to another one-sided defeat.

This bold attack was strategically sound for two reasons. First, speed was Israel's ally. A smaller force can maneuver more quickly and strike unexpectedly. "For in matters of war, speed is often more useful than courage" (Vegetius 110). Additionally, advantages tend to compound when pursued resolutely. Their sudden assault threw the massive Canaanite army into chaos. "For part of the victory consists in throwing the enemy into disorder before you engage them" (Vegetius 75). Israel had both the fluidity to launch a swift attack and the grit to carry it out until the end.

Second, Israel selected the battlefield wisely. The terrain at the Waters of Merom favored infantry. Chariots, though technologically superior, require open ground. Wetlands and uneven ground neutralized this advantage. *De Re Militari* confirms the principle: "If strongest in cavalry, we should prefer plains and open ground; if superior in infantry, we should choose a situation full of enclosures, ditches, morasses and woods" (Vegetius 62).

Joshua followed the Lord's command to the letter. "And Joshua did to them just as the Lord had told him; he hamstrung their horses and burned their chariots with fire" (Joshua 11:9).

With their technical assets destroyed, the Canaanite resistance collapsed quickly. Since Hazor was the head of the northern coalition, Joshua struck it first. He executed its king and burned the city to the ground. Though Joshua went on to conquer many royal cities, Hazor was the only one he burned. "Joshua captured all the cities of these kings, and all their kings; and he struck them with the edge of the sword and utterly destroyed them, just as Moses the servant of the Lord had commanded" (Joshua 11:12). Israel carried off livestock and valuables, growing stronger with each conquest (Joshua 11:10–14).

Joshua stayed faithful throughout. "Just as the Lord had commanded His servant Moses, so Moses commanded Joshua, and so Joshua did; he left nothing undone of all that the Lord had commanded Moses" (Joshua 11:15). The northern campaign took years, but it was nearly complete. One final task remained: the Anakites (Joshua 11:21).

Driving out the Anakites was the capstone of Joshua's military career. These were not ordinary foes—they were giants, with fortified cities and a reputation for striking fear into the hearts of Israel. Years earlier, when Moses sent twelve spies into Canaan, it was the Anakites who terrified ten of them. Though they praised the land's bounty, they faltered when they saw the giants. "Nevertheless, the people who live in the land are strong, and the cities are fortified and very large. And indeed, we saw the descendants of Anak there!" (Numbers 13:28).

Caleb silenced the crowd, urging the people to trust the Lord and take the land. "Then Caleb quieted the people before Moses and said, 'We should by all means go up and take possession of it, for we will certainly prevail over it'" (Numbers 13:30). But the others sowed the seeds of doubt. "So they brought a bad report of the land which they had spied out to the sons of Israel, saying, 'The land through which we have gone to spy out is a land that devours its inhabitants; and all the people whom we saw in it are people of great stature'" (Numbers 13:32). The ten spies placed fear above faith. Because of this, they never entered the Promised Land (Numbers 14:29–30).

The Israelites had not encountered the Anakites in forty-five years (Joshua 14:10). Much had changed since then. When they first saw these towering enemies, Israel was newly freed from Egypt, underequipped, inexperienced, and easily shaken. Now, after decades of warfare, the Israelites were seasoned. They had conquered thirty-one kingdoms (Joshua 12:24). With each victory, Israel gained both resources and real-world experience.

De Re Militari emphasizes the value of such experience: "The well-trained soldier is eager for action, and the untrained fear it. In war, discipline is superior to strength; but if that discipline is neglected, there is no longer any difference between the soldier and the peasant" (Vegetius 43). In the past, the Anakites had inspired such dread that many Israelites wanted to return to slavery in Egypt (Numbers 14:2–4). Only Joshua and Caleb had stood firm. But now, Israel had matured. It had learned to trust the Lord through repeated victories. In other words, "A soldier who has proper confidence in his own skill and strength, entertains no thought of mutiny" (Vegetius 52). They wouldn't rebel under Joshua's faithful leadership.

The final military campaign of Joshua's life was to defeat the Anakites once and for all. "Then Joshua came at that time and eliminated the Anakim from the hill country, from Hebron, Debir, Anab, and from all the hill country of Judah and all the hill country of Israel. Joshua utterly destroyed them with their cities. There were no Anakim left in the land of the sons of Israel; only in Gaza, Gath, and Ashdod some remained" (Joshua 11:21–22).

This passage foreshadows the later battle between David and Goliath. Goliath was one of the descendants who had fled to Gath. "Then a champion came forward from the army encampment of the Philistines, named Goliath, from Gath. His height was six cubits and a span" (1 Samuel 17:4). Joshua had slain most of these giants, but Israel would once again forget this legacy of courage. In the time of Saul, fear returned where faith should have remained.

Scripture gives few tactical details about Joshua's battle against the Anakites. Yet we can infer the challenges: steep terrain, fortified cities, and massive enemies (Joshua 14:12). Joshua's troops may have used ambushes or rapid maneuvers to outmatch their size. Regardless of the exact method, the outcome was decisive. What once filled Israel with panic now lay in ruin. By God's strength, the giants were removed. Their defeat was not just a military victory, but it was also a spiritual triumph of faith over fear.

This final battle offers us little in battlefield strategy but much in spiritual insight. Israel's journey under Joshua reveals how faith is forged. They began as fearful people. But step by step, battle by battle, they learned to trust God's promises. Their strength and courage grew through obedience. The same path is open to us. If we walk faithfully and pray for strength, we too can overcome the giants in our lives.

THE LATER YEARS

After decades of war, Israel entered a new era. Joshua, once a slave in Egypt, had led the people faithfully from the Jordan to the conquest of Canaan. Now the land had rest from war (Joshua 11:23). Joshua could have retired, but he remained a dutiful servant of the Lord, enacting reforms to secure Israel's future.

Following God's command, Joshua established cities of refuge. These towns ensured justice for those who accidentally killed another, protecting them from personal vengeance while awaiting trial (Joshua 20:2–6). He also fulfilled Moses' command to provide the Levites with towns and pasturelands scattered throughout the territory (Joshua 21:3). These decisions ensured both spiritual leadership and unity across Israel.

With peace secured, Joshua released the eastern tribes—Reuben, Gad, and the half-tribe of Manasseh—who had honored their pledge to fight alongside their brothers. Before they departed, Joshua gave them a charge: to love and obey the Lord wholeheartedly (Joshua 22:5). They left with great wealth, but their construction of an altar near the Jordan nearly sparked civil war. Misunderstood as rebellion, the western tribes prepared for battle. Thankfully, diplomacy prevailed. The altar was a symbol of unity, not defiance (Joshua 22:28–34).

In his old age, Joshua summoned the people for a final address. He reminded them that it was the Lord—not their own strength—who had delivered victory (Joshua 23:3, 10). He warned of the dangers of idolatry and urged them to obey God's law (Joshua 23:6).

At Shechem, Joshua renewed the covenant. He recounted Israel's journey from Abraham to the present, emphasizing God's faithfulness.

The people pledged loyalty to the Lord. Yet Joshua remained cautious. "If you abandon the Lord and serve foreign gods, then He will turn and do you harm and destroy you after He has done good to you" (Joshua 24:20). He set up a large stone as a witness to their promise (Joshua 24:27).

Joshua died at the age of 110. He was buried in the hill country of Ephraim. Israel remained faithful during his lifetime and that of the elders who followed him (Joshua 24:31). Joshua's life exemplified faithful service, courageous leadership, and disciplined preparation. As Vegetius said, "Few men are born brave; many become so through care and force of discipline" (Vegetius 88). Joshua became great not by birth, but through obedience to the Lord from beginning to end.

JOSHUA CONCLUSION

The life of Joshua stands as one of the clearest scriptural examples of how disciplined leadership, unwavering faith, and strategic thinking can advance the will of God. Once a slave in Egypt, Joshua rose to become Moses' assistant and eventually the commander of Israel. He led with humility, not ego, and gave glory to God for every victory. His success was rooted in obedience. From the Jordan river to the Anakite giants, Joshua proved what a godly warrior–leader could accomplish. In fact, the Apostle Peter later highlights the importance of this style of leadership. "shepherd the flock of God among you, exercising oversight, not under compulsion but voluntarily, according to the will of God; and not with greed but with eagerness; nor yet as domineering over those assigned to your care, but by proving to be examples to the flock. And when the Chief Shepherd appears, you will receive the unfading crown of glory" (1 Peter 5:2–4).

The command to be *strong and courageous* frames the entire book. God repeats it three times in the opening chapter, setting the tone for Joshua's leadership. Additionally, this serves as an example to follow. This courage was not based on self-confidence, but on trust in the Lord's presence and promises. His troops understood that by following Joshua, they were following the Lord. Early in the campaign, Joshua demonstrated tactical wisdom by sending spies to Jericho. One of the most unexpected spiritual lessons came through Rahab, a prostitute who aided the spies. Despite her background, she recognized God's power and risked everything to align with His people. Her bravery echoes the divine charge given to Joshua and reminds us that God uses unlikely people in pivotal roles. Her faith saved her family, and her lineage leads directly to King David and Christ. Never forget to treat people well. We never know what role they play in God's plan.

Crossing the Jordan river marked a turning point. The ark of the covenant, symbolizing God's law and presence, led the way. As the waters parted, Israel crossed on dry ground, just as in the days of Moses. Joshua then built a memorial to ensure future generations remembered God's faithfulness. With each step, Joshua ensured that Israel moved forward in faith and discipline.

The fall of Jericho was a masterclass in faithful obedience. The Lord gave unconventional instructions, yet Joshua followed without hesitation. In response, God delivered a miraculous victory. This demonstrates how God's infinite wisdom can lead to salvation, even when the situation appears hopeless. However, Chapter 7 revealed the cost of disobedience. Achan's hidden sin led to a humiliating defeat at Ai, resulting in the deaths of Israelite soldiers. Joshua responded with swift discipline. The guilty were identified and punished, restoring holiness to the camp. Once the sin was purged, Joshua regrouped and led a successful ambush against Ai. This showed not only battlefield intelligence, but also the redemptive power of repentance and renewed obedience. Repentance requires genuine action, not empty words.

Leadership often requires discernment under pressure. It's what separates true leaders from tyrants or cowards. The Gibeonites deceived Israel into making a peace treaty. Though Joshua was misled, he honored the vow made before God. He chose de-escalation and discipline over vengeance, assigning the Gibeonites to servanthood rather than destruction. When enemies later attacked Gibeon, Joshua came to their defense without hesitation. He was a good shepherd indeed. Marching overnight, Israel struck with speed and precision. God responded with overwhelming power: throwing the enemy into confusion, halting the sun in the sky, and hurling hailstones upon the retreating armies. This moment, one of the most dramatic divine interventions in all of Scripture, echoed the plagues of Egypt and reaffirmed God's active role in Israel's victories.

As campaigns continued, Joshua executed decisive strikes across the southern and northern coalitions. The latter was vast, with chariots and cavalry as numerous as sand on the shore. Yet Joshua's faith did not falter. He leveraged the terrain, obeyed God's instructions, and routed the enemy. His battle-hardened troops utilized ambush tactics for a one-sided victory. Things were very different now. The cowardly had fallen by the wayside many years ago in the desert. Those who remained leveraged their faith as a strength and emulated the courage of their leader.

Joshua's final battle was against the Anakites, giants who had once terrified Israel's spies. Joshua defeated them, achieving what earlier generations had feared to attempt. His life reminds us: with God, we too can face and conquer giants in our own lives. Yet, he did not stop with battlefield victories. Joshua ensured lasting stability by dividing the land according to Moses' instruction, establishing cities of refuge to preserve justice, and distributing land to the Levites to provide spiritual guidance. Near the end of his life, Joshua renewed the covenant with Israel at Shechem. He issued a clear warning to remain true to the Lord. As a final reminder, he set up a stone to stand as a witness to this covenant.

Joshua's story closes with discipline, unity, and reverence. He did not seek personal glory. He prepared the next generation to walk in righteousness. He led not just as a warrior, but also as a faithful servant. The lessons of his life remain timely: obey God, be strong and courageous, and act decisively.

PART TWO:
THE JUDGES—THE DARKEST
DAYS IN ISRAEL

JUDGES INTRODUCTION

Fundamentally, the Book of Judges is about duality. Throughout this book, we see the darkest depths of humanity contrasted with true heroism and God's grace. It is often considered the darkest book in the Bible, with good reason. This collection of stories is one of the clearest pictures of humanity's fallen nature—with nearly every act of depravity on display. The Book of Judges puts nearly every form of human depravity on full display. Murder, mutilation, human sacrifice, slavery, and rape are recurring themes throughout this book. Yet it is not all doom and gloom. The faithful rose up to fight the dark forces. The need for faith, courage, and a fighting spirit remained a rallying cry in those dark days. Let's examine further.

The era of the judges spans roughly 350 years; from Joshua's death to the rise of Israel's first king, Saul. This was arguably the most unstable period of Israel's long history. A recurring phrase is repeated several times in Judges. "In those days there was no king in Israel; everyone did what was right in his own eyes" (Judges 17:6). Similar sentiments are found in Judges 18:1, 19:1, and 21:25. Yet this does not refer to Earthly authorities. Rather, it refers to the abandonment of God's ways and holy covenant. Adam and Eve did as they saw fit in the Garden of Eden (Genesis 3). Moreover, this was Satan's fatal flaw as well (Isaiah 14:12–14, Revelation 12:7–9) Rebranding sin and vice as independence or empowerment is one of Satan's favorite schemes.

Following these examples, the Israelites placed their desires above the Lord's by repeatedly violating their sworn covenant agreement. This disobedience and darkness are foreshadowed in the second chapter of Judges. The Lord spoke through an angel to warn the Israelites again: "I brought you up out of Egypt and led you into the land which I have sworn to your fathers; and I said, 'I will never break My covenant with you, and as for you, you shall not make a covenant with the inhabitants of this land; you shall tear down their altars.' But you have not obeyed Me; what is this thing that you have done?" (Judges 2:1–2). In His patience, He gave them a final warning that echoed back to Moses and Joshua. "Therefore I also said, 'I will not drive them out from you; but they will become like thorns in your sides, and their gods will be a snare to you'" (Judges 2:3).

Another theme of Judges is the cyclical nature of humanity. The author G. Michael Hopf is credited with the now famous saying. "Hard times create strong men. Strong men create good times. Good times create weak men. And, weak men create hard times" (Hopf 18). This quote is Judges in a nutshell. Indeed, Joshua was the strong man who soldiered through the hard times. By embracing God's wisdom, Joshua prevailed against countless deadly challenges! In the previous chapter, we examined all the good things Joshua did to ensure the stability of Israel into the future. His attempts were noble, but he was still a mortal man. During Joshua's leadership, the people remained faithful. After his death, some of the elders carried on his waning legacy and contributions (Judges 2:7). Yet nothing lasts forever: "For there is no lasting remembrance of the wise, along with the fool, since in the coming days everything will soon be forgotten. And how the wise and the fool alike die!" (Ecclesiastes 2:16).

Alas, this too came to pass. The good times Joshua fought so hard to create eventually caused a disconnect. His descendants forgot the lessons of the past. "All that generation also were gathered to their fathers; and another generation rose up after them who did not know the Lord, nor even the work which He had done for Israel" (Judges 2:10). They were insulated from the harsh reality Joshua and Caleb had faced while growing up. This insulation softened them as a society. Indeed, this softening and ignorance led them to create hard times again.

This softness quickly led to arrogance, which led to the abandonment of the Lord's ways. Much like the above quote about strong men and hard times, Judges operates on a four-part cycle. The cycle of Judges goes like this: *apostasy*, *oppression*, *crisis*, and *deliverance.* It always begins the same way. First, Israel forgets the Lord's ways and commits *apostasy*. This is usually in the form of idolatry. Second, its enemies *oppress* Israel and subjugate it for many years. Eventually things reach a breaking point. A *crisis* point happens and the Israelites repent and cry out for mercy. Finally, God *delivers* them from oppression. He does this by raising judges to carry out the task. Yet each time the cycle repeats, Israel sinks a little bit deeper into depravity. The cycle of Judges characterizes a downward spiral.

Scripture elaborates more on the cyclical pattern of Judges. Part 1: apostasy. "Then the sons of Israel did evil in the sight of the Lord and served the Baals, and they abandoned the Lord, the God of their fathers, who had brought them out of the land of Egypt, and they followed other gods from the gods of the peoples who were around them, and bowed down to them; so they provoked the Lord to anger" (Judges 2:11–12). This blatant betrayal angered the Lord. Recall Joshua's encounter with an angel in Joshua 5. The angel fought on the side of the Lord, not on either side of humanity. At that time, Joshua and Israel fought for the Lord. Yet in most of Judges, they stopped fighting the good fight and were treated accordingly. They served false gods and felt the repercussions of their idolatry. "Wherever they went, the hand of the Lord was against them for evil, as the Lord had spoken and just as the Lord had sworn to them, so that they were severely distressed" (Judges 2:15).

Due to this, Israel entered Part 2: oppression. "Then the anger of the Lord burned against Israel, and He handed them over to plunderers, and they plundered them; and He sold them into the hands of their enemies around them, so that they could no longer stand against their enemies" (Judges 2:14). After many years of brutal oppression, a faithful few stood out. Some of the Israelites remembered the ways of the past. This led to the boiling point of Part 3: crisis. Eventually some of the Israelites cried out and repented to the Lord. True to His word, He remained faithful to their holy covenant. In His infinite grace and patience, the Lord answered their prayers.

There was a glimmer of hope flickering upon the backdrop of despair. The hard times created some of God's strongest soldiers. We now see the final phase of the cycle: deliverance. "And when the Lord raised up judges for them, the Lord was with the judge and saved them from the hand of their enemies all the days of the judge; for the Lord was moved to pity by their groaning because of those who tormented and oppressed them" (Judges 2:18). These judges served well. They were flawed, yet faithful. They weren't perfect; they made mistakes. Yet all of the judges had the following characteristics: a fighting spirit, faith, grit, and adaptability. They offer many lessons for us to follow today. Sometimes the lesson is what *not* to do.

The judges shared Joshua's desire for a brighter future. They displayed valiant effort in this endeavor. Unfortunately, some things never change. "But it came about, when the judge died, that they would turn back and act more corruptly than their fathers, in following other gods to serve them and bow down to them; they did not abandon their practices or their obstinate ways" (Judges 2:19). This stoked the Lord's anger. Due to this repeated failure of the covenant, God used this period as a trial. This primarily served as a test of faith (Judges 2:20–22). Additionally, this tumultuous period taught Israel combat skills and warfare (Judges 3:2).

The Book of Judges offers many practical combat lessons. There is a theme throughout the Bible, and it is especially pronounced in this book. With the good graces of the Lord, anything is possible. If we are fighting on the side of the Lord, all we need are modest tools, a fighting spirit, and above all, faith. In the time of Judges, Israel was always the underdog. Throughout the book, it was dramatically outclassed in terms of troop numbers and technology. In spite of the odds, Israel prevailed. Due to these disadvantages, we see many examples of unconventional warfare at play. This includes assassinations, clandestine operations, psychological operations (PSYOPs), improvised weaponry, guerilla warfare, and sabotage. We see more traditional warfare as well, though it is less pronounced. In addition to battlefield wisdom, there are many spiritual lessons to glean. With all this in mind, let's take a closer look.

EHUD—SLY AS A FOX, BRAVE AS A LION: JUDGES 3

The story of Ehud offers profound insights into leadership, strategy, courage, and faith. Though his account spans just one chapter in the Bible, it holds timeless lessons with both spiritual and tactical weight. It's unlikely you ever heard this bold story in Sunday School.

During Joshua's time, the Israelites witnessed great sacrifices and acts of valor. Joshua served as a clear example of a faithful warrior–leader, devoted to the Lord. But as that generation passed away, a new one rose up; one that had not seen the miracles of Exodus or the conquest of Canaan. It was unfamiliar with the Lord's ways (Judges 2:10). Lacking guidance and strong leadership, it drifted. Jericho, once conquered by Joshua, eventually fell under the control of King Eglon of Moab. During this dark chapter, the Moabites subjugated Israel. For eighteen long years, the people suffered under his tyrannical rule (Judges 3:14).

Then, from an unlikely place, deliverance came.

Long before David slew Goliath, Ehud toppled a kingdom with nothing but a honed blade and even sharper wits. His story reflects a familiar pattern in Scripture: With the good graces of the Lord, we are capable of incredible feats. All we need are modest tools, a fighting spirit, and faith. After nearly two decades of brutal oppression, the Israelites cried out to God. In response, He raised up Ehud (Judges 3:15). One of the first judges, Ehud executed what today would be called a special operation. As Sun Tzu wrote, "All warfare is based on deception" (Tzu 4). Ehud's mission is a textbook example of that principle. Through cunning, foresight, and stealth, he infiltrated the enemy court and assassinated King Eglon.

In preparation, Ehud forged a double-edged short sword, about one cubit long (~18 in.) At first glance, this seems counterintuitive for a covert mission. A smaller blade would be easier to conceal. Moreover, A small blade can slash a man's throat as easily as a larger one. Why risk detection with a weapon of that size? Because it was all a part of the plan.

Ehud was left-handed. This gave him a tactical advantage. During that time, weapons were typically carried on the left hip and drawn with the right hand. A guard checking for blades would naturally search that side. Ehud, however, strapped his dagger to his right thigh, exploiting the enemy's blind spot. This allowed him to bypass security and strike from where he was least expected. Their complacency, and failure to attend to detail, opened the door for their undoing.

After delivering the tribute, Ehud seized his moment. Turning back, he said to the king, "I have a secret message for you, O king" (Judges 3:19). The words caught Eglon's attention. Intrigued, the king dismissed his attendants. As the doors closed behind them, silence filled the chamber. Ehud's gambit had worked. He was now alone with the tyrant. As Sun Tzu would say, "Hold out baits to entice the enemy. Feign disorder, and crush him" (Tzu 4). King Eglon took the bait. But the mission was far from over.

Imagine Ehud's tension in that moment: a lone man standing before a monarch, the fate of his people hanging in the balance. One mistake, and the result would be torture and death. *The Art of War* warns, "He who exercises no forethought but makes light of his opponents is sure to be captured by them" (Tzu 19). But Ehud had planned thoroughly. More than that, he was guided by faith. An operation of this scale required not just strategy, but also extraordinary courage and resolve. He had reached the king's chamber, but his mission was just beginning.

Ehud stepped forward and said, "I have a message from God for you" (Judges 3:20). Without hesitation, he drove the dagger deep into the king's massive belly. His strike was swift and decisive, fueled by duty and divine purpose. The blow pierced through flesh and into his gut. The king convulsed, partly from the excruciating pain, partly from the sheer shock of such an audacious attack. His bowels discharged. The fat engulfed the vicious blade, even covering the handle (Judges 3:21–23). As the saying goes: *The mouse only realizes why the cheese was free after it's too late.*

The foot-and-a-half blade disappeared like a man swallowed in quicksand. The contrast of the moment was jarring; regal perfumes now mingled with the stench of filth and death. The immaculate chamber, once draped in wealth and opulence, was now smeared with blood and waste. The thin veneer of luxury betrayed by rot and corruption from the inside. Yet in truth, it had always been tainted. Now, the aesthetics of the king's chamber reflected the character of its occupant.

Ehud's method was as strategic as it was brutal. He immediately locked the chamber doors behind him and slipped away (Judges 3:23). A noxious odor permeated the regal palace. Shortly after the assassination, the king's attendants approached. Smelling the foul odor, they assumed Eglon was relieving himself. Not wanting to disturb their gluttonous master, they delayed (Judges 3:17, 24). It is never wise to embarrass a king. Ehud had accounted for this. He knew their hesitation would buy him precious time. While the guards stood outside, unsure and unwilling, he vanished into the night, quiet and lethal as a black cat.

As *The Art of War* observes, "What the ancients called a clever fighter is one who not only wins, but excels in winning with ease" (Tzu 8). Ehud was such a fighter. Yet the mission was only half complete. Eglon was dead, but Moab still held the advantage. Ehud had seized the initiative, now he had to exploit it

His plan had worked because it was tailored for this moment. Only a long dagger could pierce through the king's thick belly and deliver a fatal internal wound. If Ehud had gone for the throat, the kill might have been cleaner, but the guards would have been alerted too soon. The smell of waste became its own diversion, giving him a plausible cover for escape. It delayed pursuit just long enough to turn a covert operation into a full-scale uprising.

As Sun Tzu said, "If you know the enemy and know yourself, you need not fear the result of a hundred battles" (Tzu 7). Ehud knew his enemy's weaknesses, not just Eglon, but the attendants who served him.

The Israelites were vastly outnumbered and technologically inferior. But Ehud found a gap in their armor, and he drove the blade straight through it.

CASE STUDY

King Eglon was not the last time evil sat on a throne unchallenged; far from it. A remarkably similar assassination took place in World War II. Operation Anthropoid was the codename for the mission to eliminate Reinhard Heydrich, better known as "The Butcher of Prague." A high-ranking Nazi official, Heydrich rapidly ascended the party ranks due to his ruthlessness and efficiency. Personally appointed by Hitler to lead the newly created Reich Main Security Office (RSHA), he became the primary architect of Nazi oppression and genocide.

According to Burian, "In 1939 he was asked to prepare a so-called 'final solution to the Jewish problem.' From 1941 he personally supervised the creation of a system of extermination concentration camps. On January 20, 1942, he presided over a conference in Wannsee, where the 'final solution' was adopted" (Burian, Michal, et al 25). In September 1941, Heydrich was appointed Reichsprotektor of Bohemia and Moravia (modern-day Czech Republic). He immediately imposed martial law. Mass executions followed (Burian, Michal, et al 24–25). Out of this darkness came a daring mission to stop him.

Two Czechoslovakian soldiers, Jozef Gabčík and Jan Kubiš, were recruited by the British Special Operations Executive (SOE). On October 2, 1941, they began preparations (Burian, Michal, et al 31). Shortly after, they were flown to Scotland to undergo intensive training. "The aim of this Special Course was accurate shooting and grenade and bomb throwing... This demanding training took place in a variety of situations the two men might encounter in the occupied land. They practiced on fixed and moving targets in open terrain as well as in enclosed buildings and rooms" (Burian, Michal, et al 37). They were also trained as paratroopers and became proficient with improvised explosives.

After months of preparation, Gabčík and Kubiš parachuted into enemy territory. Their target, Heydrich, had grown arrogant. He routinely traveled with minimal security in an open-top Mercedes, a fatal oversight. After surveying the terrain, the soldiers strategically selected a sharp curve in the road, a wise choice. The bend forced Heydrich's car to slow significantly, exposing him at a vulnerable moment. All of the training the paratroopers had received led to this brief encounter. Victory loves preparation.

Choice of location, timing, and surprise, the same ingredients that defined Ehud's success were all in play. According to *Principles of War*, "One of the most important principles of offensive war is to surprise the enemy with a fast attack. The more unexpected the attack, the more successful it will be" (Clausewitz 8). The two paratroopers clearly understood this timeless principle. Boldly staring down the Butcher, Gabčík stepped into the road and raised his STEN submachine gun to unleash a vicious hailstorm of bullets into the open vehicle. Everything was going according to plan. until *CLICK*. At the critical moment, his weapon jammed.

As the old adage goes, "If you fail to plan, you plan to fail." This is where their specialized training kicked in. Without skipping a beat, Gabčík's partner, Kubiš, leapt into action. From his briefcase, he pulled out an improvised bomb. Much like David facing Goliath, his throw had been honed through countless repetitions in training. With a practiced motion, he hurled the explosive at the target (Burian, Michal, et al 64). The modified grenade tore through the rear fender and blew open the passenger-side door. With their target struck, the two paratroopers made their escape. Heydrich initially survived the attack but died from his injuries several days later; long after the two assassins had vanished (Burian, Michal, et al 68). Just like Ehud, they slipped away undetected, but the war was far from over. Now back to Ehud.

STRIKING WHILE THE IRON IS HOT

A fter his daring escape, Ehud rallied the Israelites and proclaimed, "Pursue them, for the Lord has handed your enemies the Moabites over to you" (Judges 3:28). The sudden death of King Eglon threw the Moabites into immediate disarray; an opportunity Ehud was quick to exploit. As Sun Tzu said, "Rapidity is the essence of war: take advantage of the enemy's unreadiness, make your way by unexpected routes, and attack unguarded spots" (Tzu 22). Ehud and his troops clearly understood this timeless principle. *De Re Militari* underscores this point further: "in matters of war speed is often more useful than courage" (Vegetius 110). Fortunately, the Israelites possessed both.

With Ehud leading the fight, the Israelites surged toward a key battleground chokepoint. Swiftly, they seized the fords of the Jordan river, denying their enemy an escape route (Judges 3:28). This was a wise strategy. Rivers are excellent natural lines of defense. Controlling the crossing neutralized the Moabites' numerical advantage. First, it trapped any forces attempting to flee from Jericho. Second, it prevented reinforcements from reaching them across the river. This single move shifted the balance decisively. *Principles of War* elaborates on the value of this strategy: "Forces should not be evenly distributed along the river to directly interfere with the crossing — too dangerous. Instead, we should try to observe the enemy, and once the enemy starts to cross, attack it on all sides while its full forces have not arrived yet, and it is limited by the narrow perimeter by the river" (Clausewitz 22). Such a choke point allows a small fighting force to neutralize a larger force efficiently. Ehud leveraged this to full effect.

By combining the advantages of terrain with speed and ferocity, the Israelites made short work of the enemy forces. They struck down ten thousand vigorous Moabite soldiers. None escaped (Judges 3:29). In one day, Ehud slew King Eglon and broke Moab's brutal grip on Israel. Ehud and his forces truly embodied the advice of Miyamoto Musashi: "Attack without warning where the enemy is not expecting it, and while his spirit is undecided follow up your advantage and, having the lead, defeat him" (Musashi 26). Vegetius echoes this sentiment: "But if he knows himself inferior, he must avoid general actions and endeavor to succeed by surprises, ambuscades and stratagems" (Vegetius 64).

Eighty years of peace in the land followed this daring special operation and the bloody battle for freedom (Judges 3:30).

EHUD CONCLUSION

In the incredible story of Ehud, we find not only a tale of courage and strategy, but also valuable lessons that resonate through the ages. Ehud's story highlights the importance of faith, courage and resilience in the face of apparently insurmountable odds. One courageous person can make a difference, "for it is not numbers, but bravery which carries the day" (Vegetius 9). By utilizing unconventional tools and tactics, Ehud reminds us that creativity and adaptability are paramount to success. As Sun Tzu brilliantly stated, "Water shapes its course according to the nature of the ground over which it flows; the soldier works out his victory in relation to the foe whom he is facing" (Tzu 12). Ehud didn't complain about his disadvantages; he adapted to them. He embodied the military axiom: *improvise, adapt, and overcome.*

In spite of all the challenges, Ehud's unwavering faith in the Lord prevailed. His unconventional methods and faith led him to a decisive victory. The Israelites enjoyed eighty years of peace as a result of this valiant freedom fighter. As the Founding Father Thomas Jefferson famously said, "The tree of liberty must be refreshed from time to time with the blood of patriots and tyrants." Unfortunately, another drought would fall upon the tree of liberty. The cycle of Judges continued once again.

DEBORAH AND JAEL – UNLIKELY HEROES RISE TO THE OCCASION: JUDGES 4–5

D ecades after the death of Ehud, the Israelites again did evil in the eyes of the Lord. As a consequence, they were sold into the hands of King Jabin of Canaan. His kingdom boasted a formidable army led by the ruthless commander Sisera. Sisera's nine hundred iron chariots struck terror into the hearts of the Israelites. They were severely outmatched. There wasn't a single spear or shield amongst the forty thousand of Israel (Judges 5:8). With little hope, the Israelites suffered under this oppressive reign for twenty years, until they finally cried out to the Lord for help (Judges 4:1–3).

And into this moment steps the next Judge: Deborah.

Deborah was a prophet and leader who ruled from the hill country of Ephraim. She held court under the Palm of Deborah, where Israelites came to settle their disputes (Judges 4:5). One day, filled with divine urgency, she summoned the military commander Barak. The Spirit moved through her. She relayed how the Lord had commanded this battle and declared: "I will draw out to you Sisera, the commander of Jabin's army, with his chariots and his many troops to the river Kishon, and I will hand him over to you" (Judges 4:7).

Barak was stunned. His reply was timid: "If you will go with me, then I will go; but if you will not go with me, I will not go" (Judges 4:8). Clearly, he hadn't internalized the warrior ethos. His hesitation revealed not only personal fear, but also the deeper spiritual decay of Israel. The courage that once defined its warriors had faded. If the military leader was this hesitant, the state of the nation was even worse. Leading from the rear is a recipe for disaster.

Much to his surprise, Deborah agreed. "She said, 'I will certainly go with you; however, the fame shall not be yours on the journey that you are about to take, for the Lord will sell Sisera into the hand of a woman'" (Judges 4:9). At first glance, her response may seem arrogant or boastful. But soon we will see, she wasn't speaking of herself. Another unlikely hero would soon emerge from the fray.

Deborah and Barak were outnumbered and underequipped. But they had one powerful advantage: the terrain. With courage and tactical insight, they led their ten thousand troops to seize the high ground of Mount Tabor. Elevation is a timeless battlefield asset. It doesn't matter how many troops a commander has if he can't position them properly. A plan was beginning to form.

By taking the heights of Mount Tabor, Deborah stripped Sisera of his greatest asset. The steep, rocky slopes neutralized his chariots, which were designed for speed and power on flat terrain. In this instance, the iron armor became a liability. Even worse for Sisera, his forces were positioned near the Kishon River; soft, muddy ground ideal for trapping wheels. Deborah had set a brilliant trap. Sisera now stood between a rock and a hard place.

Her strategy echoed the timeless wisdom of Vegetius: "If strongest in cavalry, we should prefer plains and open ground; if superior in infantry, we should choose a situation full of enclosures, ditches, morasses and woods, and sometimes mountains" (Vegetius 62). Deborah understood this intuitively.

Transforming an adversary's greatest strength into a weakness is nothing short of strategic genius. As the battle was staged between a river and a mountain, the options for Sisera's chariots to retreat were slim to none. Additionally, Deborah received help from an unexpected ally; Heber the Kenite, a descendant of Moses' brother-in-law (Judges 4:11). Heber played a crucial role in this mission. He informed Sisera that Barak was preparing to attack. That intelligence was all it took to lure Sisera and his forces into the Valley of Jezreel. Like all good traps, this one offered enticing bait. The plains offered great maneuverability for the chariots, but that advantage would be short lived. Just like Ehud's gambit against King Eglon, Sisera was exactly where Deborah wanted him.

With Sisera's full force gathered in one place, the trap was set. It was time for battle. With unwavering conviction, Deborah told Barak, "'Arise! For this is the day on which the Lord has handed Sisera over to you; behold, the Lord has gone out before you.' So Barak went down from Mount Tabor with ten thousand men following him" (Judges 4:14). Her words echoed Moses' assurance to the Israelites as they prepared to enter the Promised Land: "So be aware today that it is the Lord your God who is crossing over ahead of you as a consuming fire. He will destroy them and He will subdue them before you, so that you may drive them out and eliminate them quickly, just as the Lord has spoken to you" (Deuteronomy 9:3).

Invigorated by Deborah's faith and the divine promise, Barak led his ten thousand men into the fray, ready to confront the formidable enemy and reclaim Israel's freedom (Judges 4:14). Bravely, the Israelites charged head first against the iron chariots and Canaanite soldiers. They embodied Joshua's example of courage when the odds were stacked against him. And just as in Joshua's time, the Lord intervened. A torrential downpour broke loose. "The stars fought from heaven, From their paths they fought against Sisera. The torrent of Kishon swept them away" (Judges 5:20–21). The scene echoed the Red Sea overwhelming the Egyptian charioteers during Moses' escape, drowning them all (Exodus 14:26–28).

The once-favorable plains had become a nightmare. The chariots, just moments before their greatest advantage, were now their greatest liability. Heavy wheels sank into the mud, immobilizing the Canaanite forces. No matter how advanced the technology, the environment always gets a say. The Lord was surely looking out for Israel that day.

As Vegetius observed, "part of the victory consists in throwing the enemy into disorder before you engage them" (Vegetius 75). Deborah's forces understood this well and seized the momentum. Moreover, "In large-scale strategy, when the enemy starts to collapse you must pursue him without letting the chance go. If you fail to take advantage of your enemies' collapse, they may recover" (Musashi 24).

With the chariots disabled, panic spread through the Canaanite ranks like wildfire. Fleeing uphill toward Mount Tabor was suicide, so they took the only other option; retreat toward the Kishon River. But that escape was cut off as well. Many were swept away in its floods

(Judges 5:21). Deborah and Barak embodied Musashi's timeless instruction: "Chase him towards awkward places, and try to keep him with his back to awkward places. When the enemy gets into an inconvenient position, do not let him look around, but conscientiously chase him around and pin him down" (Musashi 21). They followed that advice to the letter.

The Canaanite forces' actions provide a textbook example of tactical failure. According to *Principles of War*: "We must not lose composure and resolve in these conditions; loss of these qualities is often among the first losses in a war." To prevent that: "One must get used to the idea of dying with honor, to continually nurture that idea to get used to it" (Clausewitz 2). As the legendary swordsman put it, "Generally speaking, the Way of the warrior is resolute acceptance of death" (Musashi 1). The warrior will die for a just cause.

The Canaanites began this battle with immense technological superiority. They had the ancient equivalent of tanks. Meanwhile, the Israelites didn't even possess spears or shields. This battle serves as a timeless lesson: *never over-rely on technology for victory*. While advanced tools can be advantageous, they are no substitute for a fighting spirit. We must embrace the way of the warrior. It doesn't matter how dangerous our weapons are, or how sophisticated our defenses may be; if you lose your will to fight, all the technology in the world is worthless.

The battle raged on. Barak led his ten thousand daring warriors in a one-sided clash. Exploiting the chaos and confusion, the Israelites made short work of the Canaanite forces. Muddy terrain was soon stained with blood as their enemies fell by the sword. The remaining troops fled. During the retreat, Commander Sisera dismounted his chariot and slipped away from the fray (Judges 4:15). But combat was far from over. Barak pursued the enemy until not a single man was left, save their commander (Judges 4:16).

Sisera continued his flight, abandoning his men rather than fighting alongside them. He chose the coward's path, and a coward's death soon followed. Eventually, he reached the tent of Jael, the wife of Heber the Kenite. Perhaps due to a previous alliance or arrangement, Sisera thought this was a safe place to hide (Judges 4:17). The term *Kenite* refers to a clan of metalworkers. Ironically, the very people who forged the Canaanites' war machines would now bring their downfall.

As *The Art of War* advises, "Hold out baits to entice the enemy. Feign disorder, and crush him" (Tzu 4). Jael embodied this tactic literally and with flawless precision. She welcomed the fleeing commander warmly (Judges 4:18). Gratefully, Sisera scurried into her tent and collapsed from exhaustion. On the surface, it seemed like the perfect hiding place. In that culture, only a woman's husband or father was permitted to enter her tent. Sisera's pursuers were unlikely to look there.

Jael comforted him and covered him with a blanket. Exhausted from battle and flight, Sisera asked for water. She went above and beyond. All good traps are baited well. She brought him milk served in a fine bowl fit for nobility (Judges 5:25). Sisera must have felt immense relief. He had barely escaped death, and now he was given a warm place to rest and rich milk to drink. While his soldiers were being cut down outside, he lay surrounded by comfort.

But he knew he wasn't completely safe. He instructed Jael to guard the tent door. If anyone came by, she was to say he wasn't there (Judges 4:20).

His luck was truly extraordinary, but not in the way he imagined. As Sun Tzu said, "Attack him where he is unprepared, appear where you are not expected" (Tzu 4). Sisera had escaped every armed combatant on the battlefield. Yet he ran headfirst into the most dangerous adversary of all: *the one he never saw coming*. According to *Principles of War*, "Surprise plays a much greater role in strategy than in tactics; it is the most valid starting point for victory" (Clausewitz 13). Jael understood this principle clearly.

She saw her moment and acted. Sisera had dozed off; it was now or never. With resolve in her heart, she grabbed the only tools at her disposal and set forth. "She reached out her hand for the tent peg, And her right hand for the workmen's hammer. Then she struck Sisera, she smashed his head; And she shattered and pierced his temple" (Judges 5:26). A wise choice. The temple is the weakest part of the human skull, making it a perfect target. Whether Jael knew that anatomically or not, we can't say. But what we *can* say is this: the blow was absolutely devastating. The tent peg pierced his skull not once, but twice—a clean shot, through and through. The formidable spike drove straight through his temple and out the other side of his skull. Jael swung with such brutal force that the stake pinned his head to the ground (Judges 4:21). The thirsty sand drank the Canaanite's blood as the fog of war began to clear.

The tent peg's material is not specified, but it was strong enough to pierce two layers of bone and anchor firmly in the ground. Given her clan's metalworking background, it's plausible that Jael wielded an iron spike—fitting, since iron chariots had once terrorized her people. Now, iron may have sealed their fate.

This unlikely woman delivered the final blow in a brutal campaign. Armed with nothing but faith and common household tools, she destroyed one of the most feared commanders in the region. The Israelites were poorly armed and outnumbered. Yet they had something the Canaanites could never manufacture—*the will to win*. Jael was an everyday woman in the right place at the right time. But none of that would have mattered if she had lacked the courage to take her shot.

Shortly after, Barak arrived on the scene.

Jael met him outside and invited him in. She opened her tent and revealed the fallen commander (Judges 4:22). Sisera had died a coward's death. He had deserted his troops and paid the price, not with glory, but with dishonor. His downfall was forever immortalized in Judges 5, the Song of Deborah.

When Barak saw the impaled body, Deborah's prophecy came into full focus (Judges 4:9). The Lord's word always bears fruit, even when we don't understand it at the time. The moment strengthened Barak's faith and fighting spirit. Sisera's defeat had shattered the Canaanite army; now it was time to finish the war. Fired up from the battlefield, Barak embodied a timeless principle of warfare: "With overwhelming energy, pursue our successes. Only the pursuit of a defeated enemy will give us the fruits of victory" (Clausewitz 13).

King Jabin was exposed. With Sisera's forces annihilated, there was nothing left to protect him. The Israelites seized the opportunity. They advanced without hesitation and slaughtered the Canaanite king that very day (Judges 4:23–24). Much like their predecessor Ehud, all they needed were faith, modest tools, and the fighting spirit to wield them against the odds. After this decisive victory, the land had peace for forty years (Judges 5:31).

DEBORAH AND JAEL
CONCLUSION

There are many lessons to learn from this dark and violent chapter of Israel's history. First, much of Deborah's decisive leadership came from her faith in the Lord. King David would later write, "The Lord is for me; I will not fear; What can man do to me?" (Psalm 118:6). Deborah clearly lived by this truth long before David penned it.

Barak, however, lacked that same faith. His hesitation reflected the broader moral and spiritual decay of the time. As Deborah warned, the glory would not be his. Courage and faith are essential for leadership. We will never have perfect information or ideal conditions. There is no perfect time to act. Many leaders fall into the trap of *paralysis by analysis*—but often this is just cowardice dressed up as strategy or thoroughness. Fear is one of Satan's most effective tools. He uses it to paralyze believers and lead them astray. The Apostle Peter warned us: "Be of sober spirit, be on the alert. Your adversary, the devil, prowls around like a roaring lion, seeking someone to devour. So resist him, firm in your faith, knowing that the same experiences of suffering are being accomplished by your brothers and sisters who are in the world" (1 Peter 5:8–9).

Just like many real-world battles, terrain played a decisive role. Deborah's choice to take the high ground at Mount Tabor demonstrated how natural advantages can neutralize superior technology. At the outset, resistance seemed impossible. The Israelites were poorly armed, facing iron chariots. But God was watching. The heavy rains bogged down the enemy's greatest advantage. Victory came not from superior weapons, but from *faith joined with decisive action*. That combination is always a winning formula.

A fighting spirit is more powerful than any weapon. With resolve, resourcefulness, and trust in the Lord, we can overcome what seems insurmountable. On paper, the Canaanites should have won. But their hearts failed them. Panic took root. And their flight became their downfall. The Israelites capitalized on the momentum and struck down King Jabin before the day ended. They demonstrated persistence, boldness, and the discipline to press the advantage.

Once again, we saw how God uses the unlikely to accomplish the extraordinary. Jael was not a soldier, yet she struck the killing blow. With simple tools and unshakable courage, she ended the life of a feared commander. Her story reminds us that faith, resolve, and a readiness to act are more valuable than any weapon forged by man. We see the pattern again: With the good graces of the Lord, even the lowliest warrior can rise. All that's needed is conviction, courage, and the will to act.

The Song of Deborah closes with a prayer: "May all Your enemies perish in this way, Lord; But may those who love Him be like the rising of the sun in its might" (Judges 5:31).

After this fateful day, the land had rest for four decades. But the cycle would return. Strong men—and strong women—led to peace. But peace bred comfort, and comfort bred weakness. Before long, Israel would forget what God had done for it. It would become an enemy to itself as the moral decay crept in again.

GIDEON – A JOURNEY FROM COWARDICE TO COURAGE: JUDGES 6–8

The cycle turned yet again. The soft times created weakness and moral complacency. Once again, the Israelites lost their way. The people abandoned God and troubles soon followed. The Midianites recruited Amalekites and other forces to invade and oppress the tribes of Israel for seven years. Due to the people's apostasy, the Lord handed them over to these enemies. This crushing oppression from the Midianites drove the Israelites into mountain clefts, caves and strongholds (Judges 6:1–3). Like mice fleeing hawks, they scurried into this rough shelter, disheartened. *The Art of War* teaches us that we must be ready to receive the enemy. We accomplish this by making our position "unassailable" (Tzu 16). The mountains made an excellent position for defense. But this defense came at a cost. In all things in life, there must be balance. Due to the overwhelming forces they faced, the Israelites made a costly error. Clausewitz points out: "never rely entirely on the favorable conditions of the terrain; therefore, never succumb to the temptation of passive defense" (Clausewitz 10).

This passivity led to short-term wins at the expense of long-term losses. There is a difference between surviving and thriving. The rocky terrain made for an excellent fighting position but lacked fertile soil for planting crops. Wherever the Israelites planted their crops, the enemy swooped in to destroy. The massive army either ate or destroyed everything to impoverish the Israelites (Judges 6:5–6). The Midianite leaders embodied the wisdom of Sun Tzu: "Hence a wise general makes a point of foraging on the enemy. One cartload of the enemy's provisions is equivalent to twenty of one's own" (Tzu 5).

Moving like a swarm of locusts these forces invaded and consumed everything in their path. They plundered and destroyed the land with malicious intent. They set up camps in the fields and obliterated the crops. The swarm stole the sheep and cows for food and confiscated the donkeys as beasts of burden. Not a single animal escaped the enemy's greedy hands. This is a classic strategy. According to Vegetius, "The main and principal point in war is to secure plenty of provisions and to destroy the enemy by famine" (Vegetius 49). The invaders ravaged the land, leaving it barren as a desert. They did not spare a living thing for Israel (Judges 6:4). With heavy hearts, the Israelites learned a hard truth. "Famine makes greater havoc in an army than the enemy, and is more terrible than the sword" (Vegetius 49). After this utter devastation the Israelites cried out to the Lord for help (Judges 6:6).

God answered their cries with a prophet. The prophet reminded them how God had rescued them from slavery in Egypt. How He had freed them from their oppressors and given them land. Yet instead of embracing God's love, they had turned their backs to him by worshiping false gods (Judges 6:7–10). Then, an angel of the Lord came to speak to Gideon, son of Joash. Gideon was threshing wheat while hidden in a winepress. The winepress was cramped, yet he could conceal the valuable food from the Midianites. The angel of the Lord appeared to Gideon and said, "The Lord is with you, valiant warrior" (Judges 6:12). Gideon stood perplexed by this statement. He was confused as to why Israel had fallen into the hands of the Midianites. The Lord replied, "Go in this strength of yours and save Israel from the hand of Midian. Have I not sent you?" (Judges 6:14). Gideon timidly replied to God. He told the Lord how his clan was the weakest. Furthermore, he was the weakest in his family (Judges 6:15). Coincidentally, Moses gave a similar reaction after the Lord commissioned him to free the Israelites in Egypt. "But Moses said to God, 'Who am I, that I should go to Pharaoh, and that I should bring the sons of Israel out of Egypt?'" (Exodus 3:11). God often uses unlikely characters as heroes; Gideon was no exception. Just as the Lord reassured Moses, He reassured Gideon as well. God responded to Gideon's questions resolutely. He reassured Gideon, the He would be with him as he stuck down all the Midianites.

Gideon's timid nature prevailed in spite of this divine revelation. He asked God to give him a sign by accepting his offering of meat and unleavened bread.

Our loving God patiently obliged. Gideon placed the offering on a rock. An angel of the Lord touched the offering with his staff. Flames flared from the rock and consumed the bread and meat. God accepted the offering and again reassured him, "Peace to you, do not be afraid; you shall not die" (Judges 6:23). It is unwise to question the Lord by asking him to provide a sign for you. More on that shortly. We are called to serve the Lord, not to ask Him to serve us! Yet this was a product of Gideon's lack of confidence in himself and faith in the Lord. As for Barak, this hesitation reflects the spiritual decay of Israel at the time.

Later that evening, Gideon received his first assignment as a warrior of God. In his town stood a blasphemous altar to Baal and Asherah. His task was to destroy this demonic worship site and replace it with a proper altar for the Lord. In another parallel to Moses, God instructed him to destroy the unholy altar to Baal and cut down the wooden Asherah pole (Exodus 34:13). On top of the rubble, he was to build an altar for the Lord. Using the wood from the Asherah pole he was to start a fire and consecrate one of his father's bulls as a burnt offering (Judges 6:25–26).

This marks a turning point in Gideon's transformation as a holy warrior. Armed with faith in the Lord, he gathered ten servants and set out on this mission to reclaim the land. Even though God had earlier told him he would not die, he was still frightened. Due to the fear of his family and townspeople he embarked on this mission in the dead of night (Judges 6:27). Despite his trepidation, his faith allowed him to carry out the mission. Remember, courage is not the absence of fear, but the ability to get the job done *in spite of fear*. Gideon courageously destroyed the demonic altar and rendered the burnt offering to the Lord. The mission was complete. Now, Gideon had to face his family and the town.

The following day, the residents awoke to their pagan idols in ruins. An altar to Yahweh stood victoriously upon the rubble. They investigated and caught word that Gideon was responsible. Enraged, the townspeople went straight to Joash and demanded bloodshed: "Then the men of the city said to Joash, 'Bring out your son, that he may die, for he has torn down the altar of Baal, and indeed, he has cut down the Asherah which was beside it'" (Judges 6:30). Joash boldly faced the crowd to defend his son. Roaring like a lion, he replied: "Will you contend for Baal, or will you save him? Whoever will contend for him shall be put to death by morning. If he is a god, let him contend for himself, since someone has torn down his altar!" (Judges 6:31). This challenge was enough to sway the residents of Ophrah. They relented after seeing the error of their ways. The boldness of destroying the demonic altar earned Gideon the moniker Jerub-Baal, meaning, *Let Baal contend with him* (Judges 6:32).

Gideon excelled in his first mission! However, his story was just beginning. Once again, the Midianites and their allies set forth upon the land. During this imminent invasion, Gideon started to rally troops of his own. The Spirit of the Lord came upon Gideon. He blew a trumpet to summon followers. He also sent messengers throughout the land, calling allies to arms (Judges 6:33–35). The people had faith in Gideon to protect them, yet he had doubts. Imposter syndrome crept in as he fell back to his timid ways. His faith faltered and once again he asked the Lord for a sign. He placed a wool fleece on the threshing room floor. Full of hesitation and fear, he prayed to the Lord. He said, "behold, I am putting a fleece of wool on the threshing floor. If there is dew on the fleece only, and it is dry on all the ground, then I will know that You will save Israel through me, as You have spoken" (Judges 6:37).

Surely enough, God showed loving patience and fulfilled his request. Gideon awoke the following day and saw the Lord had answered his prayer. Despite this, Gideon *still* had doubts! "Then Gideon said to God, 'Do not let Your anger burn against me, so that I may speak only one more time; please let me put You to the test only one more time with the fleece: let it now be dry only on the fleece, and let there be dew on all the ground'" (Judges 6:39). Ever so graciously, God obliged. That evening the fleece was dry, yet the ground was covered with dew. With this not so subtle encouragement from the Lord, Gideon was all in.

It's worth noting that this stemmed from Gideon's lack of faith. Testing God should not be looked upon as an example to follow. For example, Moses states in Deuteronomy 6:16: "You shall not put the Lord your God to the test, as you tested Him at Massah." Massah means testing, and the region earned that name after the Israelites doubted

God's grace. After a near mutiny, Moses prayed for guidance. The Lord instructed Moses to strike a rock with his staff and drinking water flowed freely (Exodus 17:5–7). Furthermore, Jesus denounced this practice on multiple occasions (Matthew 12:39, 16:4; Luke 11:29–30). Jesus compared testing the Lord to adultery, as it demonstrates lack of faith.

Now that Gideon was bought in, the Israelites prepared for war. At dawn the following day, Gideon and his troops made camp. His forces numbered an impressive thirty-two thousand (Judges 7:3). God again spoke to Gideon: "And the Lord said to Gideon, The people who are with you are too many for Me to hand Midian over to them, otherwise Israel would become boastful, saying, 'My own power has saved me.' Now therefore come, proclaim in the hearing of the people, saying, 'Whoever is afraid and worried, is to return and leave Mount Gilead.' So twenty-two thousand from the people returned, but ten thousand remained" (Judges 7:2–3). We can now see the wisdom of this approach. The Israelites constantly struggled with pride and idolatry during this tumultuous period of history. Should the victory come too easily, they would forget about the Lord's ways. With this instruction from the Lord, Gideon let the fearful go home. While their numbers were an asset, the liability of their apprehension outweighed the benefit. Out of the thirty-two thousand troops, only ten thousand remained. We see another parallel to Moses here. In Deuteronomy 20:8 he said to the Israelites: "Who is the man that is afraid and fainthearted? Let him go and return to his house, so that he does not make his brothers' hearts melt like his heart!"

Fear is contagious. It must be remedied before it turns into mass hysteria. It is better to let the cowards go home before battle lest their actions lead to panic and chaos during battle. Recall the samurai mentality, "the way of the warrior is resolute acceptance of death" (Musashi 1). The twenty-two thousand cowards were not cut out for the fight. The Lord again spoke to Gideon. "There are still too many men. Take them down to the water, and I will thin them out for you there." Gideon listened to God. Most of the men kneeled down to drink. However, some of them stayed on their feet and drank from their cupped hands. God instructed Gideon to keep the men who remained on their feet and release all the others. Faithfully, he obeyed. Only 300 of the battle-ready soldiers remained (Judges 7:4–7). That means Gideon's final crew was *less* than one percent of this initial army!

Gideon and his select crew moved onward. He demonstrated great faith by thinning his troops, yet some fear still remained. The Lord again spoke to Gideon that night, "Arise, go down against the camp, for I have handed it over to you" (Judges 7:9). God sensed Gideon's fear and reassured him once again. "But if you are afraid to go down, go with Purah your servant down to the camp, so that you will hear what they say; and afterward you will have the courage to go down against the camp" (Judges 7:10–11). God encouraged Gideon to gather intelligence on the enemy. According to *The Art of War*, "Spies are a most important element in war, because on them depends an army's ability to move" (Tzu 29). Faithfully, Gideon snuck into enemy territory.

At first glance, the scene was terrifying. Gideon and his forces were greatly outnumbered. "Now the Midianites, the Amalekites, and all the people of the east were lying in the valley as numerous as locusts; and their camels were without number, as numerous as the sand on the seashore" (Judges 7:12). Yet God delivered on his promise to Gideon. Shortly after he arrived, Gideon overheard two men talking. One of the men told his friend about a recent dream where the Midianite camp collapsed. The friend gave his interpretation. "This is nothing other than the sword of Gideon the son of Joash, a man of Israel; God has handed over to him Midian and all the camp" (Judges 7:14).

Gideon bowed down and worshipped the Lord as soon as he heard this. He returned to his camp and called out to his soldiers, "Arise, for the Lord has handed over to you the camp of Midian!" (Judges 7:15). He was finally embodying the advice Moses gave the Israelites shortly before they entered the Promised Land. "Be strong and courageous, do not be afraid or in dread of them, for the Lord your God is the One who is going with you. He will not desert you or abandon you" (Deuteronomy 31:6).

Gideon's faith was restored and he finally learned to trust the Lord. With this newfound faith, he formed a bold plan. Recall from Sun Tzu, "All warfare is based on deception" (Tzu 4). Each man was given a trumpet and a jar with a torch inside. Gideon called his troops to attention and instructed them to follow his lead. He said, "When I and all who are with me blow the trumpet, then you also blow the trumpets around the entire camp and say, 'For the Lord and for Gideon!'" (Judges 7:18). The Israelites embodied a maxim from *The Art of War*: "Attack him where he is unprepared, appear where you are not expected" (Tzu 4).

On Gideon's signal, they all blew their trumpets and smashed the jars with vigor. The jars loudly cracked like thunder in the quiet of the night. The trumpets howled like hungry wolves hunting their prey. The cacophony woke the enemy with shock and awe! Encouraged by their leader, the men waved their torches for an intimidating spectacle. They all roared, "A sword for the Lord and for Gideon!" (Judges 7:20). This ruse gave the appearance of a much larger army. While they were vast in number, the Midianites fled and cried out in terror. The Midianites should have been more discerning in their troop selection. As Vegetius stated, "it is not numbers, but bravery which carries the day" (Vegetius 9). The Midianites violated core tenets from *Principles of War*. Stated by Clausewitz: "We must not lose composure and resolve in these conditions; loss of these qualities is often among the first losses in a war. Without these qualities, even the most brilliant gifts of mind go to waste. One must get used to the idea of dying with honor, to continually nurture that idea to get used to it" (Clausewitz 2).

Gideon performed a PSYOP upon the enemy. This worked because people aren't afraid of the dark, per se. Rather, they are afraid of the *unknown* within the darkness. Gideon leveraged this natural human tendency to his advantage. True power isn't simply the technology or the skillset one possesses. Rather, true power lies in what the enemy *perceives* you to have! Therefore, one can create the illusion of power with smoke and mirrors. PSYOPs are rather paradoxical. They are merely illusions, yet the *effects* of these illusions are very tangible and real. Gideon exemplified this perfectly. While his forces were few in number, they elicited a massive reaction from the enemy. Gideon's team only had jars, torches, and trumpets. Yet with these modest tools they were able to drive out the vast army with ease. In this initial attack, Gideon embodied the perfect victory. According to Sun Tzu, "to fight and conquer in all your battles is not supreme excellence; supreme excellence consists in breaking the enemy's resistance without fighting" (Tzu 6). Gideon excelled indeed.

CASE STUDY

A similar operation was carried out in World War II. On June 6, 1944, the largest amphibious invasion occurred on the beaches of Normandy, France. Better known as D-Day, over 130,000 Allied troops stormed the machine-gun-infested beaches. This was the first foothold into Nazi-occupied France. While successful, the courageous mission came with heavy casualties. Machine gun fire and drowning caused most of the more than 9,000 Allied deaths on that bloody day (Lange). D-Day was the major turning point in the war that led to the Nazi downfall (Eisenhower Presidential Library). The events that occurred on that fateful day still resonate today.

Much of D-Day's success comes from a lesser known operation that began in December 1943 (Munez and Luebering). Operation Fortitude was a deception campaign (PSYOP) led by the Allied forces. Much like Gideon, the Allies created the illusion of a grand spectacle. The main objective of Fortitude was to draw attention away from the true target of Normandy. The Allies misdirected the Nazi military officials to believe the invasion would occur elsewhere. Just like a magician, they drew the prying eyes away from the real target.

The Allies utilized a two-pronged strategy to accomplish this. The first prong was Fortitude North. The Allied commanders directed attention to Nazi-occupied Norway. They leaked false communications and increased naval activity towards the North Sea (Donovan). However, Fortitude South is where we see the major parallel to Gideon's strategy. Fortitude South convinced the Nazi forces that the invasion was going to occur in the Pas-de-Calais, France. This region is the shortest distance from the English coast. Therefore, it was a rational point to launch an invasion into enemy territory. The Allied forces reinforced this perception.

To accomplish this, the Allies created a ghost army called the First US Army Group. To build the spectacle, the Allies constructed thousands of prop tanks and airplanes. By staging this fleet of vehicles with new infrastructure, the Allies gave a stunning illusion. German surveillance planes reported this massive fleet to their leaders. This gambit forced the Germans to prepare for invasion in the wrong location. Additionally, the legendary General George S. Patton was publicly declared the commander of this fictitious unit to give credibility (Munez and Luebering). Sun Tzu would certainly have applauded the lengths they went to for this mission.

The two-pronged approach of Operation Fortitude spread the German forces thinly. This gave the Allies an excellent window of opportunity to attack. They didn't merely wait on an opportunity to arise; *they created one!* Without this deception campaign, D-Day likely would have turned out very differently. In fact, this operation probably solidified the Allied victory for the entire war.

Now back to Gideon's story.

GIDEON CONTINUED

The trumpets blasted sounds of war. Clay pots shattered underneath the feet of the assaulting army. Torchlight flickered to create a claustrophobic feeling. Shadows danced as their minds filled the gaps between dark and light with unseen horrors. This strategy aligns with Miyamoto Musashi's advice. "In large-scale strategy you can frighten the enemy not by what you present to their eyes, but by shouting, making a small force seem large" (Musashi 26). In the chaos of the scene, the Midianites panicked and lashed out. In their confusion they began to turn and attack one another (Judges 7:22). Utilizing the fog of war, the Israelites began to divide and conquer. They embodied another maxim from *The Art of War*. "If he is taking his ease, give him no rest. If his forces are united, separate them" (Tzu 4).

The once mighty army was fleeing and fracturing. The Midianites forces fell under attack from multiple fronts—never ideal. Encouraged by the changing tides of battle, some of the Israelites came down from the hills to join the fight! Things continued deteriorating for the Midianites. For example, "Then Gideon sent messengers throughout the hill country of Ephraim, saying, 'Come down against Midian and take control of the waters ahead of them, as far as Beth-Barah and the Jordan'" (Judges 7:24). This order was tactically sound. This aligns with Vegetius's teachings to military leaders: "form ambuscades with the greatest secrecy to surprise the enemy at the passages of rivers" (Vegetius 66). During this counterattack, the Ephraimite troops killed two Midianite leaders. After the battle they presented their heads to Gideon (Judges 7:25).

However, this was far from a victorious celebration. When they first reunited, the Ephraimites were furious at Gideon and brashly confronted him. They probably felt slighted that Gideon had not recruited them for battle. Instead of arguing back, he put his own ego aside and complimented their battlefield performance. "God has handed over to you the leaders of Midian, Oreb and Zeeb; and what was I able to do in comparison with you?" (Judges 8:3). Their bitterness and resentment quickly subsided. This parallels a wise proverb, "A gentle answer turns away wrath, but a harsh word stirs up anger" (Proverbs 15:1).

With things patched up, Gideon's army continued its pursuit of the remaining Midianites. While passing through Sukkoth, it encountered resistance. Exhausted from the grueling battle and pursuit, Gideon asked for food for his troops. The local officials replied in a taunting manner. Apparently, Gideon had used up all his diplomacy in the last quarrel. He lashed out with a chilling response. "So Gideon said, 'For this answer, when the Lord has handed over to me Zebah and Zalmunna, I will thrash your bodies with the thorns of the wilderness and with briers'" (Judges 8:7). Gideon's crew continued. He made the same request for food to the people of Peniel. They responded in the same arrogant manner as the people of Sukkoth. Once again, Gideon let anger get the better of him. "So he said also to the men of Penuel, "When I return safely, I will tear down this tower'" (Judges 8:9). With haste they continued their pursuit.

The troops of Zebah and Zalmunna, the Midianite leaders, were greatly diminished. Approximately 120,000 swordsmen had fallen. This left them with only 15,000 fighting men remaining (Judges 8:10). The enemy stood greatly weakened; now was the time to act! As stated in *De Re Militari*: "If the enemy makes excursions or expeditions, the general should attack him after the fatigue of a long march, fall upon him unexpectedly" (Vegetius 66). It was time for the killing blow.

To set up the checkmate maneuver, Gideon took a route used by nomads of the region. This allowed him to take the enemy by surprise (Judges 8:11). Recall from *The Art of War*: "Take advantage of the enemy's unreadiness, make your way by unexpected routes, and attack unguarded spots" (Tzu 22). This is always advisable, if possible. The attack succeeded. Gideon captured the Midianite leaders and routed the remainder of their army. They were nothing but sitting ducks to the emboldened Israelites (Judges 8:12). Gideon then returned to Sukkoth, where he let his anger get the better of him. Here we see yet another parallel to Moses. As promised, Gideon inflicted his vengeance on the people who taunted him earlier. He whipped some with briar thorns, along with killing the men of Peniel and Sukkoth (Judges 8:16–17).

Moses lost control of his anger on several notable occasions. The earliest recorded was when he killed the Egyptian slave driver. "Now it came about in those days, when Moses had grown up, that he went out to his fellow Hebrews and looked at their hard labors; and he saw an Egyptian beating a Hebrew, one of his fellow Hebrews. So he looked this way and that, and when he saw that there was no one around, he struck and killed the Egyptian, and hid his body in the sand" (Exodus 2:11–12). This led him to losing credibility amongst people. The following day, he attempted to break up a fight between two of his fellow Hebrews. They called him out for this hypocritical behavior. After this, he realized the error of his ways. He later pointed out that it is God's role to avenge wrong doers, not ours. "Vengeance is Mine, and retribution; In due time their foot will slip. For the day of their disaster is near, And the impending things are hurrying to them" (Deuteronomy 32:35). In taking personal vengeance, we place our own desires before the Lord's. King Saul also committed this crucial error, to his detriment. This is expounded upon in a later chapter. However, Gideon failed to realize this mistake with the men of Sukkoth and Peniel. His personal vengeance was swift and brutal!

After this, Gideon executed the Midianite leaders Zebah and Zalmunna. The war was over now. Their deaths catalyzed an era of peace for the land. After the dust settled, the Israelites begged Gideon to be their king. "Then the men of Israel said to Gideon, 'Rule over us, both you and your son, your son's son as well, for you have saved us from the hand of Midian!'" (Judges 8:22). Gideon remained true to his faith and denied their request. "But Gideon said to them, 'I will not rule over you, nor shall my son rule over you; the Lord shall rule over you'" (Judges 8:23). Gideon had created good times for the Israelites. Soon they would forget about the Lord and lose their way yet again.

After denying the request to be their king, Gideon posed a bizarre request. He asked that everyone give him one golden earring from their share of the plunder (Judges 8:24). The Israelites gladly obliged. They spread out a cloth and the gold jewelry quickly piled up. The mass of the earrings was 1,700 shekels (approx. 43 pounds). With the plunder, Gideon cast an ephod, a type of robe, which symbolized his victory. This golden ephod served as a type of trophy for the community. In peaceful times, the Israelites turned to worshiping this golden ephod. Just as Moses had warned about in Deuteronomy 7:16, the golden ephod had captured the Israelites like a snare (Judges 8:27). They worshiped the *victory*, not the Lord who led them to it. The Israelites began the cycle of corruption over again.

GIDEON CONCLUSION

There are many lessons to be learned from Gideon's story. Once again, we see that God often chooses unlikely heroes to accomplish His purposes. Gideon began as cowardly and weak. He came from a small, insignificant clan and was the least in his family. In spite of this, through God's guidance, he assembled a sizable army—only to have God instruct him to whittle it down to a mere 300 men (Judges 7:2). God was teaching a critical lesson: it was not their strength that would bring victory, but their faith.

Gideon built his courage one step at a time. Each small act of obedience strengthened his faith, allowing him to press forward despite his fears. As Hebrews reminds us: "The Lord is my helper, I will not be afraid. What will man do to me?" (Hebrews 13:6). True courage is not the absence of fear but acting righteously in spite of it. Like Gideon, we too can grow our courage by starting small—trusting God, even when the path is unclear. Cultivating courage requires *action*, not just wishful thinking!

We also learned that modest forces, when applied with boldness and cunning, can overcome seemingly impossible odds. Gideon's army was vastly outnumbered, yet he leveraged its strengths, exploited the enemy's weaknesses, and used unconventional tactics. He did not waste time wishing for better tools. He used what he had with faith and ingenuity. Akin to David slaying Goliath, he overcame overwhelming odds with nothing more than modest tools and faith. His transformation was not instantaneous—it was forged through God's patience and Gideon's willingness to embrace faith over fear. From a doubting man into a bold warrior, Gideon's journey shows us that victory comes not by might, nor by status, but by walking faithfully with God.

Yet Gideon's story also issues a warning: winning the battle is not the end of the fight. After the victory, Israel quickly slipped back into idolatry, worshiping the golden ephod Gideon had made. Success carries hidden dangers—pride, complacency, and the temptation to forget the source of all blessings. Therefore, we must remember the following: In times of fear, trust in God's strength. In times of success, guard against pride. In times of peace, remain vigilant. Each era has its own challenges. We must adapt our strategies accordingly.

As *The Art of War* states, "no military leader can ever become great without a dash of boldness in him" (Tzu 3). But boldness alone is not enough. It must be anchored in humility, discipline, and enduring faith. True victory is not merely winning a battle. It is remaining faithful long after the dust has settled.

ABIMELECH – THE ANTI-JUDGE: JUDGES 9

The vicious cycle returned once again. The hard times had forged Gideon into a strong man who defended Israel with valor. Yet after Gideon's forty-year reign of peace and prosperity, the Israelites lost their way. "Then it came about, as soon as Gideon was dead, that the sons of Israel again committed infidelity with the Baals, and made Baal-berith their god" (Judges 8:33). They abandoned the Lord who had rescued them. The moment Gideon was gone, the people turned their backs on all the great things their leader had accomplished with God's help.

As history often shows, good times create weak men. Abimelech, one of Gideon's sons born to a concubine in Shechem, stood as a prime example (Judges 8:31). While polygamy is not expressly forbidden in the Old Testament, it always leads to serious and dramatic problems.

Abimelech's story is a cautionary tale of unchecked wrath and greed. His lust for power eclipsed the love of his dozens of half-brothers. The moment Gideon died, Abimelech scrambled for power. He turned swiftly to the political arena in a city already plagued with idolatry and demon worship.

Abimelech stirred the people of Shechem with an appeal to blood loyalty: "Which is better for you: for seventy men, all the sons of Jerubbaal, to rule over you, or for one man to rule over you? Also, remember that I am your bone and your flesh" (Judges 9:2). Jerub-Baal, Gideon's nickname after tearing down Baal's altar, now became a pawn in Abimelech's ambition. In pursuing his crown, Abimelech betrayed the Lord, dishonored his father, and violated the Fifth Commandment (Exodus 20:12). His disrespect toward Mosaic law was a grim foreshadowing of horrors to come.

Being half-Canaanite, Abimelech found fertile ground among the Shechemites. They gave him seventy shekels of silver from the temple of Baal-Berith. With this dirty money, Abimelech hired a band of reckless mercenaries (Judges 9:4). His actions fulfilled the warning of Proverbs 10:2: "Ill-gotten gains do not benefit, But righteousness rescues from death." Yet instead of righteousness, Abimelech chose a path of blood and betrayal. Some people only learn lessons the hard way. The treasure he sought would soon slip from his blood-soaked hands.

Leading his gang of thugs, Abimelech returned to his father's home in Ophrah and slaughtered his seventy brothers "on one stone" (Judges 9:5). The symbolism is chilling. Rather than offering sacrifices to the Lord on stone altars, Abimelech perverted the holy act—offering human blood to his lust for power. He embodied what Jesus warned about. "The thief comes only to steal and kill and destroy; I came so that they would have life, and have it abundantly" (John 10:10). Abimelech fit the description perfectly! This is a classic example of Satanic inversion. Satan cannot create, he can only pervert and corrupt what the good Lord has made.

Gideon's youngest son, Jotham, escaped the carnage. Refusing to cower, he climbed Mount Gerizim and issued a dire warning to the citizens of Shechem. Jotham reminded them of Gideon's sacrifices and compared their new "king" to a thornbush—a worthless, dangerous plant offering no shade or fruit. Thorns draw blood rather than sustain life. He prophesied that fire would consume both Abimelech and the Shechemites (Judges 9:16–20). Then he fled for his life.

Meanwhile, Abimelech's unstable rule began to crumble. His mercenaries, loyal only to silver, soon turned to robbing travelers (Judges 9:25). There is no honor among thieves. As Machiavelli warned in *The Prince*: "Mercenaries and auxiliaries are useless and dangerous; they are disunited, ambitious, and without discipline, unfaithful, valiant before friends, cowardly before enemies" (Machiavelli 66). Predictably, rebellion brewed. A man named Gaal son of Ebed moved into Shechem and quickly incited the people against Abimelech. Emboldened by liquid courage, the mood turned resentful. The citizens, ever fickle, sided with the newcomer and plotted revolt (Judges 9:26–29).

When Abimelech heard of the plot through the city's governor, Zebul, he responded swiftly. Zebul advised an ambush at dawn—catching the rebels off guard, following a timeless principle of warfare: "One of the most important principles of offensive war is to surprise the enemy with a fast attack" (Clausewitz 8). Sun Tzu echoes this sentiment in *The Art of War*. "Attack him where he is unprepared, appear where you are not expected" (Tzu 23). The ambush succeeded. Many of Gaal's followers were slain as they fled toward the city gates. Though Gaal himself escaped, Zebul promptly drove him and his clan from Shechem (Judges 9:39–41) As the Apostle Paul later wrote: "Do not be deceived: 'Bad company corrupts good morals'" (1 Corinthians 15:33). Gaal's drunken boastfulness led many to ruin.

Yet Abimelech's thirst for blood remained unsatisfied. The next day, as the workers of Shechem went into the fields, he launched another ambush. His men cut down the workers and captured the city (Judges 9:42–44). All day long, Abimelech and his mercenaries ransacked Shechem, slaughtering its inhabitants without mercy. Survivors fled into the stronghold of the temple of Baal-Berith. But Abimelech cornered them there. He cut branches from nearby trees and ordered his men to do the same. They piled the wood around the stronghold and set it ablaze. Over a thousand men and women perished in the flames (Judges 9:45–49).

Strategically, it was effective. As Sun Tzu wrote: "If it is possible to make an assault with fire from without, do not wait for it to break out within, but deliver your attack at a favorable moment" (Tzu 26). Symbolically, it was chilling. Abimelech had earlier shed his brothers' blood on one stone—a cruel parody of Levitical sacrifice. Now he burned an entire community alive, eerily resembling the Canaanite practice of human sacrifice via fire to Molech (sometimes spelled Molek or Moloch), which Moses had sternly forbidden (Deuteronomy 12:31). Given Abimelech's Canaanite upbringing, this may have been more than just sacking of the city. We will examine this demonic practice in greater detail in the next chapter.

Still thirsting for domination, Abimelech marched on Thebez (Judges 9:50–51). Yet this time, the defenders were ready. They fled to a fortified tower and refused to be taken by surprise. In doing so, they lived out Sun Tzu's principle: "Rely not on the enemy's not coming, but on our own readiness to receive him" (Tzu 16). Abimelech, reckless and enraged, made a fatal error. He advanced uphill against a fortified enemy—directly violating another crucial law of warfare: "not to advance uphill against the enemy" (Tzu 15). Worse still, he foolishly repeated the same tactic he had used at Shechem. As Sun Tzu warned: "Do not repeat the tactics which have gained you one victory" (Tzu 12). Yet embracing sin clouds sound judgement and reason.

As Abimelech approached the tower to set it ablaze, an anonymous woman dropped a millstone from the roof. The improvised weapon sailed true to its target. The heavy stone struck him on the head, shattering his skull. In desperation, Abimelech ordered his armor-bearer to kill him—to avoid the disgrace of dying at a woman's hand. However, the legend of his embarrassing death was known well into the future (2 Samuel 11:21). His armor-bearer complied with the morbid request, thus ending Abimelech's bloody reign (Judges 9:53–54). His humiliating death fulfilled the proverb: "Pride goes before destruction, And a haughty spirit before stumbling" (Proverbs 16:18).

ABIMELECH: CONCLUSION

In the end, Abimelech received the fate he deserved. Jotham's warning proved true. Like a thornbush, Abimelech brought only destruction. Perhaps King David reflected on this story when he wrote: "But the worthless, every one of them, are like scattered thorns, Because they cannot be taken in hand; Instead, the man who touches them Must be armed with iron and the shaft of a spear, And they will be completely burned with fire in their place" (2 Samuel 23:6–7). Good shepherds must use violence against bloodthirsty wolves. It's the only way to protect the flock.

There is a certain poetic justice to it all. Abimelech killed his brothers on one stone—and in the end, a single stone crushed him. The weapon that ended him—a humble millstone—was a domestic tool, much like Jael's tent peg, which felled Sisera (Judges 4:21). Once again, God used unlikely instruments and unexpected heroes to bring justice. With God's favor, even modest tools and a faithful spirit can achieve mighty victories.

All that is required is the courage to act.

JEPHTHAH – THE DANGER OF IMPULSIVITY: JUDGES 10–12

Decades passed after the heroism of Gideon and the atrocities by his son, Abimelech. Israel continued to decline morally. During this low point a new judge emerged, Jephthah. Like most judges, he came from an unlikely background. Before his arrival, Israel engaged in apostasy once again. "Then the sons of Israel again did evil in the sight of the Lord, and they served the Baals and the Ashtaroth, the gods of Aram, the gods of Sidon, the gods of Moab, the gods of the sons of Ammon, and the gods of the Philistines; so they abandoned the Lord and did not serve Him" (Judges 10:6). The apostasy cycle churned yet again.

One of these stands out in particular: the Ammonites. Their religious practices are relevant to this story and will be examined in detail. Indeed, Joshua's prophetic warning came true. "If you abandon the Lord and serve foreign gods, then He will turn and do you harm and destroy you after He has done good to you" (Joshua 24:20). Since Israel abandoned the Lord, He did not protect it. "And the anger of the Lord burned against Israel, and He sold them into the hands of the Philistines, and into the hands of the sons of Ammon" (Judges 10:7). The coalition of Ammonites and Philistines held an iron grip east of the Jordan river. Gilead was their focus. This brutal oppression lasted nearly two decades! (Judges 10:8).

Crushed by suffering and despair, Israel entered the next phase of the cycle and cried out to the Lord. "Then the sons of Israel cried out to the Lord, saying, 'We have sinned against You, for indeed, we have abandoned our God and served the Baals'" (Judges 10:10). They confessed their wrongdoing, but this time, God did not answer with immediate mercy. He reminded them of the seven nations He had already delivered them from (Judges 10:11–12). Then came His sharp rebuke: "Go and cry out to the gods which you have chosen; let them save you in the time of your distress" (Judges 10:14). This was a painful but just response. There is a reason "You shall have no other gods before Me" is the First Commandment (Exodus 20:3).

The Israelites acknowledged their guilt and pleaded once more. "Then the sons of Israel said to the Lord, 'We have sinned, do to us whatever seems good to You; only please save us this day'" (Judges 10:15). In an echo of Joshua's response to the sin of Achan, they acted. They removed the corruption from among them, destroyed their foreign gods, and served the Lord again. Only then did God relent. He could no longer bear to watch His people suffer. This moment serves as a critical reminder: *repentance without action is nothing more than wishful thinking*. When the heart is sincere, and obedience follows, the Lord responds with mercy. Despite Israel's repeated failures, He offered them yet another path to redemption.

Israel now prepared for war in Gilead. It assembled at Mizpah, which translates to *watch tower*, an appropriate name for overseeing military operations. Recall, "The highest ground is reckoned the best" (Vegetius 70). They knew they must fight back, but there was a glaring issue. Like a ship without a captain, Gilead was rudderless. They had people, but no true warriors. No one to lead. The leaders at Mizpah issued a feeble call to action. "And the people, the leaders of Gilead, said to one another, 'Who is the man who will begin to fight against the sons of Ammon? He shall become head over all the inhabitants of Gilead'" (Judges 10:18). This is when our unlikely hero emerges from the fray.

Jephthah was a local to the region. A mighty warrior, but also a social pariah (Judges 11:1–3). His upbringing was harsh. His father rejected him because his mother was a prostitute. The other sons, Jephthah's half-brothers, bullied him and drove him out. They made their motivation clear. "Gilead's wife bore him sons; and when his wife's sons grew up, they drove Jephthah out and said to him, 'You shall not have an inheritance in our father's house, for you are the son of another woman'" (Judges 11:2). This illustrates the tragic fallout of polygamy and broken homes. Like father, like sons; Jephthah's father stood by in silence as cruelty reigned. Cast aside like garbage, Jephthah fled to the land of Tob. In this new land, his leadership stood out. He attracted a gang of hardened men who followed him loyally (Judges 11:3).

Whether these men were mercenaries or outcasts remains unclear. But one thing is certain: Jephthah's leadership presence set him apart. Despite being the son of a prostitute, he commanded genuine authority. To lead a band of ruffians, and keep their loyalty, required a forceful, commanding presence. As tensions with the Ammonites boiled over, Gilead came looking for a savior. It approached Jephthah with a desperate offer. "Come and be our leader, that we may fight against the sons of Ammon" (Judges 11:6).

The irony was not lost on Jephthah. Gilead had rejected him and driven him out. Now it came pleading for deliverance. Could it really be true? The only man capable of rescuing them was the very man they had rejected. It echoes how the Pharisees rejected Christ, the only one who could save them. Jephthah answered with rightful skepticism: "Did you not hate me and drive me from my father's house? So why have you come to me now when you are in trouble?" (Judges 11:7). The elders replied, "For this reason we have now returned to you, that you may go with us and fight the sons of Ammon, and become our head over all the inhabitants of Gilead" (Judges 11:8). Still unconvinced, Jephthah pressed for confirmation: "If you bring me back to fight against the sons of Ammon and the Lord gives them up to me, will I become your head?" (Judges 11:9). This marked the first time his leadership was tied directly to divine deliverance. The elders affirmed both the holy mission and their sincerity. "And the elders of Gilead said to Jephthah, 'The Lord is witness between us; be assured we will do as you have said'" (Judges 11:10). With the agreement sealed, Jephthah was ratified as commander of Gilead. And here, he reaffirmed his commitment to the Lord.

Jephthah immediately got to work. Admirably, he sought to de-escalate the conflict through diplomacy. If this leader of scoundrels could set aside his ego for peace, so can we. He took the initiative to open communication with the king of Ammon. His inquiry was direct: why had the Ammonites attacked? (Judges 11:12). The king responded swiftly: "It is because Israel took my land when they came up from

Egypt, from the Arnon as far as the Jabbok and the Jordan; so return them peaceably now" (Judges 11:13).

This conflict had occurred roughly three hundred years earlier, during the time of Moses. Hardly ground for a fresh war. Still, Jephthah pursued peace. He reminded the king that Israel had requested safe passage through Edom and Moab but was denied and forced to pass through the wilderness instead (Judges 11:14–18). Israel later asked for the same from the Amorites, but King Sihon did not answer with a pen, he chose the sword. The Israelites defended themselves valiantly. The Book of Numbers recounts the event: "Then Israel struck him with the edge of the sword, and took possession of his land from the Arnon to the Jabbok, as far as the sons of Ammon; for the border of the sons of Ammon was Jazer" (Numbers 21:24). In other words, the Ammonites had *not* lost any land. Their claim rang hollow.

Jephthah then made a theological argument. The Lord had given Israel that land, and if the situation were reversed, the Ammonites would have done the same (Judges 11:23–24). Finally, Jephthah offered a simple, unassailable point: Israel had lived in that land for three centuries (Judges 11:26). Why raise the issue now? The answer was obvious. Simple opportunism and greed. Jephthah reaffirmed Israel's innocence: "So I have not sinned against you, but you are doing me wrong by making war against me. May the Lord, the Judge, judge today between the sons of Israel and the sons of Ammon" (Judges 11:27).

Jephthah's letter marked the end of polite negotiations. "But the king of the sons of Ammon disregarded the message which Jephthah sent him" (Judges 11:28). When de-escalation fails, action becomes necessary. When you are forced to act, it's best to act first with overwhelming force. There was no room left for compromise. The king of Ammon didn't provoke the pushovers of Gilead; he challenged a hardened warrior. Jephthah answered the call resolutely. The Spirit of the Lord came upon him, and he took the fight to the Ammonites (Judges 11:29).

Jephthah understood that the Lord's blessing was essential. Yet before the battle, he made a grievous mistake. In an effort to secure victory, he attempted to bargain with God. His vow was reckless: "If You will indeed hand over to me the sons of Ammon, then whatever comes out the doors of my house to meet me when I return safely from the sons of Ammon, it shall be the Lord's, and I will offer it up as a burnt offering" (Judges 11:30–31). It was a fateful misstep; one that would exact a terrible price.

The battle was one-sided. The Lord delivered the Ammonites into Jephthah's hands. His forces swept through twenty towns with ease, and Israel was finally free (Judges 11:32–33). After nearly two decades of oppression, the nation had peace. At first glance, this seems like a classic underdog victory. Jephthah's cold upbringing forged him into a warrior. Rejected by family and society, he walked a lonely path. Yet through hardship, he prevailed. He trusted the Lord, and that faith led to overwhelming triumph. The same people who cast him out now exalted him as their commander. All of that is true, but this story is no celebration. It ends in tragedy, born of a critical misunderstanding of God's will.

Jephthah had already been chosen for a holy mission. He knew the importance of honoring the Lord. But in a misguided attempt to guarantee success, he made a vow he never should have uttered. And when he returned home in triumph, that vow came back to haunt him. His daughter, his only child, came out to greet him.

The horror set in immediately. He had placed himself in a lose–lose scenario. The contrast could not have been more brutal. The army returned in glory. The townspeople rejoiced with music and dancing (Judges 11:34). But Jephthah was utterly devastated. "So when he saw her, he tore his clothes and said, 'Oh, my daughter! You have brought me disaster, and you are among those who trouble me; for I have given my word to the Lord, and I cannot take it back'" (Judges 11:35).

Jephthah placed himself in a terrible dilemma. Should he break his vow to the Lord and spare his daughter? Or honor his reckless promise and lose her? Mosaic Law takes vows seriously: "If a man makes a vow to the Lord, or takes an oath to put himself under a binding obligation, he shall not break his word; he shall act in accordance with everything that comes out of his mouth" (Numbers 30:2). Yet that same Law sternly forbids child sacrifice: "There shall not be found among you anyone who makes his son or his daughter pass through the fire" (Deuteronomy 18:10). Either path brought devastating consequences. Jephthah was trapped! Not by fate, but by his own misunderstanding of God's will.

His daughter responded with stunning resolve. "So she said to him, 'My father, you have given your word to the Lord; do to me just as you have said, since the Lord has brought you vengeance on your enemies, the sons of Ammon'" (Judges 11:36). She made one humble request: to roam the hills and grieve with her friends (Judges 11:37). Jephthah granted her wish. She went to mourn the future she would never have. Upon her return, he fulfilled his vow and sacrificed his only child (Judges 11:38).

This morbid story demands reflection. How could such a thing happen? Why would Jephthah make such a vow? And why would his daughter accept it? The answers lie in the cultural and spiritual climate of the time.

This tragedy was the bitter fruit of Israel's apostasy. Having turned from the Lord, the nation came under the yoke of Ammonite and Philistine rule (Judges 10:6–8). During that time, the demonic practices of these nations seeped into Israel's culture. Molech, the chief god of Ammon, demanded child sacrifice (1 Kings 11:7). Moses had already condemned such worship in the strongest possible terms: "You shall not behave this way toward the Lord your God, because every abominable act which the Lord hates, they have done for their gods; for they even burn their sons and daughters in the fire for their gods" (Deuteronomy 12:31).

Joshua also warned of this spiritual drift: "For if you ever go back and cling to the rest of these nations, these which remain with you, and intermarry with them, so that you associate with them and they with you, know with certainty that the Lord your God will not continue to drive these nations out from before you; but they will be a snare and a trap to you, and a whip on your sides and thorns in your eyes, until you perish from this good land which the Lord your God has given you" (Joshua 23:12–13). And so it happened. Israel had forsaken the Lord and embraced the ways of the world. Jephthah, caught in this divided culture, was a man split between two kingdoms. He knew the sovereignty of God. He invoked the Lord's name repeatedly before both the elders of Gilead and the king of Ammon (Judges 11). Yet he also carried the imprint of a culture steeped in confusion, compromise, and bloodshed.

In spite of his reverence for the Lord, Jephthah misunderstood God's will. He made an impulsive vow, one that reflected the practices of surrounding cultures more than the heart of the Lord. Jephthah had confused the dominant religious customs of his time with divine expectation. The Book of Judges reveals some of the darkest chapters in Israel's history, and Jephthah was shaped by that climate. His exile from Gilead likely meant he received less religious instruction than most. Though he was familiar with the Lord, he lacked grounding in His commands. The result was devastating.

The Apostle Paul offers guidance for this very kind of spiritual confusion. He wrote: "Therefore I urge you, brothers and sisters, by the mercies of God, to present your bodies as a living and holy sacrifice, acceptable to God, which is your spiritual service of worship. And do not be conformed to this world, but be transformed by the renewing of your mind, so that you may prove what the will of God is, that which is good and acceptable and perfect" (Romans 12:1–2). Unknowingly, Jephthah had conformed to the demonic patterns of the culture.

Christ also spoke directly to the problem of divided loyalty. He warned: "No one can serve two masters; for either he will hate the one and love the other, or he will be devoted to one and despise the other. You cannot serve God and wealth" (Matthew 6:24). While that verse specifically addresses the worship of money (Mammon), the principle applies broadly: God's will and worldly influence cannot coexist in the same heart.

Jesus also warned against making hasty vows. He said: "Again, you have heard that the ancients were told, 'You shall not make false vows, but shall fulfill your vows to the Lord.' But I say to you, take no oath at all, neither by heaven, for it is the throne of God, nor by the earth, for it is the footstool of His feet, nor by Jerusalem, for it is the city of the great King. Nor shall you take an oath by your head, for you cannot make a single hair white or black. But make sure your statement is, 'Yes, yes' or 'No, no'; anything beyond these is of evil origin" (Matthew 5:33–37). We have the gift of hindsight and history. Jephthah, however, did not. His vow was not a sign of faith. Rather, it was a tragic misstep rooted in fear, pride, and poor theology.

Even Sun Tzu, a secular strategist, warned of this kind of recklessness: "Now the general who wins a battle makes many calculations in his temple before the battle is fought. The general who loses a battle makes but few calculations beforehand. Thus do many calculations lead to victory, and few calculations to defeat: how much more no calculation at all!" (Tzu 4). Perhaps Jephthah's upbringing, shaped by rejection and hardened among scoundrels, fostered a habit of impulsive thinking. That mindset may serve in chaos, but not when entrusted with the fate of a nation. Among warriors, impulse without wisdom leads to ruin.

There was no time to grieve; another conflict awaited Jephthah. The dust had barely settled from the battle with the Ammonites. Recall that the Ammonites had also attacked Ephraim earlier in the campaign (Judges 10:9). Yet rather than offer thanks or unity, the Ephraimites confronted Jephthah with open hostility: "Now the men of Ephraim were summoned, and they crossed to Zaphon; and they said to Jephthah, 'Why did you cross over to fight against the sons of Ammon without calling us to go with you? We will burn your house down on you!'" (Judges 12:1). Instead of praise for defeating their common enemy, Jephthah only faced more hostility.

Such aggressiveness was completely unwarranted. Jephthah responded truthfully and again attempted diplomacy. "So Jephthah said to them, 'I and my people were in a major dispute with the sons of Ammon; and I did call you, but you did not save me from their hand. When I saw that you were no deliverer, I took my life in my hands and crossed over against the sons of Ammon, and the Lord handed them over to me. Why then have you come up to me this day to fight against me?'" (Judges 12:2–3). Under the circumstances, Ephraim's accusations ring hollow. Their actions appear driven by base motives, pride, entitlement, or a desire for a share in the spoils. Perhaps they felt insulted, perceiving Jephthah's success as a slight against their standing. This was not the first time Ephraim had behaved this way. A similar confrontation occurred after Gideon's victory in Judges 8.

Sadly, Ephraim's conduct further reflects the spiritual and moral decay of Israel during the time of the Judges. What should have been a moment of unity became a flashpoint for civil unrest. Unlike Gideon, Jephthah could not pacify the Ephraimites. His efforts to de-escalate were rejected. Instead of seeking peace, Ephraim doubled down with insults: "Then Jephthah gathered all the men of Gilead and fought

Ephraim; and the men of Gilead defeated Ephraim, because they said, 'You are survivors of Ephraim, you Gileadites, in the midst of Ephraim and in the midst of Manasseh'" (Judges 12:4). Internal division in Israel was more than just political—it was spiritually dangerous. Christ warned, "Every kingdom divided against itself is laid waste; and no city or house divided against itself will stand" (Matthew 12:25).

De Re Militari offers practical wisdom for this leadership dilemma. As a rule, a commander should rely more on discipline and structure than coercion: "It is better to form his troops to submission and obedience by habit and discipline than to be obliged to force them to their duty by the terror of punishment" (Vegetius 52). But Ephraim had rejected discipline. The time for gentleness had passed. *De Re Militari* continues: "But if the height of the mutiny requires violent remedies ... punish the ring-leaders only in order that, though few suffer, all may be terrified by the example" (Vegetius 52). Jephthah followed this principle. War broke out between Gilead and Ephraim, and Jephthah took the lead.

His strategic thinking proved decisive. He quickly secured the fords of the Jordan—a natural choke point and vital crossing (Judges 12:5). Rivers are formidable defensive lines when used wisely. Clausewitz advised, "we should try to observe the enemy, and once the enemy starts to cross, attack it on all sides while its full forces have not arrived yet, and it is limited by the narrow perimeter by the river" (Clausewitz 22). *De Re Militari* echoes this: "Form ambuscades with the greatest secrecy to surprise the enemy at the passages of rivers" (Vegetius 66). The Ephraimites made the same fundamental error as the Ammonites; they underestimated their adversary.

Jephthah's tactical proficiency at this stage was exceptional. But he was no madman on a rampage; his aggression was tightly focused on the mutinous Ephraimites. This is where his quick thinking came into play. He devised a clever ruse to distinguish ally from enemy. At the checkpoints, Gileadean soldiers asked travelers if they were Ephraimites. If they denied it, they were given a simple test: "Just say, 'Shibboleth.' But he said, 'Sibboleth,' for he was not prepared to pronounce it correctly. Then they seized him and slaughtered him at the crossing places of the Jordan. So at that time forty-two thousand from Ephraim fell" (Judges 12:6). Their dialect betrayed them. What seemed like a minor speech difference became a decisive tactical filter. Jephthah turned linguistic nuance into military advantage, maintaining control and preventing full-blown civil war.

Ultimately, this unnecessary conflict resulted in the deaths of 42,000 Ephraimites. It was a tragedy. But Jephthah's resolute leadership preserved broader national stability. By punishing Ephraim, he spared the other eleven tribes from further descent into chaos. Without such decisive measures, the rebellion could have escalated into widespread civil war. And the more Israel fought internally, the more vulnerable it became to external threats. Disunity weakens defenses. Israel's enemies were always watching.

JEPHTHAH CONCLUSION

Jephthah's story is one of great sorrow and tragedy. His life was bittersweet, marked by incredible heights and devastating lows. Once an outcast, he rose to leadership among scoundrels and outlaws. The very people who had driven him away later begged for his return. Overnight, he went from untouchable to uncontested commander of the region. He delivered Israel from destruction, but at the overwhelming cost of his only child. Later, he rescued the nation again, though the price was steep: 42,000 Ephraimites died to preserve unity. At first glance, this account may seem bizarre. Yet it carries profound lessons if we are willing to look closely.

One of the most important lessons is that repentance requires genuine change. Israel began this story by serving foreign gods. After years of oppression, it cried out to God, but words alone were not enough. At first, the Lord told it to seek help from its idols. Only when Israel acted and removed its false gods did the Lord respond. HE couldn't bear its suffering any longer (Judges 10:16). Real repentance is visible. It is not what we say, but what we do.

Another lesson is the danger of misinterpreting God's will. Jephthah clearly revered the Lord. He invoked God's sovereignty before both Israel's elders and Ammon's king. Yet he also absorbed the toxic practices of his time. His impulsive vow, born of zeal rather than wisdom, cost him his only daughter. His exile likely left him under-instructed in God's law. A fish is unaware that it's submerged in water until it is washed ashore or snatched by a bird from above. In this same way, Jephthah did not realize how deeply the world had shaped him. The prudent reader must ask: are we conflating any of today's cultural practices with God's will? Jephthah's mistake stands as a solemn warning. If you are unsure, return to Scripture for clarity.

Jephthah also reminds us to pursue peace before war. He tried to de-escalate with both the king of Ammon and the leaders of Ephraim. If a leader of outcasts could restrain his ego for diplomacy, so can we. But peace only works when both parties desire it. We must still be ready to defend ourselves and those under our care. History shows that capable people prevent disaster, while helpless people are left to endure it. As Teddy Roosevelt wisely said, "Speak softly and carry a big stick; you will go far."

SAMSON – THE LORD'S SUPER SOLDIER: JUDGES 13–16

The account of Samson mirrors the broader structure of Israel's history during the time of Judges. His personal life traces the same four-part cycle that defined the nation: apostasy, oppression, crisis, and deliverance. In this way, Samson serves as a microcosm of Israel itself. Though God blessed him with supernatural strength, Samson's moral resolve remained painfully human.

Samson is unique among the judges. His story is spectacular, ranging from unmatched physical feats to moments of arrogance and self-destruction. Without question, the Lord was the source of his unparalleled power. Like Ehud, Samson often acted alone and relied on unconventional tactics. But whereas Ehud resembled a stealthy assassin, Samson was more akin to Hercules or Superman, a force of nature, feared and admired. Yet like all such figures, he had a fatal weakness. Samson's personal flaws reflected the same unfaithfulness and impulsiveness that plagued the nation of Israel.

Samson was set apart for a special purpose. Though he accomplished great things in God's plan, his life is not one to imitate. The Lord gifted him with extraordinary strength and paired it with a sharp mind for tactics and strategy. He was a one-man wrecking force against the Philistines; a divine weapon forged for conflict. But his victories came in *spite* of his character, not because of it. His hubris and lack of discipline eventually led to his downfall.

Samson's divine calling began with a miraculous birth. The cycle of judges entered another dark chapter. "Now the sons of Israel again did evil in the sight of the Lord, and the Lord handed them over to the Philistines for forty years" (Judges 13:1). In the midst of this oppression, an angel of the Lord delivered an astounding message to an Israelite woman. "Behold now, you are infertile and have not given birth; but you will conceive and give birth to a son" (Judges 13:3). She was instructed to avoid fermented drinks and unclean food. Then came a prophetic announcement: "you will conceive and give birth to a son, and no razor shall come upon his head, for the boy shall be a Nazirite to God from the womb; and he will begin to save Israel from the hands of the Philistines" (Judges 13:5).

The word *Nazirite* literally means "set apart" or "dedicated," highlighting Samson's divine purpose. This wasn't a new concept. The Lord had given Moses instructions for Nazirite vows long before, detailed in Numbers 6.

After the angel's first visit, Samson's mother told her husband Manoah about the encounter. Manoah prayed to the Lord for clarity. "Lord, please let the man of God whom You have sent come to us again so that he may teach us what we are to do for the boy who is to be born" (Judges 13:8). God answered his prayer, and the angel appeared again—this time to Samson's mother while she was alone in a field. She quickly ran to summon her husband.

When Manoah arrived, he questioned the angel about the prophecy. "Then Manoah said, 'Now when your words are fulfilled, what shall be the boy's way of life and his vocation?'" (Judges 13:12). The angel reaffirmed the boy's Nazirite calling. Still unaware of the messenger's divine nature, Manoah offered him a meal. The reply was telling: "Though you detain me, I will not eat your food, but if you prepare a burnt offering, offer it to the Lord" (Judges 13:16).

Hoping to honor the mysterious visitor, Manoah asked for his name. The angel's response perplexed the couple even further. Cryptically, the divine messenger responded: "Why do you ask my name, for it is wonderful?" (Judges 13:18). The full glory of God's magnificence is simply beyond human understanding. Even the name of a heavenly messenger was too much for human minds to grasp.

Manoah obeyed and prepared a burnt offering to the Lord. Then something extraordinary occurred. "For it came about when the flame went up from the altar toward heaven, that the angel of the Lord ascended in the flame of the altar. When Manoah and his wife saw this, they fell on their faces to the ground" (Judges 13:20). Only then did they realize they had been in the presence of a divine being. Manoah panicked, fearing they would die for having seen the Lord. But his wife responded with calm wisdom: "If the Lord had desired to kill us, He would not have accepted a burnt offering and a grain offering from our hands, nor would He have shown us all these things, nor would He have let us hear things like this at this time" (Judges 13:23).

Some time later, Samson was born. As he grew, the Holy Spirit began to stir within him (Judges 13:24–25). The circumstances of Samson's birth are rich with symbolism. His origin reflects Israel's own. The Nazirite vow was more than personal; it echoed the national calling of Israel to be set apart for God's purposes. God called Abraham out of his father's household and promised him a new land for future generations. "And I will make you into a great nation, And I will bless you, And make your name great; And you shall be a blessing; And I will bless those who bless you, And the one who curses you I will curse. And in you all the families of the earth will be blessed" (Genesis 12:2–3).

That promise was reiterated at Sinai. "'You yourselves have seen what I did to the Egyptians, and how I carried you on eagles' wings, and brought you to Myself. Now then, if you will indeed obey My voice and keep My covenant, then you shall be My own possession among all the peoples, for all the earth is Mine; and you shall be to Me a kingdom of priests and a holy nation.' These are the words that you shall speak to the sons of Israel" (Exodus 19:4–6).

Centuries later, Jesus Christ fulfilled this covenant in its ultimate form. He made it clear that His followers were chosen by God and entrusted with a divine mission. "You did not choose Me but I chose you, and appointed you that you would go and bear fruit, and that your fruit would remain, so that whatever you ask of the Father in My name He may give to you" (John 15:16). Christ reiterated this truth while confronting the Pharisees, who had rejected Him. "Therefore I say to you, the kingdom of God will be taken away from you and given to a people producing its fruit" (Matthew 21:43).

In the meantime, however, we return to Samson's era, a time long before Christ fulfilled the covenant. Israel still bore the responsibility of walking faithfully in the role God had already given it. Now it was up to Israel, and to Samson, to walk faithfully in that calling.

There are more parallels between Samson's birth and the early days of Israel. Samson's mother was barren, yet the Lord sent an angel to foretell his birth. God did something similar with Sarah, Abraham's wife. "I will bless her, and indeed I will give you a son by her. Then I will bless her, and she shall be a mother of nations; kings of peoples will come from her" (Genesis 17:16).

Later, God sent three angels to reinforce the point. He spoke through them, saying, "'I will certainly return to you at this time next year; and behold, your wife Sarah will have a son.' And Sarah was listening at the tent door, which was behind him" (Genesis 18:10). At this point, Sarah was very old. She chuckled to herself at the idea of giving birth. The Lord responded with firm clarity. "But the Lord said to Abraham, 'Why did Sarah laugh, saying, "Shall I actually give birth to a child, when I am so old?"' Is anything too difficult for the Lord? At the appointed time I will return to you, at this time next year, and Sarah will have a son" (Genesis 18:13–14). The barrenness of Sarah and Samson's mother represents a human limitation that only the Lord can overcome. These miraculous births demonstrate God's grace and faithfulness to His promises.

Another striking parallel appears in the divine fire that followed the offering given by Samson's parents. Witnessing it, they collapsed in awe. This echoes an earlier scene in Israel's history. In Leviticus 9, Moses oversaw the first priestly sacrifices offered by Aaron and his sons. After Aaron blessed the people and presented the offering, something miraculous occurred. "And Moses and Aaron went into the tent of meeting. When they came out and blessed the people, the glory of the Lord appeared to all the people. Then fire went out from the Lord and consumed the burnt offering and the portions of fat on the altar; and when all the people saw it, they shouted and fell face downward" (Leviticus 9:23–24).

These events hold deep meaning. In both cases, the pattern is the same: God calls, His people obey, and their sacrifice is accepted. Fire affirms the covenant. It is a symbol of God's purity, power, and presence. From that moment forward, both Israel and Samson's household entered a new era.

Chapter 13 of Judges establishes the foundation of Samson's divine calling. Chapters 14 through 16 show how that calling played out. They are a living example of Israel's cycle in the Book of Judges. Samson had been set apart, but like Israel, he was not always faithful. Samson soon wandered into rebellion. He traveled to Timnah and became infatuated with a Philistine woman. When he returned home, he demanded his parents arrange the marriage. They were dismayed by his desire to wed an outsider.

Their concern was warranted. Both Moses and Joshua warned of the consequences of such unions. Moses said, "and you might take some of his daughters for your sons, and his daughters might prostitute themselves with their gods and cause your sons also to prostitute themselves with their gods (Exodus 34:16). Samson would soon find out *why* these prophetic voices warned of this pitfall.

Yet Samson stood headstrong and cocky. Stubbornly he pushed his parents to arrange the marriage anyway: "But his father and his mother said to him, 'Is there no woman among the daughters of your relatives, or among all our people, that you go to take a wife from the uncircumcised Philistines?' Yet Samson said to his father, 'Get her for me, because she is right for me'" (Judges 14:3). This arrogant attitude typified Israel in the time of Judges. Little did it know, however, that this marriage would become an opportunity for the Lord to confront the Philistines (Judges 14:4).

As they traveled back to Timnah, Samson passed near a vineyard. A young lion suddenly pounced. The beast came roaring toward him, but the Lord intervened. "And the Spirit of the Lord rushed upon him, so that he tore it apart as one tears apart a young goat, though he had nothing in his hand; but he did not tell his father or mother what he had done" (Judges 14:6). Again, we see a familiar pattern. With the favor of the Lord, we are capable of incredible feats. All we need are modest tools, a fighting spirit, and faith. Samson illustrates this well.

The lion also represents Israel's enemies. Israel was often the underdog, yet with God's help, it triumphed. But this moment offers more than triumph. It offers a warning. Recall the angel's instructions to Samson's mother regarding the Nazirite vow: Samson must avoid anything from the vine (Judges 13:14).

The lion emerged from the shadows of the vineyard. The danger came as Samson approached the very thing he was supposed to avoid. The message is clear: *break the vow, and danger will follow*. Yet Samson said nothing to his parents. Later, Samson returned to Timnah and pursued the Philistine woman again (Judges 14:7), ignoring the Lord's command against marrying outside the faith (Deuteronomy 7:3–4).

On the journey, Samson passed the lion's carcass. But something had changed. A swarm of bees had built a hive inside it (Judges 14:8). Instead of recoiling, Samson reached into the carcass with his bare hands and scooped out the honey. This was a blatant violation of the Nazirite vow of cleanliness (Numbers 6:6–7). Samson did not hesitate. He touched the carcass, defiled himself, and indulged. Then he gave some of the honey to his parents (Judges 14:9), never telling them where it came from.

This deception reveals a growing character flaw. Samson traded long-term holiness for short-term pleasure. He violated his sacred calling and involved others in his defilement. Eating the honey and pursuing the Philistine woman were only the beginning. This was the start of Samson's personal cycle of Judges; a cycle of impulse, rebellion, and consequences.

After this, Samson's father went to see his bride (Judges 14:10). A customary feast was held for the occasion. The Philistines assigned thirty companions to Samson, likely to serve as security or keep watch over the outsider. To pass the time, Samson offered them a wager. Should they answer correctly, and within seven days of the feast, he would give them each new garments and clothes. Should they fail, they would owe him thirty sets of garments; one from each man (Judges 14:12–13).

Once again, Samson's arrogance pulled him into conflict. He challenged the Philistines with a riddle: "So he said to them, 'Out of the eater came something to eat, And out of the strong came something sweet'" (Judges 14:14). For three days, they were stumped. But by the fourth day, frustration turned to threats. They approached Samson's wife and issued a terrifying ultimatum: "Then it came about on the fourth day that they said to Samson's wife, 'Entice your husband, so that he will tell us the riddle, or we will burn you and your father's house with fire. Have you invited us to impoverish us? Is this not so?'" (Judges 14:15).

Panicked, she turned on Samson and pleaded with him in tears. "So Samson's wife wept in front of him and said, 'You only hate me, and you do not love me; you have proposed a riddle to the sons of my people, and have not told it to me'" (Judges 14:16). Samson resisted at first. He contended that he hadn't even told his own parents; so why her? This highlights the cultural divide between Israel and the Philistines. God's law instructed, "Honor your father and your mother, so that your days may be prolonged on the land which the Lord your God gives you" (Exodus 20:12). Apparently, the Philistines had no such ethic. Samson's wife persisted, and eventually, he gave in.

As soon as she had the answer, she relayed it to the others. Before sunset on the seventh day, they approached Samson with the solution. Furious, Samson accused them of cheating through deceit (Judges 14:18). Now obligated to pay the wager, Samson took matters into his own hands. The Spirit of the Lord came upon him with power. "Then the Spirit of the Lord rushed upon him, and he went down to Ashkelon and killed thirty men of them and took what they were wearing and gave the outfits of clothes to those who told the riddle. And his anger burned, and he went up to his father's house" (Judges 14:19).

The Lord was using Samson as a tool to humble the Philistines. But Samson remained angry and bitter. After settling the debt, he returned to his father's house. In his absence, his wife was given to another man (Judges 14:20). Later, during the wheat harvest, Samson returned with a gift, a young goat, hoping to visit his wife. But her father stopped him dead in his tracks. "Her father said, 'I really thought that you hated her intensely; so I gave her to your companion. Is her younger sister not more beautiful than she? Please let her be yours instead'" (Judges 15:2).

This news was too much for Samson to bear. He was dead set to get vengeance for this (Judges 15:3). Samson not only possessed incredible strength; he also had a sharp understanding of guerrilla warfare tactics. His next move was as unconventional as it was devastating. "And Samson went and caught three hundred jackals, and took torches, and turned the jackals tail to tail and put one torch in the middle between two tails. When he had set fire to the torches, he released the jackals into the standing grain of the Philistines and set fire to both the bundled heaps and the standing grain, along with the vineyards and olive groves" (Judges 15:4–5).

At first glance, this plan seems bizarre, but it was tactically brilliant and symbolically potent. Facing a nation as a lone man, Samson embraced the jiu-jitsu principle of *minimum input for maximum output*. By targeting the Philistines' food supply, he inflicted massive damage with minimal effort. He struck during the dry harvest season when the fields were stockpiled and especially flammable. The result was widespread devastation of grain, vineyards, and olive groves.

Targeting food stores is a timeless strategy. *De Re Militari* notes: "The main and principal point in war is to secure plenty of provisions and to destroy the enemy by famine.... Famine makes greater havoc in an army than the enemy, and is more terrible than the sword" (Vegetius 49). Samson applied this instinctively. He avoided head-on conflict and struck at a weak point fulfilling Sun Tzu's maxim "to avoid what is strong and to strike at what is weak" (Tzu 12). And in doing so, he embodied another of Tzu's essentials for victory: "Prepare yourself and take the enemy unprepared" (Tzu 7). As Vegetius also affirms, "To distress the enemy more by famine than the sword is a mark of consummate skill" (Vegetius 90).

This attack carried spiritual weight as well. One of the chief deities of the Philistines was Dagon (Judges 16:23), a god whose name literally means "grain." In setting fire to their fields, Samson did more than cause physical loss; he struck at the heart of their idolatry. He humiliated their false god. This is one of the ways the Lord used Samson's personal rage to fulfill divine purpose. Moses commanded Israel to tear down pagan altars and destroy these false idols (Deuteronomy 12:2–3). Samson followed in the footsteps of Gideon as he tore through the land. Though Samson's motives were personal, his actions aligned with this divine directive. And in many ways, this destructive campaign foreshadows the final mission of his life.

This was a devastating blow to the Philistines. Enraged, they demanded to know who was responsible. Bystanders were quick to name him (Judges 15:6). In a bitter twist of fate, the very people who had manipulated Samson now paid the ultimate price. His wife and her father, who folded under pressure, were hunted down and burned alive by the Philistines. Back then, appeasement spared them. This time, it did not. Evil men rarely keep their word. Appeasement may delay violence, but it seldom prevents it.

These murders enraged Samson. He boldly proclaimed: "If this is how you act, I will certainly take revenge on you, and only after that will I stop" (Judges 15:7). Immeasurable strength combined with unbreakable resolve led to a whirlwind of violence. With the same strength he used to tear apart a lion, Samson unleashed his power on the Philistines. Little is written about this skirmish, but it was devastating. He viciously slaughtered countless Philistine foes. Afterward, he withdrew and took shelter in a cave at the rock of Etam (Judges 15:8).

The Philistines, unwilling to let the matter go, pursued him into Judah. They appeared in force, prompting concern from the local Israelites. The people of Judah feared an invasion and asked the Philistine commanders why they had come. They responded: "We have come up to bind Samson in order to do to him as he did to us" (Judges 15:10). As we can clearly see, vengeance begets more vengeance. Their goal was clear: outsource the capture to Samson's own people, thereby minimizing their own losses.

Three thousand men from Judah climbed up to confront Samson. This staggering number revealed just how seriously they viewed the threat he posed to peace and stability. Rather than stand with him, they echoed the voice of defeat. "'Do you not know that the Philistines are rulers over us? What then is this that you have done to us?' And he said to them, 'Just as they did to me, so I have done to them'" (Judges 15:11). Samson's response was as simple as it was true. His answer mirrored the Philistines' earlier justification. The cyclical nature of vengeance here is unmistakable. It underscores why Moses and Joshua instituted cities of refuge to prevent endless retaliation and the spread of blood feuds (Joshua 20).

Judah pressed on. Despite the justness of Samson's cause, the people had no intention of protecting him. They were there to capture him. Samson made only one request: Their word that they would kill Samson themselves (Judges 15:12). The men agreed. The Israelites took oaths seriously, and this one sealed Samson's fate, or so they thought.

Bound with ropes, he was handed over. But everything was unfolding exactly as Samson had planned.

As they approached the Philistine camp, the enemy shouted with excitement. Their foe was finally captured, or so they believed. "When he came to Lehi, the Philistines shouted as they met him. And the Spirit of the Lord rushed upon him so that the ropes that were on his arms were like flax that has burned with fire, and his restraints dropped from his hands" (Judges 15:14). Samson's strategy here followed a core principle of *The Art of War*: "All warfare is based on deception. Hence, when able to attack, we must seem unable" (Tzu 4). Seemingly bound and helpless, Samson lulled the Philistines into complacency. Then, using whatever was at hand, he turned the tide. "Then he found a fresh jawbone of a donkey, so he reached out with his hand and took it, and killed a thousand men with it" (Judges 15:15). After the dust settled, Samson gave a brief but biting victory speech (Judges 15:16).

His words were not mere boast; they fulfilled prophecy. Joshua had said, "For the Lord has driven out great and mighty nations from before you; and as for you, no one has stood against you to this day. One of your men puts to flight a thousand, for the Lord your God is He who fights for you, just as He promised you" (Joshua 23:9–10). Samson went above and beyond! He didn't just rout a thousand men; he permanently removed them from the battlefield. but his story was far from over.

This battle recalls the earlier judge Shamgar, who struck down six hundred Philistines with an ox goad (Judges 3:31), and it holds deep spiritual significance. A familiar pattern emerges throughout Judges and beyond; with the Lord's blessing, His people accomplish extraordinary feats using the humblest of means. Few weapons are more unassuming than an old jawbone.

This modest weapon held deeper significance. It was another way God humbled the mighty Philistines. The Apostle Paul elaborates on this recurring theme: "but God has chosen the foolish things of the world to shame the wise, and God has chosen the weak things of the world to shame the things which are strong, and the insignificant things of the world and the despised God has chosen, the things that are not, so that He may nullify the things that are, so that no human may boast before God" (1 Corinthians 1:27–29). In this same way, the Lord worked through Samson's weaknesses to confront the Philistines and reveal His power.

Exhausted but reverent, Samson cried out to the Lord. He was grateful for the victory, but not without his trademark sarcasm: "Then he became very thirsty, and he called to the Lord and said, 'You have handed this great victory over to Your servant, and now am I to die of thirst and fall into the hands of the uncircumcised?'" (Judges 15:18). The Lord responded with mercy: "But God split the hollow place that is in Lehi so that water came out of it. When he drank, his strength returned and he revived. Therefore he named it En-hakkore, which is in Lehi to this day" (Judges 15:19). En-hakkore roughly means *the spring of him who called.*

The Lord answered, just as He had in the wilderness with Moses. Then, too, the people cried out in desperation. God told Moses to strike the rock at Horeb, and water flowed out for the people to drink (Exodus 17:3–7). The prophet Isaiah captured this divine pattern beautifully: "He gives strength to the weary, And to the one who lacks might He increases power. Though youths grow weary and tired, And vigorous young men stumble badly, Yet those who wait for the Lord Will gain new strength; They will mount up with wings like eagles, They will run and not get tired, They will walk and not become weary" (Isaiah 40:29–31). The Lord works in mysterious ways for those faithful to Him.

Samson's story continues much later. One day, he went to the Philistine city of Gaza. Once again, the Lord worked through Samson in spite of his moral shortcomings. Samson visited a prostitute and stayed the night (Judges 16:1). By this point, he was well known to the Philistines. They allowed him into the city as part of a trap: "When it was reported to the Gazites, saying, 'Samson has come here,' they surrounded the place and lay in wait for him all night at the gate of the city. And they kept silent all night, saying, 'Let's wait until the morning light, then we will kill him'" (Judges 16:2). Their strategy was tactically sound; ambush him at dawn, while he was likely sluggish and distracted. Indeed, "A clever general, therefore, avoids an army when its spirit is keen, but attacks it when it is sluggish and inclined to return" (Tzu 14).

But Samson was one step ahead. City gates were heavily fortified and sealed at night to protect against invasion. In the ancient world, city gates were not just physical barriers, but also symbols of a city's strength, sovereignty, and stability. The Philistines had wisely funneled him into a bottleneck. Unfortunately for them, the Lord was with Samson. He woke in the middle of the night and sprang the trap: "Now Samson lay asleep until midnight, and at midnight he got up and took hold of the doors of the city gate and the two doorposts, and pulled them up along with the bars; then he put them on his shoulders, and carried them up to the top of the mountain which is opposite Hebron" (Judges 16:3).

Modern strongman competitions include a "yoke carry"—a bar loaded with immense weight, shouldered and carried for distance. Perhaps this divine feat of strength inspired such events. Regardless, this was a devastating blow to the Philistines. Gaza was a major city, and by removing its gate, Samson exposed it to external threats and stripped it of its symbol of strength. He carried the gate nearly 40 miles! A staggering feat, and a humiliating message. Considering that modern strongman competitors only carry their yokes about two dozen yards, this miraculous feat of strength surely befuddled the Philistines.

The damage was not only physical, but also psychological and logistical. This attack wasn't just brute strength; it was sabotage by design! Sabotage has a twofold aim. According to the OSS (precursor to the CIA): "Simple sabotage is more than malicious mischief, and it should always consist of acts whose results will be detrimental to the materials and manpower of the enemy" (OSS 12). Samson's sabotage was devastatingly effective. The Philistines were forced either to launch a grueling eighty-mile recovery mission or to rebuild the entire gate from scratch. Either choice would drain time, labor, and resources. Additionally, the city would have to post guards at a temporary entrance, compounding the manpower strain. Just as Gideon shattered his enemy's confidence with torches and jars, Samson demoralized his foes by turning their strength into vulnerability.

Samson's strategy also highlights a timeless principle found in hand-to-hand combat: the art of placing your opponent in a lose–

lose dilemma. Much like a fractal, the micro reflects the macro. This concept appears across all levels of combat, whether on the battlefield or in a one-on-one fight. In grappling, a common tactic is to threaten a chokehold while setting up an arm lock. The choke is the undisputed king of fighting. It does not rely on size, strength, or endurance. A well-placed choke renders even the toughest opponent unconscious in seconds.

Skilled fighters know this, so they defend chokes with everything they have. But this opens new opportunities. This is the essence of a true dilemma. When your opponent reaches to fight your grip or hand position, he exposes his elbows; the very opening you desired initially. While he believes he is defending, you are already transitioning to an arm lock. And if he adjusts again, you return to the choke. Either way, you win.

Striking disciplines echo this same philosophy. Say your opponent is too defensively sound for a clean knockout punch. You start low. Hard leg kicks disrupt his rhythm and footwork. He begins to worry more about his legs than protecting his chin. Eventually he grows tired or distracted. For a brief moment, his defenses slip, *then* you land the knockout blow! Just like in jiu-jitsu, your setup *forced* him to react poorly. Skilled opponents won't simply give you opportunities. Rather, you must create them.

Samson employed the same layered logic. He removed Gaza's gate not just for brute display, but with careful sabotage. And by carrying it toward Hebron, a major city of Judah, he may have been sending a message to those who previously handed him over: *do not interfere with my mission again*. It was a display of power and tactical deterrence wrapped into one. Whether to demoralize Philistines or to warn his own people, his act imposed his will on all parties. The clever combatant always does.

THE BEGINNING OF THE END

Some time later, Samson stumbled once again. He fell in love with another Philistine woman named Delilah (Judges 16:4). Unbeknownst to Samson, she was plotting against him. The rulers of the Philistines approached her with an offer of vast fortune. "So the governors of the Philistines came up to her and said to her, 'Entice him, and see where his great strength lies and how we can overpower him so that we may bind him to humble him. Then we will each give you 1,100 pieces of silver'" (Judges 16:5). This was wealth beyond belief; approximately 28 pounds of silver from each ruler! Delilah was happy to oblige.

Notice, they didn't ask her to kill Samson. Given his trust in her, she easily could have poisoned him. In terms of strategy, that would have been the most efficient and safest option. But that wasn't the goal. The Philistine rulers wanted something far more gruesome. They sought to make a public mockery of God's champion; a slow, humiliating torture that would exalt their false gods and challenge the Lord's power. From their perspective, a quick death was too merciful.

With silver glimmering in her eyes, Delilah pursued the assignment. She wasted no time with subtlety: "So Delilah said to Samson, 'Please tell me where your great strength lies, and how you can be bound to humble you'" (Judges 16:6). Instead of fleeing from this obvious trap, Samson arrogantly toyed with his enemies. He told her, "If they bind me with seven fresh animal tendons that have not been dried, then I will become weak and be like any other man" (Judges 16:7). Sure enough, the Philistines tried this. Delilah tied him with the seven bowstrings while men hid in the room, waiting to pounce. "The Philistines are upon you, Samson!" (Judges 16:9). He broke free with ease and she tried again to sway him.

Samson should have walked away then and there. He knew they were trying to trap him, yet he continued to provoke them. This time, he claimed that new ropes would sap his strength (Judges 16:11). He surely took delight in this demoralizing ruse. After all, these were the same kind of ropes Judah had used when it handed him over. The Philistines wouldn't have known that; Samson had left no survivors. Naively, Delilah attempted the same approach. Once more, she called out to him tauntingly. And once more, he snapped the ropes with ease (Judges 16:12).

Enraged and embarrassed, Delilah pressed him again. The cycle repeated itself. This time, Samson told her to weave his hair into a loom and tighten it with a pin. He again claimed this would sap his strength. The third trap failed like the others. But Delilah wasn't finished. She went on the attack, emotionally and psychologically. "Then she said to him, How can you say, 'I love you,' when your heart is not with me? You have toyed with me these three times and have not told me where your great strength is" (Judges 16:15). Much like his previous Philistine wife, she quickly resorted to emotional manipulation.

Samson's cocky attitude was catching up to him. Perhaps King Solomon thought of this very story when he later wrote, "When pride comes, then comes dishonor; But with the humble there is wisdom" (Proverbs 11:2). Samson would gain wisdom, but only after *much* suffering. Trouble was brewing for Samson, yet he was the only one who didn't see it coming.

Delilah was relentless in her pursuit. Despite being the strongest man on the planet, Samson had weaknesses as well. "And it came about, when she pressed him daily with her words and urged him, that his soul was annoyed to death" (Judges 16:16). Eventually, Samson cracked and told Delilah everything. He told her about his Nazirite covenant and that should he lose his hair, his strength would soon follow (Judges 16:17).

Delilah knew she had struck gold. With greed in her heart, she summoned the Philistine rulers. They returned with silver in hand (Judges 16:18). Much like Judas Iscariot centuries later, Delilah betrayed sacred trust for a handful of coins. Samson's arrogant disregard for his vow had finally caught up to him. "And she made him sleep on her knees, and called for a man and had him shave off the seven locks of his head. Then she began to humble him, and his strength left him" (Judges 16:19). His hair was more than just a symbol. It was the physical sign of a holy covenant. When the covenant was broken, so too was his strength. His power had never come from his hands or his victories. It came from God alone.

For the first time in his life, Samson was now defenseless. Delilah woke him again with the same warning: "Samson, the Philistines are upon you!" But the true horror of the moment had not yet dawned. "She said, 'The Philistines are upon you, Samson!' And he awoke from his sleep and said, 'I will go out as at other times and shake myself free'" (Judges 16:20) We can only imagine the terror he felt in that instant. This once-mighty warrior now stood powerless before a ruthless enemy. The Philistines pounced. "Then the Philistines seized him and gouged out his eyes; and they brought him down to Gaza and restrained him with bronze chains, and he became a grinder in the prison" (Judges 16:21). Blinded and broken, Samson was reduced to slavery behind enemy lines.

This was more than a military defeat. It was the personal collapse of a judge who forgot his mission. Samson's torment reflected the cycle repeated throughout the Book of Judges. He had placed his own desires above God's commands. The warnings of his parents had gone unheeded. So too had the wise counsel of Moses and Joshua. Now he entered the second stage of the cycle: oppression. Joshua had warned Israel exactly what would happen should they turn their backs on the Lord: "know with certainty that the Lord your God will not continue to drive these nations out from before you; but they will be a snare and a trap to you, and a whip on your sides and thorns in your eyes, until you perish from this good land which the Lord your God has given you" (Joshua 23:13). Samson, now a prisoner of the Philistines, suffered this prophecy in full. The enemy had become a trap. Their shackles were now his chains. Their vengeance was the thorn in his eyes.

This subdued version of Samson symbolizes something far deeper. His physical blindness reflected Israel's spiritual condition during the time of the Judges. Blinded by their own rebellion, they forgot the Lord's commands. In this blindness, they fell into the worship of false gods. Among these idols was Dagon, a principal deity of the Philistines who represented grain and harvest. By grinding grain in captivity, Samson was forced to serve this false god, his labor becoming an unholy tribute to the very idol he was born to oppose.

Yet there was a glimmer of hope for this fallen warrior. "However, the hair of his head began to grow again after it was shaved off" (Judges 16:22). Just as God never fully abandoned a flawed Israel, He would not abandon His flawed servant either. Israel had stumbled repeatedly in its early history, but after repentance, the Lord showed mercy (Judges 10:16). The same mercy was extended to Samson.

The Philistine captors knew nothing of the Nazirite vow or what Samson's hair truly represented. His strength had been lost not merely because of a haircut, but because his vow had been defiled long before Delilah took the blade. The cutting of his hair was simply the final, outward expression of an already broken covenant. Yet God, in His mercy, allowed a new season of consecration to begin. As Samson's hair grew, so too did the quiet return of his strength.

After some time had passed, the Philistine rulers assembled for a grand celebration in honor of Dagon. They gathered in a massive temple to sacrifice to their god. "Now the governors of the Philistines assembled to offer a great sacrifice to Dagon their god, and to celebrate, for they said, 'Our god has handed Samson our enemy over to us'" (Judges 16:23). The people echoed the leaders' cry. "When the people saw him, they praised their god, for they said, 'Our god has handed our enemy over to us, Even the destroyer of our country, Who has killed many of us'" (Judges 16:24). The crowd grew drunk as the wine fueled the people's arrogance and hatred.

The raucous crowd demanded Samson. "It so happened when they were in high spirits, that they said, 'Call for Samson, that he may amuse us'" (Judges 16:25). The guards obliged, dragging him from his cell and thrusting him into a storm of noise and mockery. Blind and humiliated, Samson stood surrounded by enemies who had once feared him. Now they laughed. From the roof alone, three thousand Philistines looked on, reveling in his fall. Many more packed the temple below.

We can only imagine how disorienting it must have been. Samson, once God's super soldier, was now paraded like a carnival freakshow. He could not see their faces, but he could hear everything. The sneers.

The jeers. The cruel laughter. The drunken shuffling of bodies and the clinking of countless cups and wine jars added to the chaos. The sound pressed in from every side, a wall of scorn that seemed to stretch on forever. Blinded, degraded, and utterly alone, Samson stood in the heart of enemy worship—mocked by a nation that thought his God had abandoned him. The enemy celebrated. The crowd laughed. But they did not see what God was quietly restoring.

This torture and humiliation pushed Samson into the next phase of the cycle: *crisis*. He knew it was time for drastic measures. He turned to his guiding servant and asked for one final favor. "Put me where I can feel the pillars that support the temple, so that I may lean against them" (Judges 16:26). In doing so, Samson embraced a principle from *The Art of War*: "Let your plans be dark and impenetrable as night, and when you move, fall like a thunderbolt" (Tzu 14).

The brutality he suffered forced Samson to confront his pride and repent. He realized he had failed the Lord, not the other way around. From the depths of his brokenness, he cried out to God: "Lord God, please remember me and please strengthen me just this time, O God, that I may at once take vengeance on the Philistines for my two eyes" (Judges 16:28).

As with the other judges, the Lord responded. With resolve in his heart, Samson set his plan in motion. He stood between the two central pillars of the massive pagan temple, placed a hand on each one, and braced himself for the final mission. Then he roared out like a lion. "'Let me die with the Philistines!' And he pushed outwards powerfully, so that the house fell on the governors and all the people who were in it. And the dead whom he killed at his death were more than those whom he killed during his lifetime" (Judges 16:30).

This moment teaches another martial lesson: *never underestimate your opponent*. The Philistines believed they had neutralized the threat. They could have executed Samson quickly and moved on. Instead, they gave in to their baser instincts; cruelty, vengeance, and pride. This led directly to their downfall. Or in the words of Scripture, "Do not rejoice when your enemy falls, And do not let your heart rejoice when he stumbles, Otherwise, the Lord will see and be displeased, And turn His anger away from him" (Proverbs 24:17–18).

The massive temple crashed down with a thunderous roar! Thousands of Philistines were crushed in the attack. Three thousand were on the roof alone, countless numbers were inside the temple walls when they collapsed. Just moments before, the temple echoed with drunken revelry. It stood opulent and fortified. Now it was nothing more than a heap of rubble. The air reeked of dust and blood. The deafening roar of the crowd was replaced by an eerie silence as the dust began to settle.

This temple was a high-value target, both symbolically and strategically. It demonstrated to all the Canaanites that even their greatest strongholds were no match for the power of God. Their false gods failed them once again. This echoed God's past acts of deliverance: when He swept Egypt's chariots into the Red Sea despite its superior strength (Exodus 14), or when the fortified walls of Jericho crumbled before Joshua (Joshua 6). Once again, the underdog prevailed. God defends His faithful.

Additionally, there were massive strategic and political implications. The Philistines had oppressed Israel for forty years (Judges 13:1). Yet in their arrogance, they gathered all their rulers in one place to celebrate Samson's humiliation. By underestimating Samson, they placed themselves right where God wanted them. With a single strike, Samson wiped out their leadership. Israel entered the final phase of the cycle: *deliverance*. Though he died in the exchange, his people were set free. With one blow, he defanged the snake. The Philistines, leaderless, were dead in the water. Israel's oppressors were undone in a single act of divine judgment. Or in the words of Proverbs, "The Lord has made everything for its own purpose, Even the wicked for the day of evil" (Proverbs 16:4).

Some time later, Samson's family came for his body (Judges 16:31). His twenty-year reign ended not in disgrace, but in deliverance. He fulfilled the mission given to him by the Lord. He died a warrior. He died victorious. His downfall echoed a saying from the Old West days: *If you're seeking vengeance, dig two graves.*

SAMSON CONCLUSION

The curious case of Samson is a story of extremes. In it, we see some of the highest peaks of victory and the deepest valleys of despair. While Samson's personal life leaves much to be desired, there are important lessons to take from his journey. He achieved victory in *spite* of his character flaws, not because of them. His questionable personal choices mirrored the condition of early Israel. The tumultuous period between Joshua and David highlights the desperate need for a sovereign. In the days when everyone did as they saw fit, chaos and violence reigned. Only returning to the Lord's ways restored peace to the land.

Even in these dark days, we learn a vital truth. In His infinite wisdom, God can redirect human sin and weakness for righteous ends. Scripture shows a consistent pattern: God uses unlikely people to fulfill His purposes. Samson was no exception. Recall that the Lord used Samson's moral weakness as an opportunity to confront the Philistines. "His parents did not know that this was from the Lord, who was seeking an occasion to confront the Philistines" (Judges 14:4).

This one-of-a-kind warrior offers more lessons. Samson's strength was a gift from God, yet his moral failures brought him low. Strength alone is not enough. Without humility and discipline, strength becomes a liability. That lesson cuts both ways. Never underestimate the enemy. Both Samson and the Philistines became arrogant in their power. Samson ignored repeated warnings about associating with the Philistines, and it led to his capture. He was the embodiment of Sun Tzu's warning: "He who exercises no forethought but makes light of his opponents is sure to be captured by them" (Tzu 19). Long before they gouged out his eyes, he was blinded by pride. Likewise, the Philistines reveled in his defeat and used him for sport, until it cost them their lives.

The Lord's super soldier offers tactical lessons as well. Samson excelled in unconventional warfare. He was not a mindless brute. He was clever and daring. His use of jackals to burn enemy crops, and his improvised jawbone weapon, reveal ingenuity under pressure. He didn't complain about lack of support or resources. He met adversity head on. He embodied the jiu jitsu philosophy of *minimum input, maximum output.*

It was famously said, "No military leader can ever become great without a dash of boldness in him" (Clausewitz 3). In this regard, Samson thrived. His targets were not chosen at random. They had psychological and spiritual significance. Burning the grain was not just about food; it was an attack on Dagon himself. He obeyed Moses' charge: "And you shall tear down their altars and smash their memorial stones to pieces, and burn their Asherim in the fire, and cut to pieces the carved images of their gods; and you shall eliminate their name from that place" (Deuteronomy 12:3). Carrying off the gates of Gaza had a similar impact. Gates represented strength, law, and order. By removing them, Samson sent a message: *you are not safe.* Strategic warfare involves undermining morale. In this, Samson was a master.

Finally, we see that repentance creates resilience. Even after being enslaved and blinded, Samson was not finished doing the Lord's work. Everyone stumbles. What matters is how we respond. Just as Israel triumphed after repenting from Achan's sin, Samson redeemed himself through faith. Repentance gave him the strength to complete his holy mission. True leadership often requires great personal sacrifice. Samson's final act embodied that sacrifice; he gave everything to set his people free. As Jesus Christ later said, "Greater love has no one than this, that a person will lay down his life for his friends" (John 15:13).

THE LEVITE AND HIS CONCUBINE; HOW ONE MAN'S COWARDICE LED TO A BLOODY WAR: JUDGES 19

The Book of Judges is widely considered the darkest book in the Bible. Fittingly, it closes with what may be the most disturbing story in all of Scripture. Judges spans roughly 350 years of moral and spiritual decay. Over and over, Israel repeats the four-part cycle of rebellion, oppression, repentance, and deliverance. But each time, it sinks a little deeper, drawing ever closer to the ways of the Canaanites. The book ends not with the rise of a righteous leader, but with a haunting summary of the era: "In those days there was no king in Israel; everyone did what was right in his own eyes" (Judges 21:25). This isn't referring to a political king or head of state. It's talking about the one Sovereign King—God.

The phrase *everyone did what was right in his own eyes* speaks to the insidious danger of moral relativism. Moral relativism is nothing more than barbarism, masquerading as an enlightened virtue. The reason is simple; when *everyone* does as they see fit, naturally only the most vicious and brutal seize control to subjugate those weaker or more vulnerable. This is what anarchy really looks like, not freedom, but chaos. Not peace, but the law of the jungle. In a society without shared values or higher authority, the strongest or most tyrannical gain power. It is not a liberation, but a slow descent into madness, where even the victors are eventually consumed. Swift claws and sharp fangs get you to the top; but no one sticks around for too long.

Like Jephthah, the protagonist of this story had unknowingly embraced the ways of the world. These final chapters highlight a nation fractured by moral compromise. With clear echoes of Achan, we're reminded that one man's sin can endanger the entire group. When cowardice goes unchecked, it leads to crisis. The Levite's story is a cautionary tale. It shows how spineless leadership can lead to atrocity and civil war. Recall that the Levites had been a priestly class since the days of Moses (Exodus 28–29). One would expect a functional moral compass from such a distinguished group. This story demonstrates just how far Israel had fallen.

The story opens with instability. "Now it came about in those days, when there was no king in Israel, that there was a certain Levite staying in the remote part of the hill country of Ephraim, who took a concubine for himself from Bethlehem in Judah" (Judges 19:1). Their relationship was troubled. For reasons unknown, the concubine left and returned to her father's house. After four months, the Levite traveled to persuade her to return. Her father was glad to see him and insisted he stay. He treated the Levite with great hospitality, providing food, drink, and rest for three days (Judges 19:4).

This dragged on for several more days. Each time the Levite prepared to leave, the father-in-law urged him to stay. But the Levite had heard this too many times. Determined to leave, he packed up his servant, concubine, and two donkeys, and departed shortly before dusk. They set out toward Jebus (later known as Jerusalem), but at that time it was still under Jebusite control. As night approached, the servant urged his master to stop there and rest. The Levite refused. "We will not turn aside into a city of foreigners who are not of the sons of Israel; instead, we will go on as far as Gibeah" (Judges 19:12). The Levite feared his group would not be safe among the Jebusites. Ironically, the true threat came from within his own people. By nightfall, they reached the city square of Gibeah, a Benjamite-controlled area. To their dismay, none of their fellow Israelites offered them shelter (Judges 19:15).

Later, they encountered an old man returning from work in the fields. Like the Levite, he also hailed from the hill country of Ephraim. Seeing the group in the square, he approached and asked where they were from and where they were headed (Judges 19:17). The Levite explained their journey and made clear they lacked nothing, save only shelter for the night. Perhaps sensing kinship or out of concern for their safety, the old man insisted they come under his roof. He warned them not to spend the night in the open square (Judges 19:20).

At first, things seemed to turn around. The travelers were relieved. They washed their feet, shared food and drink, and began to relax. But the peace was short lived; it often is. The Levite had shown discernment in avoiding the pagan city of Jebus, yet he had failed to see that the residents of Gibeah had already embraced Canaanite corruption. They had long since abandoned the ways of the Lord.

Without warning, the illusion shattered. As the travelers settled in, wicked men surrounded the house. Like wolves corralling their prey, they pounded on the door with primal fury. They hollered at the old man with a shocking demand: "While they were celebrating, behold, the men of the city, certain worthless men, surrounded the house, pushing one another at the door; and they spoke to the owner of the house, the old man, saying, 'Bring out the man who entered your house that we may have relations with him'" (Judges 19:22). No negotiation, no pretense; just raw, predatory violence.

The host pleaded with them. "Then the man, the owner of the house, went out to them and said to them, 'No, my brothers, please do not act so wickedly. Since this man has come into my house, do not commit this vile sin'" (Judges 19:23). His initial attempt to protect his guest was commendable, but it quickly crumbled into cowardice. In a shocking betrayal, he offered his own daughter, and the Levite's concubine, as a bargaining chip. "Here is my virgin daughter and the man's concubine. Please let me bring them out, then rape them and do to them whatever you wish. But do not commit this act of vile sin against this man" (Judges 19:24). The mob ignored him. And the Levite, standing by, still sought to save himself.

In a shocking act of betrayal, the Levite handed over his concubine to the frenzied mob. These wicked men committed unspeakable atrocities. They beat her and gang raped her throughout the night. As dawn approached, the predators finally fled the scene (Judges 19:25). Broken and bloodied, the woman stumbled back to the house. She collapsed at the doorstep, desperately reaching for the threshold. She died just outside the door, mere feet from safety.

The next morning, the Levite still failed to grasp the horror of what had taken place. Finding her lifeless body at the door, he said, "Get up and let's go" (Judges 19:28). But there was no answer. She was gone. He lifted her body onto one of his donkeys and began the journey home.

This is undoubtedly one of the lowest moments in Israel's history. The behavior of both the old man and the Levite was utterly repugnant. These men abandoned their most basic duties as protectors and leaders. Like Achan before them, their personal sin carried devastating consequences for the entire nation. What began as a cowardly act would soon ignite a brutal civil war. Their blasphemous betrayal did not end with one death; it eventually led to the deaths of tens of thousands of fellow Israelites.

Before analyzing the warfare which followed, we must pause to understand how Israel descended to this point. These cowardly atrocities present an important paradox: *weak men are far more damaging than capable men.* At first, this seems counterintuitive. But upon closer inspection, the danger becomes clear. Weak men will sacrifice anything or anyone to save themselves. Even their own wives and daughters are expendable if it means a shot at saving their own skin. Strong men are dangerous to bad actors; weak men are dangerous to those closest to them.

The coward refuses to fight for himself or his loved ones, while the warrior is willing to die for a righteous cause. Weak men often *appear* nice. But don't let their camouflage fool you. They may speak politely. They may appear harmless. But beneath the veneer of civility lies something far more insidious: moral rot and cowardice. And when the moment of truth arrives, when danger presses in, the mask comes off, and the coward chooses betrayal over battle without fail.

The book *Wild at Heart* describes this phenomena perfectly. "Satan is trying to appeal to the traitor's commitment to self-preservation when he uses fear and intimidation.... But the opposite is also true. When a man resolves to become a warrior, when his life is given over to a transcendent cause, then he can't be cowed by the Big Bad Wolf threatening to blow his house down" (Eldredge 153). These cowardly men had no greater purpose or higher ideals. They didn't even *try* to protect their loved ones!

As legendary swordsman Miyamoto Musashi famously said, "Generally speaking, the way of the warrior is resolute acceptance of death" (Musashi 1). Indeed, *Principles of War* echoes this timeless sentiment. "One must get used to the idea of dying with honor, to continually nurture that idea to get used to it" (Clausewitz 2). These men possessed neither duty nor honor. Their weakness drove them down the path of least resistance. They lacked the grit to stand, or the courage to even try. In every way, they stood in opposition to the warrior's code. To sum things up, "a soldier who has proper confidence in his own skill and strength, entertains no thought of mutiny" (Vegetius 52). There is a major distinction between confidence and delusion. Simply put, delusion is confidence *without* cultivated skills. Therefore, do whatever you can to build your skillsets, so you don't end up like these pathetic cowards.

These men didn't just lack courage; they were also morally bankrupt. Scripture offers no tolerance for such cowardice. Both Old and New Testaments are filled with warnings and exhortations on this point.

Let us not forget: we serve a warrior God. "The Lord is a warrior; The Lord is His name. Pharaoh's chariots and his army He has thrown into the sea; And the choicest of his officers are drowned in the Red Sea" (Exodus 15:3–4). We are made in His image. And throughout Scripture, the Lord makes it clear, He values courage and despises cowardice. This is not a call for needless violence. It is a call to protect one's family, to defend the helpless, and to serve with strength. As the renowned author and Christian apologist G. K. Chesterton wrote, "The true soldier fights not because he hates what is in front of him, but because he loves what is behind him" (Chesterton, 44).

Generations after the Exodus, God reinforced these values to Joshua before he entered the Promised Land. In the opening chapter alone, He commands Joshua three times to *be strong and courageous* (Joshua 1:1–9). These are the virtues God honors. But during the time of Judges, Israel had forgotten. The cowardice in Gibeah shows just how far it had fallen.

Scripture gives no ambiguity about what is required. "Rescue the weak and needy; Save them from the hand of the wicked" (Psalm 82:4). Moreover, "Open your mouth for the people who cannot speak, For the rights of all the unfortunate. Open your mouth, judge righteously, And defend the rights of the poor and needy" (Proverbs 31:8–9). We are not called to flee. We are called to stand and fight. As Proverbs additionally declares, "The wicked flee when no one is pursuing, But the righteous are bold as a lion" (Proverbs 28:1). Lions do not shrink from battle. They fight to the death to protect their pride. Christ understood this. The Lion of Judah gave His life for His people. He is our example, not the cowards of Gibeah.

The New Testament continues this charge. Paul, writing to Timothy, declared: "For God has not given us a spirit of timidity, but of power and love and discipline" (2 Timothy 1:7). Like Samson after his imprisonment, true strength comes from faith. In his letter to the church in Corinth, Paul reiterated the need for stoicism and resiliency amongst believers. "Be on the alert, stand firm in the faith, act like men, be strong" (1 Corinthians 16:13). These are the same traits God instilled in Joshua, and they are needed just as much today. We are called to protect the flock, not prey upon it.

Paul also gave clear commands for men in the home. "Husbands, love your wives, just as Christ also loved the church and gave Himself up for her…. So husbands also ought to love their own wives as their own bodies. He who loves his own wife loves himself" (Ephesians 5:25, 28). This is the Biblical model of leadership and sacrifice. It stands in complete contrast to the Levite who threw his concubine to the mob to save himself.

We've seen ample evidence condemning the actions of the Levite and the old man. But the most damning indictment of cowardice comes from the very end of Scripture. In the Book of Revelation, the Lord reveals to John a vision of the new Heaven and new Earth: "It is done. I am the Alpha and the Omega, the beginning and the end. I will give water to the one who thirsts from the spring of the water of life, without cost. The one who overcomes will inherit these things, and I will be his God and he will be My son" (Revelation 21:6–7).

This life-giving water is without cost not because it is cheap, but because Christ paid the full price *for* us! The offer is clear and generous.

Yet many will still reject it. And the Lord is unambiguous about the consequences: "But for the cowardly, and unbelieving, and abominable, and murderers, and sexually immoral persons, and sorcerers, and idolaters, and all liars, their part will be in the lake that burns with fire and brimstone, which is the second death" (Revelation 21:8).

The first sin listed is cowardice. That is no accident. Cowardice is a seed from which other sins grow. The coward chooses self-preservation over truth, comfort over conviction. At its root, cowardice is a rejection of faith. It denies the power and protection of God. It exalts fear above obedience. Throughout Scripture, we've seen that with God's blessing, ordinary people can achieve seemingly impossible victories. All that is required is a willing heart, humble tools, and unshakable faith. But when fear rules the heart, it robs God of glory and cripples His mission.

Courage is not the absence of fear; *it is the resolve to move forward despite it*. The coward retreats. The faithful press on.

FROM COWARDICE TO CRISIS

E ventually, the Levite made it back to his home in Ephraim. Yet, his role in the civil war was just beginning. He desired retribution, but what was one man to do against a frenzied mob? The Levite devised a gruesome, yet effective recruitment tactic. "When he entered his house, he took a knife and seized his concubine, and cut her in twelve pieces, limb by limb. Then he sent her throughout the territory of Israel" (Judges 19:29). In his groundbreaking work *How to Win Friends and Influence People*, Dale Carnegie advises, "This is the day of dramatization. Merely stating a truth isn't enough. The truth has to be made vivid, interesting, dramatic. You have to use showmanship. The movies do it. Television does it. And you will have to do it if you want attention" (Carnegie 223). Despite his moral bankruptcy, the Levite knew how to sway a crowd. This morbid message had its intended effect. The people of Israel were shocked! Public sentiment rose and people felt compelled to act against this barbarism.

The community quickly coalesced to fight back. The response was overwhelming, roughly four-hundred thousand armed men answered the call for justice. Once they were assembled, the leadership asked for the Levite's account of things. His response again highlights his character. "So the Levite, the husband of the woman who was murdered, answered and said, 'I came with my concubine to spend the night at Gibeah which belongs to Benjamin. But the citizens of Gibeah rose up against me and surrounded the house at night, threatening me. They intended to kill me; instead, they raped my concubine so that she died'" (Judges 20:4–5). After this he explained how he was the one to mutilate and distribute her body parts to each tribe of Israel. He then challenged the leadership. "Behold, all you sons of Israel, give your response and advice here" (Judges 20:7).

Notice what the cowardly Levite did with his testimony? He lied by omission. He pulled on the heartstrings of his sympathetic countrymen. Yet, he conveniently left out the fact that *he* was the one who threw his concubine to the rabid mob. This is a principal danger of weak men. Their forked tongues have no shortage of misleading half-truths. They feign righteousness while their actions betray others. Their hands feed you to the wolves while their mouths extol their own virtuous character.

The Apostle Paul comments on this effect in his second letter to the Corinthians. "No wonder, for even Satan disguises himself as an angel of light. Therefore it is not surprising if his servants also disguise themselves as servants of righteousness, whose end will be according to their deeds" (2 Corinthians 11:14–15). Jesus also referred to Satan as "The Father of Lies." (John 8:44) We can see clearly which path the Levite chose to pursue. While nothing he said was false, he deliberately framed the events to conceal his misdeeds. If he had told the whole truth, he would not have received the massive support from the fighting men. Instead, they bought the lie wholeheartedly. They rose up together to fight back.

Charged with a fighting spirit, the unified Israelites continued preparations at Mizpah. "So all the men of Israel were gathered against the city, united as one man" (Judges 20:11). The unified Israelites had a large size advantage. It was eleven tribes versus one. They could have steamrolled the Benjaminites by sheer force of will. Yet they chose diplomacy first. The unified forces wanted to minimize the bloodshed of their Benjamite brothers. To accomplish this, they sent messengers throughout the tribe of Benjamin. This was a last ditch attempt to avoid one of the worst scourges any nation can experience: civil war. The envoys spread a message calling for justice. "Then the tribes of Israel sent men through the entire tribe of Benjamin, saying, 'What is this wickedness that has taken place among you? Now then, turn over the men, the worthless men who are in Gibeah, so that we may put them to death and remove this wickedness from Israel.' But the sons of Benjamin would not listen to the voice of their brothers, the sons of Israel" (Judges 20:12–13).

The tribe of Benjamin had ample opportunity to do the right thing. Foolishly, it refused. When diplomacy fails, violence soon follows. Instead of turning over the sexual predators and murders of Gibeah, it rallied its own troops to defend the city. To be clear, this wasn't simply the people of Gibeah looking out for themselves. "Instead, the sons of Benjamin gathered from the cities to Gibeah, to go out to battle against the sons of Israel" (Judges 20:14). They were quick to fight. Indeed, the residents of Gibeah made only a small fraction of the combatants. "From the cities on that day the sons of Benjamin were counted, twenty-six thousand men who drew the sword, besides the inhabitants of Gibeah who were counted, seven hundred choice men" (Judges 20:15). Among these Benjaminites soldiers was a detachment of skilled sharp shooters. These select troops had lethal accuracy. "Out of all these people seven hundred choice men were left-handed; each one could sling a stone at a hair and not miss" (Judges 20:16). Benjamin wasn't seeking compromise; it was preparing for war.

Meanwhile, unified Israel mustered four-hundred thousand swordsmen, all fit for battle (Judges 20:17). Unlike their enemies, they sought the Lord's guidance. "Now the sons of Israel set out, went up to Bethel, and inquired of God and said, 'Who shall go up first for us to battle against the sons of Benjamin?' Then the Lord said, 'Judah shall go up first.'" Without delay, they marched to Gibeah and camped outside the city (Judges 20:19). The unified forces then took up positions and launched their initial assault.

It ended in disaster.

"Then the sons of Benjamin came out of Gibeah and struck to the ground on that day twenty-two thousand men of Israel" (Judges 20:21). For a force of twenty-seven thousand, this was a stunning success. Nearly every Benjamite struck down one of their fellow Israelites. The elite slingers likely played a key role, raining death from a distance with terrifying precision. These men weren't simply throwing stones; they were trained killers wielding a deadly weapon. Long before David slayed Goliath, the Benjamite slingers brought down thousands of their own kin with the same weapon.

We can imagine the scene. Stones flew through the air like a deadly hail storm. The clash of swords and cries of the wounded filled the sky. The cacophony concealed the faint whirring sound of the flying stones.

Then the *crunch*. Bone cracked under the force of the lethal projectile. Blood soaked the sand. Teeth scattered like shattered pottery. Before a fallen soldier hit the ground, the next projectile was already in flight. Gibeah was a slaughterhouse.

This crushing defeat was not just a tragedy; it was a failure of strategy. The brute force attack lacked any strategy or finesse. While the ambition was noble, openly confronting a highly fortified structure is costly and reckless. *The Art of War* makes this abundantly clear: "the worst policy of all is to besiege walled cities.... The rule is, not to besiege walled cities if it can possibly be avoided" (Tzu 6). Sun Tzu explains this principle more broadly. "Military tactics are like unto water; for water in its natural course runs away from high places and hastens downwards. So in war, the way is to avoid what is strong and to strike at what is weak" (Tzu 12). While water *can* break down stone, it does so slowly and at great cost. The Israelites chose the hardest path, the frontal assault, and paid for it dearly.

This bloody defeat left the unified Israelites with some soul searching to do. With heavy hearts, they went back to Bethel and prayed for guidance. They asked the Lord if they should fight again. He answered resolutely to go up against them again (Judges 20:23). Their work was far from over. Faithfully, they regrouped the next day. Unfortunately, they learned little in terms of tactics. It was another one sided victory for the Benjaminites. The second day was nearly as bad as the first. "And Benjamin went out against them from Gibeah the second day and struck to the ground again eighteen thousand men of the sons of Israel; all of these drew the sword" (Judges 20:25).

In a mere two days, forty-thousand Israelites had died with little to show. The unified forces stood forlorn and distraught. But this time, their mourning was different. It was joined by sacrifice, repentance, and complete dependence on God. "Then all the sons of Israel and all the people went up and came to Bethel, and they wept and remained there before the Lord, and fasted that day until evening. And they offered burnt offerings and peace offerings before the Lord" (Judges 20:26). Once again they sought God's wisdom on how to handle this civil war. Just as God encouraged Joshua and Moses, He encouraged Israel. "Go up, for tomorrow I will hand them over to you" (Judges 20:28).

The initial losses were devastating for the unified forces. Yet, "in war, faith in success should not die until the very last minute" (Clausewitz 37). Their faith in the Lord's blessing shone through. On the third day, they tried again. They understood the enemy always gets a say in the fight. It's not the misfortunes that define a warrior, but how he *responds* to them. In fact, we can often determine who the superior fighters are by observing "who bears up against his misfortunes with greatest resolution" (Vegetius 86).

Reinvigorated by their faith, they formed a new plan. Finally, they embraced the maxim, "Attack him where he is unprepared, appear where you are not expected" (Tzu 4). On the third day, they set an ambush. The unified forces took a page out of *The Book of Five Rings*. "There may be no help but to do something twice, but do not try it a third time. If you once make an attack and fail, there is little chance of success if you use the same approach again. If you attempt a technique which you have previously tried unsuccessfully and fail yet again, then you must change your attacking method" (Musashi 28). Embodying the legendary swordsman, they set a trap for the Benjaminites.

The unified forces set the ruse by starting in the same battle formation as the previous two days (Judges 20:31). The Benjaminites confronted them just outside the city gates. The action remained close to Gibeah. As on the days before, they began inflicting casualties on the unified forces. After a few dozen casualties, they began to yield to the Benjaminites. Yet, the Benjaminites had grown cocky due to their overwhelming success in the days previous. Unknown to the Benjaminites, the unified forces personified another maxim from Sun Tzu. "If your opponent is of choleric temper, seek to irritate him. Pretend to be weak, that he may grow arrogant" (Tzu 4). They executed this with precision. "And the sons of Benjamin said, 'They are defeated before us, like the first time.' But the sons of Israel said, 'Let's flee, so that we may draw them away from the city to the roads'" (Judges 20:32). The plan was textbook *Art of War*, "Hold out baits to entice the enemy. Feign disorder, and crush him" (Tzu 4).

The plan was working. The Benjaminites, overconfident from their previous victories, charged straight into the trap. As Musashi wrote, "When the enemy attacks, remain undisturbed but feign weakness" (Musashi 22). With their backs to the city, they failed to realize the true threat was behind them. Ten thousand Israelite soldiers were lying in wait (Judges 20:34). As the Benjaminites pushed forward, the ambush force struck. They charged into the now unprotected Gibeah, putting the entire city to the sword (Judges 20:37).

Once the city fell, the ambushers ignited it. Smoke billowed into the sky, dark and thick. This was a predetermined signal for the main force to stop retreating and counterattack. It also triggered panic in the Benjamite ranks and robbed them of their fallback position. Their fortifications were useless and unrecoverable.

Before, the Benjaminites had been like a fierce snapping turtle. Sheltered behind the stone walls of Gibeah, the hardened shell stood impervious to attack. Their slings and swords were the snapping jaws of a ruthless beast. But what good is a snapping turtle without its shell? Enticed by the lure of easy prey, they abandoned their defenses and left themselves exposed. They fell into the very trap described in *The Art of War*: "So in war, the way is to avoid what is strong and to strike at what is weak" (Tzu 12). The wise warrior strips the turtle of its shell. It is better still when the turtle removes it for you.

The smoke plumes darkened the sky. The trap was sprung. "But when the cloud began to rise from the city in a column of smoke, Benjamin looked behind them; and behold, the entire city was going up in smoke to heaven. Then the men of Israel turned, and the men of Benjamin were terrified; for they saw that disaster was close to them" (Judges 20:40–41). The timing was perfect. As Vegetius reminds us, "Part of the victory consists in throwing the enemy into disorder before you engage them" (Vegetius 75). The sight of their burning stronghold shattered the Benjaminites' morale. Fear overtook them. "Therefore, they turned their backs before the men of Israel to flee in the direction of the wilderness, but the battle overtook them while those who attacked from the cities were annihilating them in the midst of them" (Judges 20:42).

The retreat turned into a rout. The unified forces encircled them and struck from all sides. Though the Benjaminites fought valiantly, there was little they could do once the noose tightened. "So eighteen thousand men of Benjamin fell; all of these were valiant men" (Judges 20:44).

But the battle did not end there. The unified forces pressed on, relentless and disciplined. They had learned another crucial lesson of war: "If he is taking his ease, give him no rest. If his forces are united, separate them" (Tzu 4). With the advantage fully theirs, they refused to let up. Momentum, once seized, must never be surrendered.

The unified forces pressed the Benjaminites into rough terrain. "The rest turned and fled toward the wilderness to the rock of Rimmon, but they caught five thousand of them on the roads and overtook them at Gidom, and killed two thousand of them" (Judges 20:45). It was a textbook pursuit, as Musashi advised: "Chase him towards awkward places, and try to keep him with his back to awkward places. When the enemy gets into an inconvenient position, do not let him look around, but conscientiously chase him around and pin him down" (Musashi 21). Clausewitz echoed the same principle: "With overwhelming energy, pursue our successes. Only the pursuit of a defeated enemy will give us the fruits of victory" (Clausewitz 13). The Israelites were fruitful indeed!

The third day was a decisive victory. "So all those of Benjamin who fell that day were twenty-five thousand men who drew the sword; all of these were valiant men" (Judges 20:46). Only about six hundred men escaped, fleeing into the wilderness where they would hide for several months (Judges 20:47). Had they not escaped, the tribe of Benjamin would have been completely wiped out. But the fighting wasn't over. The unified Israelites pressed their advantage to extinguish any lingering resistance. Should they relent, the sparks of rebellion might rekindle old conflicts.

The Israelites did not hesitate. "The men of Israel then turned back against the sons of Benjamin and struck them with the edge of the sword, both the entire city with the cattle and all that they found; they also set on fire all the cities which they found" (Judges 20:48). Musashi's wisdom regarding large-scale strategy rings true here: "When the enemy has few men ... we knock the hat over his eyes, crushing him utterly. If we crush lightly, he may recover" (Musashi 28).

This entire assault mirrored Joshua's ambush at Ai. Like Joshua, the Israelites lured their enemy out to expose a vulnerable center (Joshua 8). But this was more than just a well-executed tactic; it marked a shift in Israel's posture. On the first two days, they attacked without a coherent strategy. Their disunity and lack of vision reflected a deeper spiritual blindness. Only after prayer, fasting, and realignment with the Lord did they begin to fight like the people of God.

During that time of reflection, perhaps they recalled what made Israel victorious in the past. Though the text doesn't state this outright, the strategic turnaround suggests a rediscovery of purpose. The Benjaminites, by contrast, showed no such humility. They fell for the trap completely; blinded by arrogance and ignorant of their heritage. Their failure wasn't just tactical, it was generational. Judges 2 explains where it all began: "The people served the Lord all the days of Joshua, and all the days of the elders who survived Joshua, who had seen all the great work of the Lord which He had done for Israel.... All that generation also were gathered to their fathers; and another generation rose up after them who did not know the Lord, nor even the work which He had done for Israel" (Judges 2:7,10).

Israel succeeded only after returning to its roots. Their first two defeats were crushing, but by emulating the example of Joshua, both spiritually and strategically, it turned the tide. The Benjaminites, too proud to bend the knee, were humbled. As Christ brilliantly said, "For everyone who exalts himself will be humbled, and the one who humbles himself will be exalted" (Luke 14:11).

THE AFTERMATH

The fighting subsided before the gravity of the situation fully sank in. Civil war had revealed its true cost; Israel had destroyed part of itself. Nearly the entire tribe of Benjamin was gone. With heavy hearts, the people cried out to the Lord. "And they said, 'Why, Lord, God of Israel, has this happened in Israel, that one tribe is missing today from Israel?'" (Judges 21:3).

The nation grieved for Benjamin, yet no one knew how to restore what had been lost. During the fog of war, it had made a hasty vow. Another ill-considered oath like Jephthah's. At Mizpah, the Israelites had sworn, "None of us shall give his daughter to Benjamin in marriage" (Judges 21:1). Only after the dust settled did they grasp the consequences. Much like Jephthah, this impulsive decision would place them in a terrible dilemma.

Their impulsive oath left them trapped. They had cornered themselves with no righteous path forward. They wanted to preserve the tribe of Benjamin, but their own words barred any peaceful path forward. Amongst the fray, one question arose: "And they said, 'What one is there of the tribes of Israel that did not go up to the Lord at Mizpah?' And behold, no one had come to the camp from Jabesh-gilead to the assembly" (Judges 21:8). Previously they had bound themselves to another solemn vow: that any tribe refusing to join the assembly would face destruction (Judges 21:5). Acting on this, the assembly dispatched twelve thousand soldiers to Jabesh Gilead. Their mission was twofold: execute those who had broken the oath, and find wives for the surviving Benjaminites. Nearly the entire city was destroyed, and four hundred young women were taken (Judges 21:12). These were offered in peace to the survivors at the rock of Rimmon (Judges 21:13).

But it wasn't enough.

Hundreds of Benjaminites remained without wives, and the tribe could not be restored without them. Once again, the people turned to compromise. Their next solution plunged even deeper into moral decay. They instructed the remaining Benjaminites to kidnap women from the festival at Shiloh (Judges 21:20). And this time, the assembly not only tolerated the act—it also helped to plan it.

It even coached Benjamin on how to manage the fallout: "And when their fathers or their brothers come to complain to us, we shall say to them, 'Give them to us voluntarily, because we did not take for each man of Benjamin a wife in battle, nor did you give them to them, otherwise you would now be guilty'" (Judges 21:22). This was Israel's solution: abduction, coercion, and collective dishonesty. Their attempts to clean up one broken vow led to more moral depravity. The story ended with the same betrayal and cowardice and callousness that sparked it all! The final chapter of Judges offers no triumph; only tragedy. The people were not restored. They were not cleansed. Judges closes with the same barbarism it began with. The final verse says it all: "In those days there was no king in Israel; everyone did what was right in his own eyes" (Judges 21:25).

THE LEVITE AND HIS CONCUBINE CONCLUSION

T he story of the Levite and his concubine offers many lessons for today. One of its clearest themes is the hidden danger of cowardice. Like a snake in the grass, the weak man cloaks himself in politeness and false virtue. He speaks with a forked tongue, hiding behind flattering half-truths and performative civility. This may work in times of peace, but under pressure, the mask always cracks. When adversity strikes, the coward sacrifices anyone or anything to protect himself.

Scripture repeatedly warns us about such people. "Better is open rebuke than hidden love. Wounds from a friend can be trusted, but an enemy multiplies kisses" (Proverbs 27:5–6). And again: "One who rebukes a person will afterward find more favor Than one who flatters with the tongue" (Proverbs 28:23). True friends speak the hard truth, not comfortable lies. As Paul wrote, "So have I become your enemy by telling you the truth?" (Galatians 4:16).

The Levite epitomized moral failure. He made no effort to resist, reason, or flee. He simply handed his concubine to the mob. His betrayal echoes forward to the trial of Jesus Christ. When Pilate had the power to release Jesus, he instead yielded to the mob. Though he recognized Jesus' innocence, he lacked the courage to do what was right. "Pilate said to them, 'Then what shall I do with Jesus who is called Christ?' They all said, 'Crucify Him!' But he said, 'Why, what evil has He done?' Yet they kept shouting all the more, saying, 'Crucify Him!' Now when Pilate saw that he was accomplishing nothing, but rather that a riot was starting, he took water and washed his hands in front of the crowd, saying, 'I am innocent of this Man's blood; you yourselves shall see'" (Matthew 27:22–24).

Like many political cowards, Pilate preserved his position by sacrificing the innocent. Paul later gave us the model of the opposite spirit: "Do nothing from selfishness or empty conceit, but with humility consider one another as more important than yourselves; do not merely look out for your own personal interests, but also for the interests of others" (Philippians 2:3–4).

The Levite's cowardice mirrored that of Achan. His selfishness sparked catastrophe. Achan's sin led to the deaths of fellow soldiers (Joshua 8). The Levite's betrayal went further. He not only abandoned his concubine, but also triggered a civil war. In just three days of battle, over sixty-five thousand Israelites died (Judges 20:21, 25, 35).

Yet this story is not only a warning—it is also a call. A call for justice. For leadership. For courage. In contrast to the Levite, the unified forces of Israel eventually rose to defend what was right. They embraced the warrior ethos. They were willing to die for justice, and when defeat came, they repented, prayed, and changed course. Like Joshua, they sought the Lord's guidance and adapted their strategy. In doing so, they remind us: *resilience is forged through humility and repentance.*

PART THREE:
SAMUEL—THE FINAL JUDGE

SAMUEL—A MIRACULOUS BIRTH AMID CORRUPTION

S amuel was a one-of-a-kind figure in early Israel. He held many roles: prophet, priest, and judge. As the final judge of Israel, he played a pivotal role in transitioning the nation from the time of the judges to the establishment of a monarchy. His story parallels those of Moses and Samson in striking ways, particularly in their miraculous births and divine appointments. Samuel's journey begins during a worship ceremony at Shiloh (1 Samuel 1:1–3).

Much like Samson's mother, Samuel's mother, Hannah, was barren. Tormented by her husband's other wife, she endured cruel bullying years. "Her rival, moreover, would provoke her bitterly to irritate her, because the Lord had closed her womb. And it happened year after year, as often as she went up to the house of the Lord, that she would provoke her; so she wept and would not eat" (1 Samuel 1:6–7). This again highlights the pitfalls of polygamous relationships. In deep anguish, Hannah cried out to the Lord: "Lord of armies, if You will indeed look on the affliction of Your bond-servant and remember me, and not forget Your bond-servant, but will give Your bond-servant a son, then I will give him to the Lord all the days of his life, and a razor shall never come on his head" (1 Samuel 1:11). Like Samson, Samuel was consecrated as a Nazirite before his birth.

Hannah poured her soul into that prayer. This was when the priest Eli noticed her. Hannah was so overcome with emotion that Eli thought she was drunk at first glance (1 Samuel 1:12–14). But once he heard her story, he blessed her: "Go in peace; and may the God of Israel grant your request that you have asked of Him" (1 Samuel 1:17). Soon after, the Lord answered her prayer. She conceived and gave birth to a son, whom she named Samuel, saying, "Because I have asked for him of the Lord" (1 Samuel 1:20). The name Samuel sounds like the Hebrew for "heard by God."

The next year, Hannah stayed home from the annual sacrifice, choosing instead to honor her vow. Once the child was ready, she brought him to Shiloh with offerings for the tabernacle. After presenting a burnt offering, she fulfilled her promise and presented her son for the Lord's service (1 Samuel 1:24–28).

Samuel now began his work in the house of the Lord. While he was faithful to the Lord, his peers were anything but. The priest Eli had two sons named Hophni and Phinehas who helped him at the tabernacle (1 Samuel 2:34). "Now the sons of Eli were useless men; they did not know the Lord" (1 Samuel 2:12). Eli's sons made a mockery of the Lord and constantly put their own selfish desires above their Godly roles. While it was customary for the Levitical priests to live off food presented in the tabernacle (Deuteronomy 18:1–5), the two wicked sons abused this system. "And this was the custom of the priests with the people: when anyone was offering a sacrifice, the priest's servant would come while the meat was cooking, with a three-pronged fork in his hand. And he would thrust it into the pan, or kettle, or caldron, or pot; everything that the fork brought up, the priest would take for himself. They did so in Shiloh to all the Israelites who came there" (1 Samuel 2:13–14). The purpose of this was to acknowledge divine providence. Yet Hophni and Phinehas rejected God's will to sate their baser desires.

Their greed led them selfishly to take the Lord's offerings for themselves. They lied to the worshippers, bypassed the protocol, and took the best pieces for themselves (1 Samuel 2:15). This was likely enough to trick most of the worshippers. However, if the person rendering the sacrifice questioned their motives, things quickly escalated. "And if the man said to him, 'They must burn the fat first, then take as much as you desire,' then he would say, 'No, but you must give it to me now; and if not, I am taking it by force!'" (1 Samuel 2:16).

This blatant disrespect of the Lord's house was terrible. But what can one expect in the time when *everyone did as they saw fit*? "And so the sin of the young men was very great before the Lord, for the men treated the offering of the Lord disrespectfully" (1 Samuel 2:17). In other words, "He who keeps the Law is a discerning son, But he who is a companion of gluttons humiliates his father" (Proverbs 28:7). This disgraceful behavior carried on for some time.

Unfortunately, their misdeeds didn't end with thievery and intimidation. Eli's sons continued to mock the tabernacle by sleeping with the female servants from the Tent of Meeting. (1 Samuel 2:22). In their arrogance, they didn't even attempt to hide or deny these claims. Eventually, Eli confronted his two sons. "So he said to them, 'Why are you doing such things as these, the evil things that I hear from all these people?'" (1 Samuel 2:23). Eli continued to plead to his sons to repent of their ways. "If one person sins against another, God will mediate for him; but if a person sins against the Lord, who can intercede for him? But they would not listen to the voice of their father, for the Lord desired to put them to death" (1 Samuel 2:25). Sinning against your fellow man is bad enough. Yet sinning directly against the Lord leads to destruction. Unfortunately Eli's pleas were too little, too late. Hophni and Phinehas disregarded their father's rebuke. Eli's sons would soon learn a valuable lesson. "A leader who is a great oppressor lacks understanding, But a person who hates unjust gain will prolong his days" (Proverbs 28:16).

After Eli's feeble attempt to rein in his sons, a mysterious prophet came to Eli. The Lord spoke through this him. He began with an account of Israel's ancestors. Additionally, he commented on how the tribe of Levi held a special role in the divinity of the Lord's house (1 Samuel 2:27–28). This historical account quickly led to biting criticism from the Lord. "Why are you showing contempt for My sacrifice and My offering which I have commanded for My dwelling, and why are you honoring your sons above Me, by making yourselves fat with the choicest of every offering of My people Israel?" (1 Samuel 2:29). Indeed, "Bread obtained by a lie is sweet to a person, But afterward his mouth will be filled with gravel" (Proverbs 20:17).

The Lord gave His criticisms—now a dire warning of His judgment. "Yet I will not cut off every man of yours from My altar, so that your eyes will fail from weeping and your soul grieve, and all the increase of your house will die in the prime of life. And this will be the sign to you which will come in regard to your two sons, Hophni and Phinehas: on the same day both of them will die" (1 Samuel 2:33–34). Fortunately, this prophet was not all doom and gloom. There was still hope in Israel. The Lord had not rescinded His holy covenant. "But I will raise up for Myself a faithful priest who will do according to what is in My heart and My soul; and I will build him an enduring house, and he will walk before My anointed always" (1 Samuel 2:35). At this time, this faithful priest was still just a little boy. His name was Samuel.

Additionally, this served as foreshadowing to Jesus Christ. The Book of Hebrews expands upon this important idea. "Therefore, since we have a great high priest who has passed through the heavens, Jesus the Son of God, let's hold firmly to our confession. For we do not have a high priest who cannot sympathize with our weaknesses, but One who has been tempted in all things just as we are, yet without sin" (Hebrews 4:14–15). Additionally, Hebrews explains why this was necessary. Jesus is simultaneously both the good shepherd and the sacrificial lamb; He paid our debt of original sin. "For it was fitting for us to have such a high priest, holy, innocent, undefiled, separated from sinners, and exalted above the heavens; who has no daily need, like those high priests, to offer up sacrifices, first for His own sins and then for the sins of the people, because He did this once for all time when He offered up Himself. For the Law appoints men as high priests who are weak, but the word of the oath, which came after the Law, appoints a Son, who has been made perfect forever" (Hebrews 7:26–28).

The contrast between Samuel and Eli's sons could not be greater. While Hophni and Phinehas cheated the worshipers, Samuel was faithful to his duties. Picture the scene. Eli's sons treated the Lord's offering as a joke. They stole offerings, threatened worshipers with violence, and flirted with the female staff in plain view of everyone. Meanwhile, Samuel was an innocent child. "Now Samuel was ministering before the Lord, as a boy wearing a linen ephod. And his mother would make for him a little robe and bring it up to him from year to year when she would come up with her husband to offer the yearly sacrifice" (1 Samuel 2:18–19).

The prophet's warning bore fruit. "Therefore the Lord God of Israel declares, 'I did indeed say that your house and the house of your father was to walk before Me forever'; but now the Lord declares, 'Far be it from Me—for those who honor Me I will honor, and those who despise Me will be insignificant'" (1 Samuel 2:30). Indeed, God blessed Samuel's family with more children to the once barren woman (1 Samuel 2:21). While Eli's sons continued their downward spiral, Samuel took a different path. "Now the boy Samuel was continuing to grow and to be in favor both with the Lord and with people" (1 Samuel 2:26).

SAMUEL'S CALLING AND RISE TO LEADERSHIP

Samuel's mother had called on the Lord for a son. Soon, God would be calling on that very child. Samuel became a beacon of hope in a dark time. "Now the boy Samuel was attending to the service of the Lord before Eli. And word from the Lord was rare in those days; visions were infrequent" (1 Samuel 3:1).

One night, everything changed. Samuel and Eli were sleeping in the tabernacle—the same holy place that held the Ark of the Covenant. The Lord called out to Samuel, but with childlike innocence, he mistook the voice for Eli's. "Then he ran to Eli and said, 'Here I am, for you called me.' But he said, 'I did not call, go back and lie down'" (1 Samuel 3:5). Eli, confused, told him to go back to sleep. This happened two more times, and each time Samuel responded with the same earnest confusion. "Now Samuel did not yet know the Lord, nor had the word of the Lord yet been revealed to him" (1 Samuel 3:7).

On the third occasion, Eli realized something greater was happening. It was no mistake; the Lord was calling on the boy. Despite his failures, Eli still feared the Lord. He gave Samuel wise counsel: "And Eli said to Samuel, 'Go lie down, and it shall be if He calls you, that you shall say, "Speak, Lord, for Your servant is listening."' So Samuel went and lay down in his place" (1 Samuel 3:9).

Samuel obeyed. When the Lord called again, Samuel responded as instructed. The message he received was devastating. The Lord told Samuel that He would soon carry out judgement against the house of Eli. The sons for their blasphemy and their father for failure to restrain them (1 Samuel 3:11–13). God concluded with finality: "Therefore I have

145

sworn to the house of Eli that the wrongdoing of Eli's house shall never be atoned for by sacrifice or offering" (1 Samuel 3:14).

This revelation weighed heavily on Samuel. As a child, he was naturally afraid to share such grim news. But when pressed by Eli, Samuel told him everything. Eli responded with solemn acceptance. "He is the Lord; let Him do what seems good to Him" (1 Samuel 3:18). This exchange marked a spiritual turning point. Israel was no longer rudderless. Unlike the days of the judges, when *everyone did as they saw fit*, there was now a prophet through whom God would speak again.

Samuel's calling was just the beginning. "And the Lord appeared again at Shiloh, because the Lord revealed Himself to Samuel at Shiloh by the word of the Lord" (1 Samuel 3:21). As Samuel grew, so did his reputation and influence. The silence had been broken. Through Samuel, God's voice was heard once more.

FALSE CONFIDENCE AND NATIONAL DEFEAT

Some time later, Israel went to war with the Philistines. The initial battle was a disaster. "Then the Philistines drew up in battle formation to meet Israel. When the battle spread, Israel was defeated by the Philistines, who killed about four thousand men on the battlefield" (1 Samuel 4:2). After this defeat, the elders of Israel gathered to reflect. They recognized a spiritual component to the loss—but their solution was shallow. "When the people came into the camp, the elders of Israel said, 'Why has the Lord defeated us today before the Philistines? Let's take the ark of the covenant of the Lord from Shiloh, so that He may come among us and save us from the power of our enemies'" (1 Samuel 4:3).

Perhaps they remembered Joshua's victories—crossing the Jordan, the fall of Jericho—when the Ark led the way (Joshua 3:3, 6:7). But they missed the deeper truth: Joshua's success came from obedience and discipline, not superstition. Still, the elders pressed forward. They sent for the Ark, carried by Eli's sons, Hophni and Phinehas, men already condemned for their wickedness (1 Samuel 2:34).

The Ark's arrival electrified the camp. "And as the ark of the covenant of the Lord was coming into the camp, all Israel shouted with a great shout, so that the earth resounded" (1 Samuel 4:5). It was a psychological tactic as much as a spiritual one. As Musashi notes: "In large-scale strategy you can frighten the enemy not by what you present to their eyes, but by shouting, making a small force seem large, or by threatening them from the flank without warning. These things all frighten. You can win by making best use of the enemy's frightened

147

rhythm" (Musashi 26). Israel's war cry echoed like a rattlesnake shaking its tail; loud, fearsome, and full of warning.

The Philistines were indeed shaken. Though they misunderstood the Ark as a pagan idol, they knew its reputation. "Woe to us! Who will save us from the hand of these mighty gods? These are the gods who struck the Egyptians with all kinds of plagues in the wilderness" (1 Samuel 4:8). But fear became iron resolve. "Take courage and be men, Philistines, or you will become slaves to the Hebrews, as they have been slaves to you; so be men and fight!" (1 Samuel 4:9).

And they did. "So the Philistines fought and Israel was defeated, and every man fled to his tent; and the defeat was very great, for thirty thousand foot soldiers of Israel fell" (1 Samuel 4:10). The Ark was captured. Both of Eli's sons were killed (1 Samuel 4:11). It was a military catastrophe, and a spiritual one. The people had treated the Ark like a charm, forgetting that God's presence demands obedience, not theatrics. The bitter prophecy spoken to Eli had now come to pass (1 Samuel 2:34, 3:12–14).

A lone survivor fled the battlefield to Shiloh, his clothes torn, his face smeared with dust. "When he came, behold, Eli was sitting on his seat by the road keeping watch, because his heart was anxious about the ark of God. And the man came to give a report in the city, and all the city cried out" (1 Samuel 4:13). Though Eli's eyes were failing, his heart still longed for the Lord. Hearing the uproar, he demanded an explanation. The man delivered grim news: Israel was defeated, Eli's sons were dead, and the Ark of God had been captured (1 Samuel 4:17).

It was not the death of his sons that broke Eli; but the capture of the Ark. "When he mentioned the ark of God, Eli fell off the seat backward beside the gate, and his neck was broken and he died, for he was old and heavy. And so he judged Israel for forty years" (1 Samuel 4:18). In a tragic irony, Eli's final act showed devotion to God over family. Yet that devotion had come too late. He failed to discipline his sons when it counted most. As Solomon later wrote, "Do not withhold discipline from a child; Though you strike him with the rod, he will not die. You shall strike him with the rod And rescue his soul from Sheol" (Proverbs 23:13–14). Sheol roughly means *underworld* or *realm of the dead*.

How did Israel fall so far? Hadn't it followed Joshua's example by bringing the Ark to battle? Sadly, no. Its approach was far closer to the

Philistines' than Joshua's. The polytheistic pagans mistook the Ark for an idol; one god among many. Israel, despite countless warnings, acted with the same ignorance. It treated the Ark like a magical relic instead of a holy symbol of God's covenant.

The Ark was not the source of Israel's power; obedience and faith were. Joshua succeeded because he honored the Lord's word. The elders meant well, but without true repentance or spiritual clarity, their gesture became a hollow act. That gap in understanding cost Israel thirty thousand men and the Ark itself.

This echoes the sin of Achan. Both he and Eli's sons placed themselves above God's commands. Phinehas and Hophni were trained priests. They should have known better. Yet sometimes, decay must be cleared before new life can flourish. Israel's priesthood had to fall before Samuel could rise.

THE WRATH OF GOD

What happened to the Ark after the devastating defeat? The Philistines took it as a trophy and brought it to Ashdod, placing it beside their idol Dagon in his temple. But the Lord would not be mocked. The next morning, Dagon had toppled before the Ark. Undeterred, they set the statue upright again. Yet the following day, Dagon lay face down once more; his head and hands broken off (1 Samuel 5:1–4). The message was clear: Dagon was no match for the Lord. The destruction of pagan idols echo back to Gideon and Samson.

What began with a broken idol quickly turned into full-blown judgment. The people of Ashdod were struck with tumors, and rats plagued their land (1 Samuel 6:5). Panicked, the leaders sent the Ark to Gath, only for the same divine punishment to strike there (1 Samuel 5:9). Gath passed it along to Ekron, but by then, the entire Philistine coalition was terrified. Death and destruction followed the Ark everywhere it went. The people demanded action: "Therefore they sent word and gathered all the governors of the Philistines, and said, 'Send away the ark of the God of Israel and let it return to its own place, so that it will not kill us and our people!' For there was a deadly panic throughout the city; the hand of God was very heavy there" (1 Samuel 5:11).

After seven months of suffering, the Philistines finally relented. They consulted their priests and diviners, who proposed a guilt offering to the God of Israel: five golden tumors and five golden rats, symbols of their affliction and guilt. They recognized their mistake like the Pharaoh in Egypt (1 Samuel 6:5–6).

Still, they couldn't fully let go of their pride. They set up a test: untrained nursing cows would pull a cart carrying the Ark. If the cows turned toward Israelite territory, they would know the plagues were from God (1 Samuel 6:8–9). Even though they bellowed for their calves, the cows marched straight to Beth Shemesh, a priestly town. The Philistine rulers followed from a distance, witnessing firsthand the sovereignty of Israel's God (1 Samuel 6:12).

The Israelites rejoiced at the Ark's return. In awe, they offered sacrifices, and the Levites handled the Ark appropriately (1 Samuel 6:13–15). But the celebration was short lived. Some of the men of Beth Shemesh, overcome with curiosity or pride, looked into the Ark. This blatant violation of God's law proved fatal as HE struck them down (1 Samuel 6:19). The mourning townspeople recognized their error. This holy fear led them to move the Ark to Kiriath Jearim, where it would remain for twenty years (1 Samuel 7:1–2).

What caused this tragedy? Pride. Just like Adam and Eve in the garden, the people disregarded God's clear instructions in favor of their own desires. The Lord had warned them long ago: even priests must consecrate themselves before approaching the holy things of God (Exodus 19:20–22). Beth Shemesh was a Levitical town; its people knew better.

The Ark's journey reminded both the Philistines and the Israelites that God's presence is not to be taken lightly. The Ark is not a weapon to be wielded or a relic to be admired. The Lord is holy, powerful, and just. And whether in judgment or in grace, His will always prevails. Put simply, "The fear of the Lord is the beginning of wisdom, And the knowledge of the Holy One is understanding" (Proverbs 9:10).

NATIONAL REPENTANCE AND HOLY VICTORY

After this bittersweet experience, Israel began to repent earnestly. In a speech echoing Moses and Joshua, Samuel gave a pathway forward. "Then Samuel spoke to all the house of Israel, saying, 'If you are returning to the Lord with all your heart, then remove the foreign gods and the Ashtaroth from among you, and direct your hearts to the Lord and serve Him alone; and He will save you from the hand of the Philistines'" (1 Samuel 7:3). Samuel wisely echoed the words his predecessors Joshua and Moses. Indeed, there is a reason the first two commandments are to honor the one true God and cast away false idols (Exodus 20:1–6). If one fails to honor these two commandments, what hope does one have of honoring the rest of them?

With a renewed spirit, Israel set forth. The people renewed their commitment to the Lord and Samuel assumed command of Israel. "Then Samuel said, 'Gather all Israel to Mizpah and I will pray to the Lord for you'" (1 Samuel 7:5). Israel followed his example dutifully, as it fulfilled a purification ritual. "So they gathered to Mizpah, and drew water and poured it out before the Lord, and fasted on that day and said there, 'We have sinned against the Lord.' And Samuel judged the sons of Israel at Mizpah" (1 Samuel 7:6).

At a glance, pouring out water may seem a simple gesture. But this act was deeply symbolic. The water represented their very hearts; emptied before God in full surrender. They offered not just their words, but also their will. Recall Hannah, Samuel's mother. She prayed with such anguish that Eli thought she was drunk. "But Hannah answered and said, 'No, my lord, I am a woman despairing in spirit; I have drunk neither wine nor strong drink, but I have poured out my soul before the Lord'" (1 Samuel 1:15). Like his mother before him, Samuel led Israel in heartfelt repentance. As King David later wrote, "Trust in Him at all times, you people; Pour out your hearts before Him; God is a refuge for us" (Psalm 62:8).

While Samuel rendered the burnt offering, the Philistines moved in. A massive force advanced on Israel's position near Mizpah. But before the Philistines could engage, the Lord thundered from the heavens. Cracking lightning and a fearsome storm rolled over the battlefield. Panic swept through the enemy ranks, and their formations dissolved into chaos (1 Samuel 7:10). Disorder set in before a single Israelite raised a sword. This was not the first time the Lord had used nature as a weapon of war. At the Red Sea, He drove back Pharaoh's army. At Gibeon, He hurled hailstones. In Deborah's time, the storm-swollen river swept away the enemy. Once again, the Lord broke the will of a superior force without Israel lifting a hand.

But divine intervention is only the first half of the victory. It was up to Samuel and the Israelites to finish the fight. They did not hesitate. "And the men of Israel came out of Mizpah and pursued the Philistines, and killed them as far as below Beth-car" (1 Samuel 7:11). This moment illustrates a key principle of warfare: when your enemy falters, strike without mercy. "When the enemy starts to collapse you must pursue him without letting the chance go. If you fail to take advantage of your enemies' collapse, they may recover" (Musashi 24). This principle applies equally well to large-scale combat and one-on-one fighting.

The Israelites understood this well. According to *De Re Militari*: "The pursuers can be in no danger when the vanquished have thrown away their arms for greater haste. In this case, the greater the number of the flying army, the greater the slaughter" (Vegetius 81). This is precisely what happened. The Philistines, thrown into confusion by the Lord's storm, had no hope of regrouping. Israel pressed the advantage with discipline and courage.

Only with God's favor could such a turnaround occur. The Lord disrupted the enemy's strength; Israel responded with strength of its own. This was not just a tactical victory, it was also a spiritual one. After the battle, Samuel commemorated the occasion with a stone memorial. True to his faithful nature, he did not credit himself or his people. He gave glory where it was due, to God. He named the stone *Ebenezer*, meaning "stone of help" (1 Samuel 7:12).

This victory marked a turning point in Israel's conflict with the Philistines. "So the Philistines were subdued, and they did not come anymore within the border of Israel. And the hand of the Lord was against the Philistines all the days of Samuel" (1 Samuel 7:13). From this point on, Samuel entered a new chapter of leadership. He established a traveling circuit of cities—Bethel, Gilgal, Mizpah—where he judged Israel and offered sacrifices. He always returned home to Ramah, where he continued to serve the Lord (1 Samuel 7:15–17).

This battle at Mizpah was more than just a military triumph. It was a confirmation that faith, not numbers, secures the victory. Israel had gathered, confessing its sins, pouring itself out before the Lord. It was vulnerable. Yet in that posture of humility, it found strength. The Philistines may have had greater weapons and numbers, but Israel had something better—the favor of the Lord and the courage to act when the moment came.

THE RISE OF THE MONARCHY

Peace is a precious commodity. Its value lies in its rarity. Samuel accomplished incredible things for Israel, but nothing lasts forever. As he grew old, his leadership began to fade. To maintain order, he appointed his two sons, Joel and Abijah, as judges. Sadly, they did not follow their father's path. "His sons, however, did not walk in his ways but turned aside after dishonest gain, and they took bribes and perverted justice" (1 Samuel 8:3).

Once again, we see human fallibility on full display. Scripture doesn't offer much about Samuel's sons, but some speculation is reasonable. Perhaps Samuel unknowingly picked up certain traits from his predecessor, Eli. As the saying goes: *Give me the boy, and I'll show you the man.* Eli raised Samuel from a young age—surely, some of his habits left a mark, for better or worse. Or perhaps the era of peace itself dulled Israel's spiritual edge. Prosperity often breeds complacency. As the brilliant Sun Tzu wrote, "If, however, you are indulgent, but unable to make your authority felt; kind-hearted, but unable to enforce your commands ... your soldiers must be likened to spoilt children; they are useless for any practical purpose" (Tzu 21). Compared to their father, Samuel's sons were useless indeed.

Whatever the cause, one thing was clear: the people had lost all faith in their leadership. The elders of Israel came to Samuel with a demand. "Then all the elders of Israel gathered together and came to Samuel at Ramah; and they said to him, 'Behold, you have grown old, and your sons do not walk in your ways. Now appoint us a king to judge us like all the nations'" (1 Samuel 8:4–5). This request displeased Samuel, and he turned to the Lord in prayer.

God's response was sobering. "And the Lord said to Samuel, 'Listen to the voice of the people regarding all that they say to you, because they have not rejected you, but they have rejected Me from being King over them. Like all the deeds which they have done since the day that I brought them up from Egypt even to this day—in that they have abandoned Me and served other gods—so they are doing to you as well. Now then, listen to their voice; however, you shall warn them strongly and tell them of the practice of the king who will reign over them'" (1 Samuel 8:7–9).

Once again, Israel had forsaken its true King.

Samuel, ever faithful, obeyed. He delivered a solemn warning. Earthly kings bring earthly burdens. There is no such thing as a free lunch. Samuel outlined the tradeoff: a king would draft their sons for war and their daughters for labor, seize the best of their land, and impose heavy taxes—taking a tenth of their grain, wine, and livestock (1 Samuel 8:11–17). In short, the people would trade their freedom for the illusion of stability. Of course, this would happen. Kings may be powerful, but they are still mortal. And with mortality comes fallibility. One cannot exist without the other.

After this massive list of downfalls, Samuel issued a chilling prophetic warning. "Then you will cry out on that day because of your king whom you have chosen for yourselves, but the Lord will not answer you on that day" (1 Samuel 8:18). While the Israelites had cast aside their Baals and Ashtoreths, they were quick to adopt another vice. They merely swapped one false idol for another.

People are imperfectible creatures dwelling in a fallen kingdom. Placing your full trust in any mortal man will eventually lead to disaster. In no time, the people began to speak of this future king with an idolatrous fervor. With arrogant hearts, they completely rejected Samuel's sage warning. "Yet the people refused to listen to the voice of Samuel, and they said, 'No, but there shall be a king over us, so that we also may be like all the nations, and our king may judge us and go out before us and fight our battles'" (1 Samuel 8:19–20).

By demanding a king, Israel was rejecting God's grace. It was not merely asking for leadership—it was abandoning its spiritual inheritance. It sought to imitate the ways of the world, forgetting that it was not meant to be like the other nations. Israel was chosen to be different, to reflect God's holiness. "So you are to be holy to Me, for I the Lord am holy; and I have singled you out from the peoples to be Mine" (Leviticus 20:26).

But how was it set apart? And why?

Let's take a look back at how it all started.

THE ABRAHAMIC COVENANT AND THE COST OF KINGS

The significance of this governmental transition cannot be overstated. The elders believed they were rejecting Samuel's corrupt sons, but in truth, they were rejecting the Lord's will. Israel had begun as a nation guided by prophetic judges—leaders raised up by God Himself. Yet now, the people demanded a monarchy. This was more than a political shift; it was a spiritual one. Israel was choosing the ways of the world over God's design.

To understand what was at stake, we must go back to the very roots of Israel's identity, back to one man: Abraham. Originally named Abram, he received a divine call that would change the course of history. "Now the Lord said to Abram, 'Go from your country, and from your relatives and from your father's house, to the land which I will show you; and I will make you into a great nation, and I will bless you, and make your name great; and you shall be a blessing; and I will bless those who bless you, and the one who curses you I will curse. And in you all the families of the earth will be blessed'" (Genesis 12:1–3).

At seventy-five years old, Abraham answered the call with courage and faith. In return, God gave him a lasting promise. "Indeed I will greatly bless you, and I will greatly multiply your seed as the stars of the heavens and as the sand, which is on the seashore; and your seed shall possess the gate of their enemies. And in your seed all the nations of the earth shall be blessed, because you have obeyed My voice" (Genesis 22:17–18). Despite their advanced age, God granted Abraham and Sarah a son, Isaac (Genesis 21). In time, Isaac married Rebekah. When she struggled to conceive, he prayed on her behalf.

158

The Lord's answer to Rebekah was clear and prophetic: "Two nations are in your womb; and two peoples will be separated from your body; and one people will be stronger than the other; and the older will serve the younger" (Genesis 25:23). Rebekah gave birth to Esau and Jacob. Though Esau was the firstborn, Jacob received the covenant blessing (Genesis 27:28–29).

Jacob prospered and had many sons through his wives Leah, Rachel, Bilhah, and Zilpah. During a moment of crisis, he found himself alone, and it was then that God met him in a mysterious and powerful way. He wrestled with an angel of the Lord until daybreak, refusing to let go.

In recognition of Jacob's struggle and persistence, God gave him a new name, Israel, which means "he struggles with God" (Genesis 32). It was a fitting name for the man and for the nation his family would become. Despite repeated rebellion, the Lord's covenant promise endured. Jacob's wives had a total of twelve sons (Genesis 35:23–26). And so, the twelve tribes of Israel were born. Each rooted in covenant, each belonging to a nation that was supposed to be set apart.

Hark back to the Lord's promise to Abraham, after he passed his final test: "And in your seed all the nations of the earth shall be blessed, because you have obeyed My voice" (Genesis 22:18). This divine promise ultimately found its fulfillment in the arrival of Jesus Christ. David, from the tribe of Judah, was part of the Messiah's ancestral line (Matthew 1). The arrival of Christ changed everything. Both the Apostle Paul and the Apostle Peter made this connection clear—Jesus was the promised offspring through whom all nations would be blessed. Through faith in Him, both Jew and Gentile could now become heirs of the same covenant that began with Abraham (Galatians 3:7–9, Acts 3:25–26). After the sacrifice of Christ, it was no longer about bloodlines, but faith.

One thing is clear throughout this story: God cherishes obedience and faithfulness. Moses underscores this theme many times, reminding Israel of its divine appointment. "For you are a holy people to the Lord your God; the Lord your God has chosen you to be a people for His personal possession out of all the peoples who are on the face of the earth" (Deuteronomy 7:6). There were likely many others God could have chosen. But Abraham had the courage to say *yes*. Despite his old age, he answered the call to adventure, duty and faith.

In his time, advanced civilizations like Egypt dominated the region with power and wealth. Yet God was not seeking earthly strength—He was seeking fidelity of the heart. Moses put it plainly: "The Lord did not make you His beloved nor choose you because you were greater in number than any of the peoples, since you were the fewest of all peoples, but because the Lord loved you and kept the oath which He swore to your forefathers, the Lord brought you out by a mighty hand and redeemed you from the house of slavery, from the hand of Pharaoh king of Egypt" (Deuteronomy 7:7–8).

The Israelites were not chosen for their power or technology. They were chosen to be different—set apart. "And the Lord has today declared you to be His people, His personal possession, just as He promised you, and that you are to keep all His commandments; and that He will put you high above all the nations which He has made, for glory, fame, and honor; and that you shall be a consecrated people to the Lord your God, just as He has spoken" (Deuteronomy 26:18–19).

All of this brings us back to Samuel. Now an old man, he stood at a crossroads that would forever change the trajectory of Israel. Would he continue the divinely led system of judges? Or would he give in to the people's demand for a king?

After warning them of the consequences, Samuel prayed for guidance. God gave him a surprising answer: "Now after Samuel had heard all the words of the people, he repeated them in the Lord's hearing. And the Lord said to Samuel, 'Listen to their voice and appoint a king for them.' So Samuel said to the men of Israel, 'Go, every man to his city'" (1 Samuel 8:21–22). And so, the era of judges was ending. A new chapter was about to begin; one that would test the heart of Israel all over again.

This message from God likely left Samuel confused. However, with the clarity of hindsight, we can glean several spiritual lessons. First, just like Adam and Eve, God respects humanity's free will. We are free to act as we please, but we are not free from the consequences.

Scripture gives us countless examples. We've examined Achan's sin in depth (Joshua 7). That story leads to the second point. By granting Israel a king, God allowed it to experience a clear contrast in leadership.

Despite always providing for the faithful, the people demanded more. During the forty-year pilgrimage to the Promised Land, God cared for Moses and the Israelites through brutal conditions. HE ensured they had food and water (Exodus 16, 17). HE even gave them manna, a miraculous provision during a fragile transition. Early Israel was always outmatched in manpower, technology, and weaponry. Yet, with God's favor, it prevailed. Earthly armies are no match for the Lord's power. As Samuel warned, the flaws of human rulers would soon become painfully clear.

Last, in God's infinite wisdom, He can redirect human folly for divine purposes. We saw this with Samson; God used his mistakes as tools of judgment against the Philistines. Likewise, God would use Israel's monarchy as a stepping stone toward the arrival of the Messiah, Jesus Christ.

Israel was about to learn a vital truth: no mortal man can save us. Only the Lord can. It misunderstood the real issue, the fallibility of the human heart. Humans are incapable of perfection. The danger of idolizing people is that they will inevitably fail. When they do, idolization often turns to demonization. The once-praised figure becomes the scapegoat for every problem. People demand change and latch onto another false idol. The cycle repeats in perpetuity. Only God can break that cycle. Only He can save us from ourselves.

Like a loving father, God tried to protect Israel. Yet out of love, He refused to trample its free will. Picture the scene: a father warns his son to avoid a hot stove. The curious boy wants to know why. The patient father explains the danger, again and again. The boy doesn't plan to touch it, he thinks to himself. He just wants a closer look. But in his folly, he gets too close. He gets burned. The pain teaches what words could not. From that day forward, the boy never disobeys that warning again.

In the same way, God allowed Israel to learn the hard way. "My son, do not reject the discipline of the Lord or loathe His rebuke, for whom the Lord loves He disciplines, just as a father disciplines the son in whom he delights" (Proverbs 3:11–12). Sometimes pain truly is the best teacher.

SAMUEL CONCLUSION

L et's recap some important takeaway lessons from Samuel's story. His very birth was miraculous—born to a barren woman who cried out to the Lord. Once again, we see that God often raises up the least likely among us to accomplish His work. Samuel shared several parallels with Samson: both had miracle births and lifelong Nazirite vows. Yet their paths could not have been more different. While Samson struggled mightily in his personal life, Samuel remained steadfast in his devotion to the Lord.

This is even more impressive given his environment. Samuel was surrounded by poor role models. Often, the Bible gives us clear examples of what *not* to do. Eli and his sons illustrate this vividly. Though Eli revered God, he failed as both a father and spiritual leader. His passivity and indulgence proved disastrous. Eli traded short-term comfort for long-term ruin. There is a well-known saying in the military: *complacency kills*. Death by a thousand cuts is not just a metaphor, it's a reality.

The true tragedy is that people often fail to recognize the consequences of their inaction until it is too late. Eli's sons died because he refused to discipline them. He loved the Lord, but when the time came to correct his household, he let his sons pursue wickedness unchecked. This failure underscores an important point: *people-pleasing is not a Christian virtue.*

There is a vital distinction between niceness and kindness. Niceness is shallow. It's about making people feel good, no matter what the cost. Kindness, by contrast, is rooted in concern for the other's true wellbeing, even if it stings. Eli was *nice* to his sons; he spared their feelings. But he was far from *kind*; his leniency enabled their destruction.

Niceness is handing a heroin addict a clean syringe with a smile. Kindness is checking them into rehab, forcibly if needed. The *nice* man pats you on the back. The *kind* man watches your back. Our culture is filled with hollow platitudes: "Be nice," "Be tolerant," "Spread positivity." These slogans are often used to shame Christians into silence. But Christ and His apostles were clear: people-pleasing is not part of the Gospel.

Jesus said, "And do not be afraid of those who kill the body but are unable to kill the soul; but rather fear Him who is able to destroy both soul and body in hell.... Therefore, everyone who confesses Me before people, I will also confess him before My Father who is in heaven. But whoever denies Me before people, I will also deny him before My Father who is in heaven" (Matthew 10:28, 32–33).

In John's Gospel, Jesus confronted the idolatry of human validation. "I have come in My Father's name, and you do not receive Me; if another comes in his own name, you will receive him. How can you believe, when you accept glory from one another and you do not seek the glory that is from the one and only God?" (John 5:43–44).

The Apostle Paul doubled down on this in his letter to the Galatians: "For am I now seeking the favor of people, or of God? Or am I striving to please people? If I were still trying to please people, I would not be a bond-servant of Christ" (Galatians 1:10). And again, in his letter to the Thessalonians: "For our exhortation does not come from error or impurity or by way of deceit; but just as we have been approved by God to be entrusted with the gospel, so we speak, not intending to please people, but to please God, who examines our hearts. For we never came with flattering speech, as you know, nor with a pretext for greed—God is our witness— nor did we seek honor from people, either from you or from others, though we could have asserted our authority as apostles of Christ" (1 Thessalonians 2:3–6).

This is not a license for cruelty or rudeness, but a reminder to stand firm when cultural norms contradict the will of God. If you find yourself questioning whether a behavior is sinful, it probably is. When in doubt, seek the Word. Stay true to God's path, even when it's lonely. You may feel like an outsider, but do not fear. As Scripture assures us, "we confidently say, 'The Lord is my helper, I will not be afraid. What will man do to me?'" (Hebrews 13:6).

In a world filled with Elis, choose to be a Samuel.

Samuel called Israel back to repentance at a time of spiritual decline. He also served as a capable military leader, strengthening the troops through faith and discipline. At Mizpah, the Lord responded to Samuel's intercession with thunderous judgment on the Philistines, leading to a great victory (1 Samuel 7:11).

When the people later demanded a king, Samuel warned them of the consequences. Though their demands grieved him, Samuel obeyed the Lord and let the people learn through experience. Just as God used Samson's flawed strength for divine judgment, He used Israel's monarchy to bring forth the Messiah. David, whom Samuel anointed, became the forefather of Christ.

All of this was possible because Samuel trusted the Lord more than his own instincts. Only in the darkness does faith truly shine.

PART FOUR:
SAUL'S RISE TO KINGSHIP

THE RISE OF SAUL—PROMISE AND POTENTIAL

Samuel understood the Lord's message loud and clear. Though his time as Israel's judge was ending, his calling had not ended. He still served as both prophet and priest; God's chosen intermediary for the people. Now he was tasked with ushering in Israel's first king.

God revealed the future king to Samuel the day before they met: "About this time tomorrow I will send you a man from the land of Benjamin, and you shall anoint him as ruler over My people Israel; and he will save My people from the hand of the Philistines. For I have considered My people, because their outcry has come to Me" (1 Samuel 9:16). The man God chose was Saul, son of Kish, a Benjamite from a family of standing. Saul made an immediate impression. He was tall, striking, and had a regal bearing; everything the people imagined in a king. But Saul was still unaware of the path laid before him.

When Kish's donkeys went missing, he sent Saul and a servant to retrieve them. The pair searched high and low across multiple territories. After days with no success, Saul grew concerned. "When they came to the land of Zuph, Saul said to his servant who was with him, 'Come, and let's return, or else my father will stop being concerned about the donkeys and will become anxious about us'" (1 Samuel 9:5). His servant urged another course as he said: "Behold now, there is a man of God in this city, and the man is held in honor; everything that he says definitely comes true. Now let's go there, perhaps he can tell us about our journey on which we have set out" (1 Samuel 9:6).

Saul was open to this spiritual appeal, but they had little left to offer the prophet. "Then Saul said to his servant, 'But look, if we go, what

166

shall we bring the man? For the bread is gone from our sacks and there is no gift to bring to the man of God. What do we have?'" (1 Samuel 9:7). Fortunately, the servant had a silver coin, just enough for a respectful offering. Encouraged, they pressed on.

When they arrived in town, they asked where the prophet could be found. Their timing was providential. One local told them, "As soon as you enter the city you will find him before he goes up to the high place to eat, for the people will not eat until he comes, because he must bless the sacrifice; afterward those who are invited will eat. Now then, go up, for you will find him about this time" (1 Samuel 9:13). Saul and his servant hurried up to the high place, and there, they crossed paths with Samuel. "When Samuel saw Saul, the Lord said to him, 'Behold, the man of whom I spoke to you! This one shall rule over My people'" (1 Samuel 9:17).

Still unaware of what was unfolding, Saul approached Samuel with an innocent request. "Please tell me where the seer's house is" (1 Samuel 9:18). In those days, prophets were also called seers. Samuel revealed his identity, then prepared to deliver news that would change Saul's life forever.

Saul was about to receive far more than the livestock he was searching for. "And Samuel answered Saul and said, 'I am the seer. Go up ahead of me to the high place, for you shall eat with me today; and in the morning I will let you go, and will tell you everything that is on your mind. And as for your donkeys that wandered off three days ago, do not be concerned about them, for they have been found. And for whom is everything that is desirable in Israel? Is it not for you and for all your father's household?'" (1 Samuel 9:19–20).

Imagine what Saul must have been thinking at that moment. He hadn't even explained why he had come. Yet Samuel already knew about the missing donkeys—and how long they had been gone. Saul quickly realized he was dealing with a prophet whose vision far surpassed normal eyesight. Surely, this was divine providence.

With a humble heart, Saul responded to Samuel: "Am I not a Benjaminite, of the smallest of the tribes of Israel, and my family the least of all the families of the tribe of Benjamin? Why then have you spoken to me in this way?" (1 Samuel 9:21). This moment echoes the story of Gideon. He too hesitated upon hearing his calling (Judges 6:15). Despite Saul's apprehension, Samuel pressed forward with assurance.

After the ceremonial sacrifice, Samuel led Saul into the dining hall reserved for honored guests. He seated the future king at the head of the table and signaled to the cook to bring out a special portion that had been set aside (1 Samuel 9:23). The cook returned with a choice cut of thigh—traditionally reserved for consecrated priests (Exodus 29:27; Leviticus 7:32). This was a sacred gesture, signifying honor and divine purpose. "Then the cook took up the leg with what was on it and placed it before Saul. And Samuel said, 'Here is what has been reserved! Place it before you and eat, because it has been kept for you until the appointed time, since I said I have invited the people.' So Saul ate with Samuel that day" (1 Samuel 9:24),

They spent the evening talking, and the next morning Samuel prepared to send Saul on his way. But before parting, there was one final act of significance. "Then Samuel took the flask of oil, poured it on Saul's head, kissed him, and said, 'Has the Lord not anointed you as ruler over His inheritance?'" (1 Samuel 10:1).

After the anointing, Samuel gave Saul a sequence of detailed prophecies. First, he would meet two men near Rachel's tomb who would confirm that the donkeys had been found (1 Samuel 10:2). Then he would encounter three men at the great tree of Tabor: "three men going up to God at Bethel will meet you: one carrying three young goats, another carrying three loaves of bread, and another carrying a jug of wine. And they will greet you and give you two loaves of bread, which you will accept from their hand" (1 Samuel 10:3–4). As a final confirmation, Saul would meet a group of prophets near Gibeah playing harps, tambourines, pipes, and lyres. This was to be a transformative encounter; one that would change Saul's heart and confirm that God's spirit was upon him (1 Samuel 10:5).

This final sign is where things truly began to change. Samuel told Saul: "Then the Spirit of the Lord will rush upon you, and you will prophesy with them and be changed into a different man. And it shall be when these signs come to you, do for yourself what the occasion requires, because God is with you" (1 Samuel 10:6–7).

Samuel then gave Saul one final directive: "And you shall go down ahead of me to Gilgal; and behold, I will be coming down to you to offer burnt offerings and sacrifice peace offerings. You shall wait seven days until I come to you and inform you of what you should do" (1 Samuel 10:8). Samuel would continue to serve in his priestly and prophetic capacity alongside Israel's new monarchy.

As surely as the Lord lives, every one of Samuel's prophecies came to pass. Saul had left home searching for donkeys. Now he had been anointed the first king of Israel. "Then it happened, when he turned his back to leave Samuel, that God changed his heart; and all those signs came about on that day" (1 Samuel 10:9). When he encountered the prophets at Gibeah, the Holy Spirit filled Saul, and he joined them in ecstatic worship. The people were shocked: "And it came about, when all who previously knew him saw that he was indeed prophesying with the prophets, that the people said to one another, 'What is this that has happened to the son of Kish? Is Saul also among the prophets?'" (1 Samuel 10:11).

Later, Saul's uncle inquired about his travels. Saul simply said he had been looking for the donkeys. He did not mention Samuel's anointing. But Saul's divine appointment would not remain secret for long.

Samuel soon gathered all Israel at Mizpah. He began with a sharp rebuke: "This is what the Lord, the God of Israel says: I brought Israel up from Egypt, and I rescued you from the hand of the Egyptians and from the power of all the kingdoms that were oppressing you. But today you have rejected your God, who saves you from all your catastrophes and your distresses; yet you have said, 'No, but put a king over us!' Now then, present yourselves before the Lord by your tribes and by your groups of thousands'" (1 Samuel 10:18–19).

The tribes were assembled and the process of selection by lot began. It narrowed to the tribe of Benjamin, then to Saul's family, and finally to Saul himself. But he was nowhere to be found. "Therefore they inquired further of the Lord: 'Has the man come here yet?' And the Lord said, 'Behold, he is hiding himself among the baggage'" (1 Samuel 10:22). Saul had concealed himself, likely overwhelmed by the gravity of what lay ahead. This note of insecurity would reappear throughout his reign.

Nevertheless, when Saul was brought forward, the people marveled at his stature. "Samuel said to all the people, 'Do you see him whom the Lord has chosen? Surely there is no one like him among all the people.'

So all the people shouted and said, 'Long live the king!'" (1 Samuel 10:24).

With the inaugural King of Israel presented, Samuel turned back to his prophetic role. He explained the responsibilities and limitations of kingship, in line with what Moses had written centuries before: "Now it shall come about, when he sits on the throne of his kingdom, that he shall write for himself a copy of this Law on a scroll in the presence of the Levitical priests. And it shall be with him, and he shall read it all the days of his life, so that he will learn to fear the Lord his God, by carefully following all the words of this Law and these statutes, so that his heart will not be haughty toward his countrymen, and that he will not turn away from the commandment to the right or the left, so that he and his sons may live long in his kingdom in the midst of Israel" (Deuteronomy 17:18–20).

Kingship in Israel came with great authority, but even greater accountability. The king was not to exalt himself above the law, but to submit to it.

Samuel recorded the duties of kingship on a scroll and laid it before the Lord. Then he dismissed the assembly. But not everyone was impressed. Some scorned Saul and refused to show him the honor due a king. Saul, for the time being, kept silent (1 Samuel 10:27).

SAUL'S EARLY VICTORIES— OPPORTUNITY AND WARNING

I n no time, a test of leadership imposed itself on Saul. An Israelite town came under siege, not by the Philistines this time, but by the Ammonites. Nahash, their ruthless leader, stormed the city of Jabesh Gilead. The Israelites pleaded for mercy via a peace treaty in exchange for subjugation (1 Samuel 11:1). But Nahash had no interest in treaties. Only wrath motivated him. "But Nahash the Ammonite said to them, 'I will make it with you on this condition, that I will gouge out the right eye of every one of you, and thereby I will inflict a disgrace on all Israel'" (1 Samuel 11:2).

This horrific threat had more than just physical consequences. It was a targeted strategy meant to cripple Israel's military capabilities. For years, the Philistines had systematically disarmed Israel—a hallmark tactic of oppressive regimes. They banned Israelite blacksmiths to prevent the manufacture of swords or spears (1 Samuel 13:19). The Israelites were left to improvise, relying on slings, bows, and farm tools turned into weapons. They paid high fees just to sharpen plows and sickles (1 Samuel 13:20–21). This weakened their army in three key ways: tactically, financially, and psychologically. This was a three-pronged trident that pinned Israel down for quite some time.

As Clausewitz explained, the objectives of war include "Gain victory and destroy the armed forces of the enemy" and "Obtain the material means of combat and other resources of the hostile army" (Clausewitz 11). The Philistines achieved both. But their dominance also served another purpose: humiliation. According to *De Re Militari*: "Fidelity is seldom found in troops disheartened by misfortunes" (Vegetius 63). Every blacksmith payment reminded Israel who was in charge.

Nahash's threat struck right at the heart of Israel's remaining strength. With so many troops relying on slings and bows, losing an eye would neutralize them entirely. Jabesh's elders knew they were cornered. So they stalled for time. "So the elders of Jabesh said to him, 'Allow us seven days to send messengers throughout the territory of Israel. Then, if there is no one to save us, we will come out to you'" (1 Samuel 11:3). Messengers fled to Saul's hometown of Gibeah. As the news spread, people wept in the streets. Saul arrived from the fields with his oxen and was stunned to find the town in mourning. He asked what had happened, and they told him about the siege.

The news of raiders usurping his kingdom sent a righteous anger through Saul. He wasn't going to lose his kingship before it even started. "Then the Spirit of God rushed upon Saul when he heard these words, and he became very angry" (1 Samuel 11:6). In a maneuver that harkened back to the days of the judges, he sent a dramatic message to rally the nation's fighting men. "He then took a yoke of oxen and cut them in pieces, and sent them throughout the territory of Israel by the hand of messengers, saying, 'Whoever does not come out after Saul and after Samuel, the same shall be done to his oxen.' Then the dread of the Lord fell on the people, and they came out as one person" (1 Samuel 11:7). Some modern readers may recoil at the imagery. But no one can question the results. Saul mustered an impressive three hundred and thirty thousand troops. Over a *quarter million* warriors, unified in purpose.

With the army assembled, Saul sent a bold message to the citizens of Jabesh. Their new leader would not let them fall. He promised their rescue by midday (1 Samuel 11:9). The people of Jabesh rejoiced at the news! With renewed hope, they issued a statement to the Ammonites, one that bought valuable time and lulled the enemy into overconfidence. "Tomorrow we will come out to you, and you may do to us whatever seems good to you" (1 Samuel 11:10). Their quick thinking mirrored the wisdom of Sun Tzu: "When able to attack, we must seem unable; when using our forces, we must seem inactive; when we are near, we must make the enemy believe we are far away; when far away, we must make him believe we are near" (Tzu 4).

The trap was set. The enemy let down its guard, completely unaware of the hammer about to fall. Saul understood how to seize the initiative and strike with precision. As stated in *The Art of War*: "A clever general, therefore, avoids an army when its spirit is keen, but attacks it when it is sluggish and inclined to return" (Tzu 14). That night, Saul lived up to this principle as he orchestrated a predawn raid. "The next morning Saul put the people in three companies; and they came into the midst of the camp at the morning watch, and struck and killed the Ammonites until the heat of the day. And those who survived scattered, so that no two of them were left together" (1 Samuel 11:11).

The overwhelming victory was decisive and total. Saul embodied the wisdom of the legendary samurai, Miyamoto Musashi: "In large-scale strategy it is important to cause loss of balance. Attack without warning where the enemy is not expecting it, and while his spirit is undecided follow up your advantage and, having the lead, defeat him" (Musashi 26). The rescue operation was a triumph. Saul had delivered, and the people knew it.

Caught up in the excitement, some among the crowd called for retribution. They demanded execution for the men who had once questioned Saul's worthiness. "Then the people said to Samuel, 'Who is he that said, "Shall Saul reign over us?" Bring the men, so that we may put them to death!'" (1 Samuel 11:12). But Saul took the higher path. He de-escalated the fervor and redirected their focus back to the true source of the victory. "But Saul said, 'Not a single person shall be put to death this day, for today the Lord has brought about victory in Israel'" (1 Samuel 11:13).

Saul wisely emphasized the victorious battle as the Lord's accomplishment, not his own. He began his reign as King of Israel with an air of humility and gratitude. As we will see shortly, this proved to be a fleeting moment in his career. In the meantime, however, jubilation filled the air. Capitalizing on the high spirits, Samuel summoned the Israelites to the holy site of Gilgal to confirm Saul's kingship formally. "So all the people went to Gilgal, and there they made Saul king before the Lord in Gilgal. There they also offered sacrifices of peace offerings before the Lord; and there Saul and all the men of Israel rejoiced greatly" (1 Samuel 11:15).

Saul was now officially inaugurated as King of Israel. Though Samuel stepped down from his public-facing role, he continued faithfully serving as priest and prophet. Even as he metaphorically passed the torch to Saul, he had one final address for Israel. His farewell speech, however, sharply contrasted with the joy of the crowning ceremony. It was not a celebration—it was a solemn warning.

Samuel began by appealing to his record. He reminded them of his faithfulness and integrity. The people acknowledged his honesty and devotion to the Lord. Then Samuel raised the stakes further, calling the Lord as his witness (1 Samuel 12:1–5). With a humble heart, Samuel shifted the focus back to God.

He recounted the Lord's deliverance from Egypt, crediting Moses and Aaron. Then he offered a condensed version of Israel's pattern during the time of the judges. He told of the brutal horrors of their oppression contrasted with the ever flowing mercy of the Lord (1 Samuel 12:9–11). This portion of the speech drew a clear line between their historic idolatry and their current political idolatry. Samuel laid bare the spiritual failure behind their demand for a king. "But when you saw that Nahash the king of the sons of Ammon was coming against you, you said to me, 'No, but a king shall reign over us!' Yet the Lord your God was your king. And now, behold, the king whom you have chosen, whom you have asked for, and behold, the Lord has put a king over you" (1 Samuel 12:12–13). Israel was getting *exactly* what it asked for. The good, bad, and the ugly.

Yet, it was not all doom and gloom. Ever faithful to the Lord, Samuel gave Israel the roadmap to success and prosperity. "If you will fear the Lord and serve Him, and listen to His voice and not rebel against the command of the Lord, then both you and the king who reigns over you will follow the Lord your God. But if you do not listen to the voice of the Lord, but rebel against the command of the Lord, then the hand of the Lord will be against you, even as it was against your fathers" (1 Samuel 12:14–15). The above statement makes one thing abundantly clear. The new monarchical governance was subject to the Lord's commands. Not just the people of the kingdom. In fact, the authority of Saul's new kingship was dependent on his fidelity to the Lord.

To emphasize the Lord's power, Samuel called upon Him once again. "Is it not the wheat harvest today? I will call to the Lord, that He will send thunder and rain. Then you will know and see that your wickedness is great which you have done in the sight of the Lord, by asking for yourselves a king" (1 Samuel 12:17). Soon after, the heavens erupted. Rain came pouring down upon Israel, ushered in by mighty thunder and lightning. The darkness of the rain clouds contrasted with the pure luminescence of the lightning bolts. Static electricity filled the air, raising the hair on the back of many anxious necks. While the Lord had used weather for destruction at other times, He was simply sending a message of His omnipotence for Samuel that day. The mighty scene left the crowd awestruck (1 Samuel 12:18). Perhaps a few of them realized the frailty of their new human king, compared to the one Sovereign King.

The electric spectacle elicited quite the reaction from the crowd. The Israelites now understood the Lord's power on a visceral level and begged Samuel for forgiveness. "Then all the people said to Samuel, 'Pray to the Lord your God for your servants, so that we do not die; for we have added to all our sins this evil, by asking for ourselves a king'" (1 Samuel 12:19). Yet again, Samuel stood as a beacon of hope for Israel. He reassured the crowd. Even though they had fallen short by idolizing Saul, there was a path to redemption. Samuel told them to serve the Lord with all their hearts (1 Samuel 12:20). Samuel then echoed Moses and Joshua while he concluded with a final warning against idolatry (1 Samuel 12:21).

Samuel again reiterated that the Lord would not reject His people, as the Lord was pleased to set them apart and make them His own. Samuel continued to reassure the crowd. While he was stepping down as judge, he would continue to pray for Israel and guide it towards a righteous path. Samuel concluded his farewell address with a bittersweet message; one we should all take to heart. "Only fear the Lord and serve Him in truth with all your heart; for consider what great things He has done for you. But if you still do evil, both you and your king will be swept away" (1 Samuel 12:24–25).

THE FIRST KING OF ISRAEL

I t was now official. Saul was King of Israel. By his new authority, Saul chose a company of three thousand fighting men for himself and his son, Jonathan. After selecting these troops, Saul dismissed the rest of the people back to their homes. While Saul made quick work of the Ammonite raiders, he wouldn't rest on his laurels. He soon found himself in a much larger conflict (1 Samuel 13:2).

In no time, Jonathan took the initiative. Echoing back to Samson, Jonathan brought the fight to the Philistines. The new king rallied around his son's vigor and ambition. "And Jonathan attacked the garrison of the Philistines that was in Geba, and the Philistines heard about it. Then Saul blew the trumpet throughout the land, saying, 'Let the Hebrews hear!' And all Israel heard the news that Saul had attacked the garrison of the Philistines, and also that Israel had become repulsive to the Philistines. Then the people were summoned to Saul at Gilgal" (1 Samuel 13:3–4). The messengers summoned the troops to Gilgal to aid Saul.

While provoking a much larger army can be risky, there is a firm logic to this strategy. *The Art of War* sheds light on this maneuver. "Whoever is first in the field and awaits the coming of the enemy, will be fresh for the fight; whoever is second in the field and has to hasten to battle will arrive exhausted. Therefore the clever combatant imposes his will on the enemy, but does not allow the enemy's will to be imposed on him" (Tzu 10). In other words, "Security against defeat implies defensive tactics; ability to defeat the enemy means taking the offensive" (Tzu 7).

The Rise of Saul

Was Saul taking decisive action against his adversaries? Or was he metaphorically kicking the hornet's nest? No one at the time knew. What they did know, however, was that this fight was going to be different. The Philistine forces dwarfed those of Nahash the Ammonite. Indeed, they showed up to the fight prepared. "Now the Philistines assembled to fight with Israel, thirty thousand chariots and six thousand horsemen, and people like the sand which is on the seashore in abundance; and they came up and camped in Michmash, east of Beth-aven" (1 Samuel 13:5). The odds were overwhelming. The Philistines had as many chariots as Saul had soldiers!

The seemingly endless Philistine army sent shockwaves through Israel. While numbers aren't everything, they do play an important factor in warfare. It is true that, "Achieving victory in war does not depend entirely upon numbers or simple courage; only skill and discipline will ensure it" (Vegetius 3). But did this newly established kingdom have what it takes to succeed against the odds?

Things quickly began to look grim for Israel. "When the men of Israel saw that they were in trouble (for the people were hard-pressed), then the people kept themselves hidden in caves, in crevices, in cliffs, in crypts, and in pits" (1 Samuel 13:6). In fact, the situation was so dire that some of the Israelites fled across the Jordan river. Meanwhile, Saul remained at Gilgal with his cowering compatriots (1 Samuel 13:7).

This is where Saul's character began to bear fruit. Recall from earlier that when Samuel summoned Saul, he was cowering away in a supply room (1 Samuel 10:20–22). This insecurity remained in him. Samuel had given Saul strict instructions about when they would meet again at Gilgal. Yet the deadline crept up and Samuel was nowhere to be found. Saul's insecurity spilled over into his troops. Saul failed to lead by example and his troops knew it. They began to scatter away from their newfound king.

A RECIPE FOR DESTRUCTION— FEAR OVER FAITH

Saul floundered as he lost control of the situation. Panicked, he rushed into a desperate maneuver. Rendering sacrifices was strictly a priestly duty to perform. Despite lacking the authority, Saul decided to perform these rituals himself. The anxious king told his men to bring him the sacrificial offerings (1 Samuel 13:9). With the haste known only to terrified men, Saul rendered these offerings in a misguided attempt to court the Lord's favor. Yet two wrongs don't make a right. "But as soon as he finished offering the burnt offering, behold, Samuel came; and Saul went out to meet him and to greet him" (1 Samuel 13:10).

Samuel immediately pressed Saul for answers. "And Saul said, 'Since I saw that the people were scattering from me, and that you did not come at the appointed time, and that the Philistines were assembling at Michmash, I thought, 'Now the Philistines will come down against me at Gilgal, and I have not asked the favor of the Lord.' So I worked up the courage and offered the burnt offering" (1 Samuel 13:11–12). Saul failed to see the error of his ways. It is true that, "In war discipline is superior to strength; but if that discipline is neglected, there is no longer any difference between the soldier and the peasant" (Vegetius 43). In an ironic twist, King Saul revealed he is more akin to a lowly peasant than a warrior king. This irony did not slip past Samuel, however.

The elderly priest did not sugar coat his criticism: "But Samuel said to Saul, 'You have acted foolishly! You have not kept the commandment of the Lord your God, which He commanded you, for the Lord would now have established your kingdom over Israel forever. But now your kingdom shall not endure. The Lord has sought for Himself a man after His own heart, and the Lord has appointed him ruler over His people, because you have not kept what the Lord commanded you'" (1 Samuel 13:13–14). Ironically, by rendering the offerings, Saul did not display his faith. Rather, his rash actions stemmed from his insecurity and flimsy faith in the Lord.

Saul didn't only step on the toes of Samuel. Rather, he defied the Lord's commands and invalidated his own conditional leadership of Israel. After the sharp criticism, Samuel left Gilgal, after which, Saul tallied up the remaining troops. Only six hundred remained (1 Samuel 13:15). Over two thirds of his men abandoned the same king they had begged for only a few days prior. According to *Principles of War*, "in war, faith in success should not die until the very last minute" (Clausewitz 37). Saul clearly didn't understand this timeless principle.

While Saul may have been cowardly and hesitant, his son Jonathan was anything but. Encouraged by his previous victory, Jonathan launched another attack on the Philistines. Recall from earlier that nearly all of Israel was disarmed by the Philistines. "So it came about on the day of battle that neither sword nor spear was found in the hands of any of the people who were with Saul and Jonathan, but they were found with Saul and his son Jonathan" (1 Samuel 13:22). Perhaps this exception imbued a sense of confidence in Jonathan. Undoubtedly, his faith in the Lord drove him forward. With no sense of caution, he brought the fight to the Philistines once again.

The Philistines stationed a detachment at a mountain pass. Seizing the opportunity, Jonathan told his armor bearer that they were going to the enemy outpost on the opposite side (1 Samuel 14:1). This was a daring mission for just two men. In fact, no one in Saul's court was aware of their departure. Not even the King of Israel himself knew that his son was provoking another conflict (1 Samuel 14:3).

The mountain pass rested between two cliffs, with a stretch of no man's land in between; an ideal location for ambush. Yet this dangerous terrain didn't fluster Jonathan in the slightest. "Then Jonathan said to the young man who was carrying his armor, 'Come, and let's cross over to the garrison of these uncircumcised men; perhaps the Lord will work for us, because the Lord is not limited to saving by many or by few!'" (1

Samuel 14:6). This was not blind risk, it was bold faith. Jonathan's trust in divine providence eclipsed any fear of numbers. Perhaps David thought of this incident when he wrote, "Even though I walk through the valley of the shadow of death, I fear no evil, for You are with me; Your rod and Your staff, they comfort me" (Psalm 23:4).

Jonathan's courage proved contagious; it typically is. His armor-bearer responded with loyalty and resolve: "His armor bearer said to him, 'Do everything that is in your heart; turn yourself to it, and here I am with you, as your heart desires'" (1 Samuel 14:7). Jonathan then proposed a quick test. The pair would walk openly in view of the Philistine guards. If the enemy invited them upward, it would be a divine sign that God had delivered them into their hands. "If they say to us, 'Wait until we come to you'; then we will stand in our place and not go up to them. But if they say, 'Come up to us,' then we will go up, for the Lord has handed them over to us; and this shall be the sign to us" (1 Samuel 14:9–10).

This bold strategy came with serious risk. As we have examined previously, "It is a military axiom not to advance uphill against the enemy" (Tzu 15).

Nonetheless, the two Israelites carried out the plan. The Philistines greeted them with mockery and arrogance. "When the two of them revealed themselves to the garrison of the Philistines, the Philistines said, 'Behold, Hebrews are coming out of the holes where they have kept themselves hidden.' So the men of the garrison responded to Jonathan and his armor bearer and said, 'Come up to us and we will inform you of something.' And Jonathan said to his armor bearer, 'Come up after me, for the Lord has handed them over to Israel'" (1 Samuel 14:11–12). At that moment, Jonathan recognized the Lord's hand in this. The Philistines underestimated their foe. In doing so, they made a costly mistake.

Jonathan ignored one military axiom, but he embodied another from *The Art of War*: "If your opponent is of choleric temper, seek to irritate him. Pretend to be weak, that he may grow arrogant" (Tzu 4). And arrogant they were. The Philistines assumed that their elevated outpost gave them full advantage. On paper, it was a perfect station. But in their pride, they effectively gave Jonathan the key to the front door.

What danger could two men possibly pose? They realized the answer far too late. "Then Jonathan climbed up on his hands and feet, with his armor bearer behind him; and the men fell before Jonathan, and his armor bearer put some to death after him. Now that first slaughter which Jonathan and his armor bearer inflicted was about twenty men within about half a furrow in an acre of land" (1 Samuel 14:13–14). Jonathan's bold strike proved one thing: "Valor is superior to numbers" (Clausewitz 88).

Jonathan's victory was so decisive, it became the first snowball in an avalanche. His surprise assault served as a catalyst for a much larger victory. God was evidently pleased with Jonathan's courage. At this point, the Lord tipped the scales dramatically in favor of Israel. The shockwaves of Jonathan's ambush reverberated through the entire Philistine camp. Then the Lord amplified the momentum. "And there was a trembling in the camp, in the field, and among all the people. Even the garrison and the raiders trembled, and the earth quaked so that it became a great trembling" (1 Samuel 14:15). Meanwhile, Saul's watchmen stood baffled by the scene unfolding before them. The once formidable army melted away in all directions. There were no ranks or formations, only sheer chaos (1 Samuel 14:16).

Saul gave the order to muster his men and quickly discovered that Jonathan and his armor bearer were missing. Seeking divine insight, Saul called on a priest to inquire of the Ark of the Covenant. But Saul grew impatient. Instead of waiting for an answer from the Lord, he cut the inquiry short: "While Saul talked to the priest, the commotion in the camp of the Philistines continued and increased; so Saul said to the priest, 'Withdraw your hand'" (1 Samuel 14:19). Once again, Saul placed his own urgency above divine guidance.

Not wanting to lose the advantage, Saul pressed forward. "Then Saul and all the people who were with him rallied and came to the battle; and behold, every man's sword was against his fellow Philistine, and there was very great confusion" (1 Samuel 14:20). This scene echoes the success of Gideon's PSYOP, and Joshua's victory at Gibeon (Judges 7:22, Joshua 10:9–10). Throwing enemies into confusion is a notable and repeated tactic of the Lord.

We have already explored the contagious nature of fear. If not suppressed, it spreads rapidly through the ranks. But the inverse is also true. Courage can be just as infectious. Sometimes, all it takes is one righteous act of bravery to awaken hearts across a nation. Jonathan's faith and valor did just that. As with Joshua before him, Jonathan obeyed the Lord and acted boldly. In return, God magnified his courage for all of Israel to witness. The mighty Philistine host, as numerous as the sand on the seashore, began to disintegrate before the Israelites. This turnaround reignited the fighting spirit in those who had been paralyzed by fear. "When all the men of Israel who had kept themselves hidden in the hill country of Ephraim heard that the Philistines had fled, they also closely pursued them in the battle" (1 Samuel 14:22). No longer were they cowering in caves and thickets; they now pursued the enemy with determination. Israel was safe—for now.

Yet despite the victory, Saul once again clouded the moment with poor judgment. In a rash and self-serving move, he made an impulsive oath that hindered his own forces. The scenario mirrors the tragic vow of Jephthah in the Book of Judges. While out on campaign, Saul ordered a fast. His reasoning was neither spiritual nor strategic—it was vain. "Now the men of Israel were hard-pressed on that day, for Saul had put the people under oath, saying, 'Cursed be the man who eats food before evening, and before I have avenged myself on my enemies.' So none of the people tasted food" (1 Samuel 14:24). Once again, Saul placed his own ego above the wellbeing of the kingdom. *De Re Militari* offers insight on this sort of behavior: "In civil dissensions men are so intent on the destruction of their private enemies that they are entirely regardless of the public safety" (Vegetius 66). Saul's reckless decree fits this warning perfectly.

Most of the troops honored this oath, despite the temptation. They remained disciplined and faithful. "All the people of the land entered the forest, and there was honey on the ground. When the people entered the forest, behold, there was honey dripping; but no man put his hand to his mouth, because the people feared the oath" (1 Samuel 14:25–26). Imagine how alluring the scene must have been. The troops had been marching, rucking, and fighting for countless hours. While they were famished, the abundant honey coaxed them in with a seduction unique to sweets. Yet they were unable to refresh themselves with this abundant food. However, Jonathan was unaware his father had bound them to this oath. His eyes lit up as the sugary honey energized him (1 Samuel 14:27).

Dutifully, one of the soldiers warned Jonathan about Saul's oath; along with the potential ramifications (1 Samuel 14:28). Here we see a rift forming between Jonathan and Saul. Jonathan addressed the soldier with a level head. "Then Jonathan said, 'My father has troubled the land. See now that my eyes have brightened because I tasted a little of this honey. How much more, if only the people had freely eaten today of the spoils of their enemies which they found! For now the defeat among the Philistines has not been great'" (1 Samuel 14:29–30). Indeed, this reasonable approach correlates with *De Re Militari*: "to enable the soldiers to charge with greater vigor, it was customary to order them a moderate refreshment of food before an engagement, so that their strength might be ... supported during a long conflict" (Vegetius 68).

Despite their exhaustion, they continued to battle the Philistines. Famished from the desperate trekking and fighting, they ate what they could, in spite of dietary laws. "So the people loudly rushed upon the spoils, and took sheep, oxen, and calves, and slaughtered them on the ground; and the people ate them with the blood" (1 Samuel 14:32). Similar to Jephthah, Saul's impulsive oath set them up for failure. Leaders are meant to be good shepherds of the flock. Yet instead of leading his men to greener pastures, he simply swapped one poor choice for a worse choice. This mistake soon caught up with Saul. One of the men pointed out this violation to the king (1 Samuel 14:33). Unaware that his rash oath was the root cause, Saul blamed the hungry troops for their violation. He then instructed the troops to roll a large stone over, to properly handle the food preparation. The troops obeyed their king dutifully, and finally ate.

After his men broke their fast, Saul attempted to regain control of the situation. "And Saul built an altar to the Lord; it was the first altar that he built to the Lord" (1 Samuel 14:35). Saul was still obsessed with his own vengeance. He told the fighting men, "'Let's go down after the Philistines by night and take plunder among them until the morning light, and let's not leave a man among them alive.' And they said, 'Do whatever seems good to you.' So the priest said, 'Let's approach God here'" (1 Samuel 14:36). The soldiers told him to do as he saw fit. Yet one of the priests told Saul they should seek guidance from the Lord first. So the King of Israel heeded this advice. "So Saul inquired of God: 'Shall I go down after the Philistines? Will You hand them over to Israel?' But He did not answer him on that day" (1 Samuel 14:37).

Tensions quickly escalated. God's silence should have been a wake-up call for Saul. Instead, the lack of response flustered him. Desperate for answers, he searched for a scapegoat. With no sign of self-reflection, Saul scolded the troops. "Then Saul said, 'Come here, all you leaders of the people, and investigate and see how this sin has happened today. For as the Lord lives, who saves Israel, even if it is in my son Jonathan, he shall assuredly die!' But not one of all the people answered him" (1 Samuel 14:38–39). The men remained silent. Here, we see Saul making a classic leadership mistake; he disregarded his subordinate commanders.

No matter how brave, intelligent, or capable a leader may be, he cannot be everywhere at once. He must rely on others to carry out the mission. As *Principles of War* observes, "In particular, you have to trust your subordinate commanders; therefore, the most trustworthy people should be selected for these posts, and this quality should count above all others" (Clausewitz 31). But Saul's egocentrism couldn't allow this. His personal vendettas overshadowed the mission, and his troops suffered as a direct result.

Eventually, it was revealed that Jonathan had unknowingly violated Saul's oath. But was this oath truly holy? Saul's decree wasn't about delivering the Lord's justice—it was about indulging his own wrath and vengeance. In fact, it appears not only to be poor strategy, but also a violation of Mosaic Law. Saul invoked the Lord's name for prideful purposes. Recall the Ten Commandments Moses brought down from Mount Sinai: "You shall not take the name of the Lord your God in vain, for the Lord will not leave him unpunished who takes His name in vain" (Exodus 20:7). Saul would not remain guiltless for long.

His rage reached a boiling point as he confronted his son. Jonathan confessed that he had tasted a bit of honey. "And Saul said, 'May God do the same to me and more also, for you shall certainly die, Jonathan!'" (1 Samuel 14:44). Fortunately, cooler heads prevailed. Filled with moral courage and righteous justice, the fighting men backed Jonathan. "But the people said to Saul, 'Must Jonathan die, he who has brought about this great victory in Israel? Far from it! As the Lord lives, not even a hair of his head shall fall to the ground, because he has worked with God this day.' So the people rescued Jonathan and he did not die" (1 Samuel 14:45). Realizing he was the odd man out, Saul backed down. Jonathan remained unharmed. Shortly after, Saul stopped pursuing the Philistines as they withdrew to their territory (1 Samuel 14:46).

THE BALANCE OF LEADERSHIP

The above conflict with Saul and his troops highlights the most fundamental dichotomy of leadership. Picture this: *a hand grasping oil*. At first glance, the task appears simple. Yet there are many subtle nuances to this performance. Hold too loosely, and the oil drips away slowly. Drop by drop, it fades until you are left with nothing. Eli and his two wicked sons are an example of this leadership style. Eli let many things slide. It was death by a thousand cuts until eventually, Eli had no handle on the situation at all. His sons slipped away from him—ultimately, to their premature deaths.

Conversely, if one squeezes too tightly, as Saul did, one too is left with nothing. One reaches the same destination even faster. For effective leadership, one must have balance and make adjustments as required. In other words, "as circumstances are favorable, one should modify one's plans" (Tzu 16). This is the delicate balance of effective leadership. Let's examine this idea in further detail.

As we saw with Eli, one cannot treat one's subordinates too laxly. The result of such negligence leads to a lack of discipline, insubordination, and potentially desertion or destruction. It is a timeless fact that "No state can either be happy or secure that is remiss and negligent in the discipline of its troops" (Vegetius 12). Reflect back to the Book of Judges. The Israelites continued to get raided and subjugated by their enemies. Of course, this isn't surprising in a time when everyone placed selfish desires above the Lord's (Judges 21:25). There was little unity or group cohesion. Without solid enforcement of rules and standards, things quickly descend into chaos and conflict.

Divide and conquer is one of the most fundamental concepts in combat. It is so foundational that even animals practice this strategy. Think of a pride of lions stalking a herd of gazelle. The lions stealthily assume position around the clueless herd. Once in range, they pounce, and the gazelles flee in unison. It is too much to keep up with the swift herd. Lions lose that race every time. Rather, they shift their focus to prey that has been separated from the protection of the herd. Once the lone target is identified, it's only a matter of time. The big cats surround the isolated gazelle and enjoy their hard-fought meal.

Humans are also animals of the pack. Without structure and discipline, we are all lone gazelles—sitting ducks waiting to get picked off by the first predator that passes by. The Apostle Peter invoked similar imagery in a letter to early church practitioners. He compared Satan to a prowling lion, ready to pounce on unsuspecting prey (1 Peter 5:8–9) Faith and community keep us safe from the ever-lurking beast.

Conversely, a leader cannot overtax his troops either. He may get away with it for a short time, but the approach is unsustainable in the long term. Saul treated his men like they were machines. Yet even machinery breaks down if not properly maintained. Leaders must find a balance. And this must be done gradually. *De Re Militari* sheds wisdom on this topic: "To accustom soldiers to carry burdens is ... an essential part of discipline" (Vegetius 15). The critical word above is *accustom*. This highlights the necessity of cultivating discipline over time. It must be layered in, like fine Damascus steel.

Additionally, Clausewitz comments on this as well: "A general who requires enormous exertions and the greatest sacrifices from his forces will have an army used to these hardships—and what a great advantage the army will have when compared to the opponent, and how much faster they will accomplish the goal, despite all difficulties" (Clausewitz 35). The key phrase above is *used* to hardships. Again, this implies the necessity of cultivating discipline over time.

Saul pushed his troops too far. But what if there was a remedy for this? As Sun Tzu brilliantly stated, "Unhappy is the fate of one who tries to win his battles and succeed in his attacks without cultivating the spirit of enterprise; for the result is waste of time and general stagnation" (Tzu 27). But what is this spirit of enterprise? We previously examined this concept with Joshua. It is a form of morale. Joshua was able to cultivate this because he gave his troops a unifying vision. He spoke to them and acknowledged their sacrifices. He spoke of the challenges they would soon face, and the glory of victory. Joshua even had common foot soldiers participate in rituals and ceremonies (Joshua 4:1–7). Saul failed to do any of this.

Nothing is more lethal than a man on a mission. But he must truly believe the cause he is fighting for is righteous and just. He will hike through snow-covered mountains or treacherous swamps. He will attack the enemy no matter how outnumbered or outmatched technologically. Men will move mountains for the right cause. As we examined previously, "the way of the warrior is resolute acceptance of death" (Musashi 1). Absolutely no one wants to die only to clean up someone else's dirty laundry, so to speak. Saul's men weren't fighting for a grand cause or ideal. Rather, it was the personal bidding of the egocentric king. Saul dug his own grave when he put his own petty interest above that of his fighting men. Kings are powerful, but the source of their power comes from armed men willing to enforce their agenda. His soldiers saw right through Saul's selfish reasoning.

So, what can leaders do to be more effective? *Be more like Joshua and less like Saul.* According to *De Re Militari*, a commander should: "form his troops to submission and obedience by habit and discipline [rather] than ... by the terror of punishment" (Vegetius 52). Joshua was no slouch—he ran a tight ship. The difference is that Joshua led by example. It was easy for his troops to be brave and disciplined because they had an example to emulate. It was easy for them to stay in high spirits because Joshua gave them a unifying vision. Despite his overwhelming success, Joshua remained humble by always giving glory to the Lord. He never treated his soldiers as less than. *The Art of War* summarizes this concept nicely: "Therefore soldiers must be treated in the first instance with humanity, but kept under control by means of iron discipline. This is a certain road to victory" (Tzu 19).

In spite of his many flaws, Saul did have some notable military victories. "Now when Saul had taken control of the kingdom over Israel, he fought against all his enemies on every side, against Moab, the sons of Ammon, Edom, the kings of Zobah, and the Philistines; and wherever he turned, he inflicted punishment" (1 Samuel 14:47). Yet, the King of Israel was about to lose it all. Call to mind the terms of Saul's reign.

Despite all the power and privileges, his leadership was conditional on his faithfulness to the Lord. The prophet Samuel came to Saul with a message. "This is what the Lord of armies says: 'I will punish Amalek for what he did to Israel, in that he obstructed him on the way while he was coming up from Egypt'" (1 Samuel 15:2). Reflect back to *Exodus*. Shortly after the Lord provided water for Moses and his followers, the Amalekites attacked the newly liberated Hebrews. This was Joshua's first battle, which he decisively won. It came after the Lord gave Moses an ominous prophecy. "Then the Lord said to Moses, 'Write this in a book as a memorial and recite it to Joshua, that I will utterly wipe out the memory of Amalek from under heaven'" (Exodus 17:14). Surely as the Lord lives, He called upon Saul to fulfill this prophecy. King Saul had one more chance to redeem himself.

THE FINAL STRAW

S aul mustered his troops accordingly. Nearly a quarter million fighting men answered the call (1 Samuel 15:4). He placed the troops strategically, inside a ravine for an ambush attack. This is a textbook approach for a reason. Indeed, "He should form ambuscades with the greatest secrecy to surprise the enemy at the passages of rivers, in the rugged passes of mountains, in defiles in woods and when embarrassed by morasses or difficult roads" (Vegetius 66). Saul wasn't leaving anything up to chance.

While marching into the ravine, Saul encountered some old allies—the Kenites. Surprisingly for the callous king, Saul showed them mercy. Remember back to Deborah's story in the book of Judges. Heber and his wife, Jael, were both Kenites. They played a pivotal role in the assassination of King Jabin of Hazor (Judges 4:11–21). Apparently, Saul remembered their contributions. He addressed them with kindness, "But Saul said to the Kenites, 'Go, get away, go down from among the Amalekites, so that I do not destroy you along with them; for you showed kindness to all the sons of Israel when they went up from Egypt.' So the Kenites got away from among the Amalekites" (1 Samuel 15:6).

With the potential collateral damage out of the way, it was go time. Saul had instructions to destroy the Amalekites totally. Saul's sizable army, matched with strategic positioning, led quickly to decisive victory. "Then Saul defeated the Amalekites, from Havilah going toward Shur, which is east of Egypt. He captured Agag the king of the Amalekites alive, and completely destroyed all the people with the edge of the sword" (1 Samuel 15:7–8). At first glance, this looks like a homerun victory. But yet again, Saul's selfishness led him to seize defeat from the jaws of victory. Not only did Saul spare King Agag, but he also kept plunder for himself. "But Saul and the people spared Agag and the best of the sheep, the oxen, the more valuable animals, the lambs, and everything that was good, and were unwilling to destroy them completely; but everything despicable and weak, that they completely destroyed" (1 Samuel 15:9). Here, Saul exhibited a sin which echoes back to Achan. Saul also stole what was dedicated to the Lord.

After this disobedience, God spoke to Samuel. "Then the word of the Lord came to Samuel, saying, 'I regret that I have made Saul king, because he has turned back from following Me and has not carried out My commands.' And Samuel was furious and cried out to the Lord all night" (1 Samuel 15:10–11). In a whirlwind of emotions, the prophet cried out to the Lord all night. Yet his despair was about to sink even lower. The prophet went to confront the King of Israel, to no avail. One bystander gave Samuel more bad news. "Samuel got up early in the morning to meet Saul; and it was reported to Samuel, saying, 'Saul came to Carmel, and behold, he set up a monument for himself, then turned and proceeded on down to Gilgal'" (1 Samuel 15:12). The king's judgment had fallen precipitously. It was bad enough blatantly to disregard the Lord's bidding. But to flaunt this act by glorifying himself was truly next level. In addition, this monument was likely a violation of the Second Commandment (Exodus 20:4–6).

Meanwhile, Saul was oblivious to any wrongdoing. His naivete was on full display as he greeted Samuel. "So Samuel came to Saul, and Saul said to him, 'Blessed are you of the Lord! I have carried out the command of the Lord'" (1 Samuel 15:13). The elderly priest responded with sarcasm as he pointed out the confiscated livestock: "What then is this bleating of the sheep in my ears, and the bellowing of the oxen which I hear?" (1 Samuel 15:14). Saul then shifted as he tried to negotiate and rationalize his disobedience. In the Navy, this behavior is often referred to as Sea Lawyering. Saul attempted to wiggle out of Samuel's confrontation: "They have brought them from the Amalekites, for the people spared the best of the sheep and oxen to sacrifice to the Lord your God; but the rest we have completely destroyed" (1 Samuel 15:15).

Samuel wasn't buying Saul's story. He promptly cut off the king and delivered last night's message from the Lord. Powered by the Holy Spirit, Samuel did not hold back with the Lord's rebuke. "So Samuel said, Is it not true, though you were insignificant in your own eyes, that you became the head of the tribes of Israel? For the Lord anointed you as king over Israel. And the Lord sent you on a mission, and said, 'Go and completely destroy the sinners, the Amalekites, and fight against them until they are eliminated.' Why then did you not obey the voice of the Lord? Instead, you loudly rushed upon the spoils and did what was evil in the sight of the Lord!" (1 Samuel 15:17–19).

Yet Saul still didn't get it. He doubled down on his Sea Lawyering and continued to justify his actions. He admitted to sparing King Agag and to taking livestock for sacrificial purposes.

Samuel gave Saul a wise response, one as poetic as it was prophetic. "'Does the Lord have as much delight in burnt offerings and sacrifices As in obeying the voice of the Lord? Behold, to obey is better than a sacrifice, And to pay attention is better than the fat of rams. For rebellion is as reprehensible as the sin of divination, And insubordination is as reprehensible as false religion and idolatry. Since you have rejected the word of the Lord, He has also rejected you from being king" (1 Samuel 15:22–23).

This echoes back to the arrogance and pride that plagued Israel in the days of the judges. As discussed previously, "In those days there was no king in Israel; everyone did what was right in his own eyes" (Judges 17:6). Unfortunately, not much had changed. Israel had simply traded one vice for another. Now it had a king, and people honored his rule, yet *he* did as he saw fit! Clearly, Saul had forgotten the conditional nature of his divine appointment. While Saul may have held earthly power, God remained the only one who is truly sovereign. This was not open for negotiation or bargaining. It is an eternal truth.

Saul finally realized the depth of his latest mistake and panicked. He took some responsibility, only to betray himself once more with a weak excuse: "I have sinned, for I have violated the command of the Lord and your words, because I feared the people and listened to their voice" (1 Samuel 15:24).

It is difficult to imagine a worse justification. Not only was Saul the King of Israel, but its Commander-in-Chief. How on earth was he supposed to stand against Israel's enemies if he could be trampled by his own men? We must ask: why was he so afraid of them? Likely, it started after they intervened during Jonathan's near-execution. Here we see another valuable lesson: Saul's chain of failures illustrates the slippery nature of sin.

This downward spiral began with a rash oath, born of a personal vendetta. It worsened with his attempt to kill Jonathan. From there, he lost credibility in the eyes of his soldiers, with good reason. In a feeble attempt to regain their confidence, he sinned again by flagrantly disobeying God's command.

We all stumble. Every one of us falls short of the glory of God. But we must *try* to correct our course as early as possible. Indeed, "if all other prudent measures are taken, complete destruction will not result from just one mistake" (Clausewitz 31). While this final betrayal was the last nail in Saul's coffin, he had begun digging his grave long before.

Saul's cowardice was his fatal flaw. As examined previously, God makes it abundantly clear that He esteems strength and courage. "Have I not commanded you? Be strong and courageous! Do not be terrified nor dismayed, for the Lord your God is with you wherever you go" (Joshua 1:9). Satan, by contrast, uses tactics of fear and intimidation. Saul had plenty of armor and weapons—but lacked the Armor of God. The Apostle Paul illustrated this concept with sharp clarity: "Finally, be strong in the Lord and in the strength of His might. Put on the full armor of God, so that you will be able to stand firm against the schemes of the devil. For our struggle is not against flesh and blood, but against the rulers, against the powers, against the world forces of this darkness, against the spiritual forces of wickedness in the heavenly places" (Ephesians 6:10–12).

Saul demonstrated the danger of misplaced fear. Fear itself is a useful emotion, designed to protect us from real danger. But Satan, with his serpentine cunning, corrupts this God-given mechanism. What was meant as a safeguard becomes a snare. This has been his tactic since Eden: twist what is holy into a trap. By fearing his men and forsaking the Lord, Saul painted a giant red target on his own back.

Once more, we draw strength from Paul's exhortation: "in addition to all, [take] up the shield of faith with which you will be able to extinguish all the flaming arrows of the evil one" (Ephesians 6:16). Faith is our first and greatest defense against the devil's dirty tricks. Saul's spiritual failings were compounded by martial ones. *The Art of War* lists several calamities by which a general can ruin himself. One fits Saul precisely: "When the common soldiers are too strong and their officers too weak, the result is insubordination" (Tzu 20). Saul feared his troops after they stood up for Jonathan. This breakdown in authority was disastrous.

Indeed, "no military leader can ever become great without a dash of boldness in him" (Clausewitz 3). Why? Because the warrior must be prepared to die for the cause. Fear of death breeds hesitation, and hesitation invites defeat. Worse still, it brings defeat without honor.

Prudence is a virtue. It separates wise commanders from reckless brutes. Yet, too much caution becomes its own trap; what some call paralysis by analysis. We will never have perfect intelligence. We won't always know the enemy's numbers, tactics, or secret weapons. War is not a math problem; it's chaos governed by will and risk. Delay has a cost.

Clausewitz saw this clearly: "However, to value caution at the expense of the set goal is false prudence. This is contradictory to the very nature of war; many things must be dared to accomplish great goals" (Clausewitz 5). That so-called caution is often a mask for deeper problems: lack of faith. Faith in God, in your men, in your mission. Without that, a commander is paralyzed by what ifs. Clausewitz concludes: "in war, nothing is accomplished without risk ... the very nature of war does not give an unconditional opportunity to always predict in advance where you are headed" (Clausewitz 31).

The advice from Sun Tzu dovetails perfectly with this point. "To secure ourselves against defeat lies in our own hands, but the opportunity of defeating the enemy is provided by the enemy himself" (Tzu 7). This presents warfighters with a paradox: the more cautious you are, the more predictable you become—and the more predictable you are, the easier you are to kill. This is why true warriors must prize defeating the enemy more than preserving themselves. Victory demands resolve.

Fortunately, the wise Chinese general offers a prescription: "Water shapes its course according to the nature of the ground over which it flows; the soldier works out his victory in relation to the foe whom he is facing" (Tzu 31). To flow with the battle, we must first believe in victory. Some commanders initiate this mentality with a bold, no-retreat approach: a burn the boats strategy. Risky, yes. But it sends an unmistakable message: *there is no escape.* Only forward. Once that truth is internalized, men convert fear into ferocity. A cornered animal is the most dangerous of all. Fighting men are no different: "Throw your soldiers into positions whence there is no escape, and they will prefer death to flight. If they will face death, there is nothing they may not achieve. Officers and men alike will put forth their uttermost strength" (Tzu 23).

After Samuel delivered God's judgment, Saul still clung to denial. He pleaded with the elderly prophet: "Now then, please pardon my sin and return with me, so that I may worship the Lord" (1 Samuel 15:25). But Samuel refused. His reasoning was sharp and final: "But Samuel said to Saul, 'I will not return with you; for you have rejected the word of the Lord, and the Lord has rejected you from being king over Israel'" (1 Samuel 15:26).

As Samuel turned to leave, Saul reached out and tore the prophet's robe. Ever poetic, Samuel responded with a statement which cut deeper than any sword: "So Samuel said to him, 'The Lord has torn the kingdom of Israel from you today and has given it to your neighbor, who is better than you'" (1 Samuel 15:28). He reminded Saul that God does not waver; people do. Still, Saul begged Samuel to join him in a public act of worship. Despite his frustration, Samuel complied out of reverence for the Lord.

But one last matter remained. God had ordered the total destruction of the Amalekites. Saul had failed. So Samuel, in his final act as judge, took up the sword. He commanded Saul to bring forth Agag, king of the Amalekites. The defeated monarch hobbled forward in chains. Hearing the clink of iron with each step, Agag believed the worst was over. He thought to himself: "Surely the bitterness of death is gone!" (1 Samuel 15:32). He was wrong.

God's judgment has no statute of limitations. In the Old West they used to say, "No man outruns the long arm of the law." The Hand of God reaches even further. "But Samuel said, 'As your sword has made women childless, so shall your mother be childless among women.' And Samuel cut Agag to pieces before the Lord at Gilgal" (1 Samuel 15:33).

Justice was served, not by the king, but by the prophet. Samuel's rebuke of Agag echoed the rebuke of Jesus to Peter: "Put your sword back into its place; for all those who take up the sword will perish by the sword" (Matthew 26:52).

This moment marked a turning point in Saul's reign. He and Samuel parted ways for good. The prophet returned to Ramah. The king returned to Gibeah. But Samuel grieved. Israel's first king had failed so tragically. Yet this ending prompts a question: *who was next?* Who was this mysterious neighbor who was better than Saul? What could the next king offer that Saul could not?

As we've seen time and again, God has a way of raising up the least likely among us.

SAUL BEFORE DAVID
CONCLUSION

Let's recap some of the takeaway lessons from Saul's early career. By all measures, Saul demonstrated great potential as the first King of Israel. He seemed to fit the bill perfectly. He came from a respected family which held standing amongst the community. Nearly all nobility comes from the upper class of society, historically speaking. In addition, Saul had a regal appearance about him. He was very tall and handsome; he stood out in a crowd. Despite his high social standing and good looks and wealth, Saul suffered from insecurity throughout his career. We examined a parallel with Gideon's account. Much like Gideon, Saul was perplexed by his divine appointment. However, their paths greatly diverged. Gideon learned to set aside his personal fears and embraced faith in the Lord. Saul never learned this valuable lesson.

Saul's reign started off well. He showed wisdom as a military leader by leveraging ambush tactics. He did the right thing for his people by rescuing the city of Jabesh. This victory solidified his role as the first King of Israel. Unfortunately, this success was short lived. Saul's reign was conditional on his service to the Lord. Shortly after his victory. Saul made his first major blunder. Despite all the incredible things the Lord had done for Israel, Saul still lacked real faith. This was clearly demonstrated when Saul panicked as some of his troops deserted him. This incident echoes back to Gideon. Before the battle with the Midianites, God gave Gideon a message to dismiss the cowardly from battle (Judges 7:3). Around twenty-two thousand troops left! After that, God told Gideon to dismiss even more. The success of Gideon proved one thing beyond the shadow of a doubt. If we follow God's path, we only need a faithful few. Saul clearly failed to internalize this message. In a desperate attempt to regain the lost troops, Saul sinned against the Lord. He usurped Samuel's position as priest by rendering offerings he was forbidden to perform. Samuel was quick to rebuke Saul for this gross violation of ceremonial law.

Aside from these failures, there was a beacon of hope. Saul's son, Jonathan, led by example rather than edict. While troops fled Saul, they rushed into battle following Jonathan's bold attack on a Philistine outpost. This highlights a crucial axiom of combat: *never underestimate your opponent*. The Philistines made this mistake with Jonathan and his armor bearer and paid dearly. The Hebrew duo killed around twenty of them. After this daring mission, God aided Israel. Panic spread like wildfire throughout the Philistine army. Again, we see a parallel to Gideon. Just like Gideon, a faithful few took the initiative. After they demonstrated their faith, the Hand of God reached down to help. Panic rippled thought the Philistine ranks like waves on a stormy sea! The massive army dispersed hastily, and Israel prevailed. Here we see the contagious nature of both fear and courage. When the Israelites realized they had the advantage, many of them rejoined the fight! A singular act of courage turned the tides of battle. Therefore, choose courage and lead by example. This is yet another reminder that God helps those who help themselves. Always remember, courage isn't the absence of fear. Rather, it means getting the job done in spite of fear. You never know who you may inspire or how God might assist you.

Warriors and mercenaries overlap significantly in their scope and function. However, they differ in a significant way. The warrior is willing to fight and die for just cause. Mercenaries are only in it for the money and are notoriously fickle. Saul failed to grasp this fundamental distinction. As stated previously, no one wants to die to help you to clean up your own dirty laundry. Your personal problems are exactly that: *personal*. Saul abused his authority as king. He made a rash oath that diminished not only troop morale, but also their combat effectiveness. With self-centered arrogance, Saul demanded his troops must fast until his personal vengeance was satisfied. Saul clearly prioritized his personal wrath over his theocratic duties.

This incident marked the beginning of Saul's downward spiral. The old adage that soldiers march on their bellies is a timeless truism. Yet,

Saul put his petty vendettas over the well-being of his army. While the troops obeyed the command, they lost respect for Saul. Samuel had previously explained the rights and duties of the kingship to the people of Israel. The kingship of Israel was in service to the Lord. The people understood this better than the king himself. Jonathan unknowingly violated this impulsive decree. For this transgression, Saul wanted his own son publicly executed, but the troops knew better. Jonathan was a true warrior. A no-nonsense, lead-from-the-front type of guy. Saul on the other hand, was a petty tyrant. The choice of who to obey was simple. The brave warriors risked their lives as they stood up to protect Jonathan. Ironically, this only fueled Saul's insecurity and paranoia, which led to a domino effect of poor decisions.

Some time later, the Lord tasked Saul with a holy mission. He was to wipe out the Amalekites totally. The Israelite-led attack was an overwhelming success. Yet once again, Saul's poor decisions got in the way. What should have been an easy win turned into one of Saul's greatest losses. He blatantly disregarded the Lord's commands by sparing King Agag, along with some choice livestock. Saul attempted to rationalize his disobedience to Samuel. The wise old priest wasn't pleased in the slightest. After Samuel's interrogation, Saul confessed the truth. Saul feared his own troops. He attempted to win back their favor by giving them plunder from battle.

Saul made one of the worst mistakes any leader can make: *failure of ownership*. Real leaders acknowledge their mistakes and ruthlessly shore them up. Bad leaders blame anyone or anything else. This serves as an important reminder of the slippery nature of sin. Saul kept digging his own grave deeper and deeper, until it was too late. We all sin. We all make mistakes. *Repentance requires genuine action*, not just talk. Saul's superficial understanding of religious matters led him to overlook this approach, ultimately to his downfall.

PART FIVE:
DAVID'S RISE AND SAUL'S FALL

SAUL AND DAVID

Israel was yet again at a low point. Their first king continued to indulge in self-destructive behavior. Saul's paranoia and impulsiveness worsened. Samuel mourned Saul's failure, lamenting the tragedy. Then the Lord intervened. God made it clear that He prefers action to inaction. "Now the Lord said to Samuel, 'How long are you going to mourn for Saul, since I have rejected him from being king over Israel? Fill your horn with oil and go; I will send you to Jesse the Bethlehemite, because I have chosen a king for Myself among his sons'" (1 Samuel 16:1). Samuel obeyed faithfully. Placing his faith above fear, he carried on his divine mission.

In Bethlehem, he met with the elders and consecrated Jesse and his sons. When he saw Eliab, Jesse's eldest, he assumed this must be God's chosen. But the Lord corrected him. HE gave Samuel a message we should all remember: "But the Lord said to Samuel, 'Do not look at his appearance or at the height of his stature, because I have rejected him; for God does not see as man sees, since man looks at the outward appearance, but the Lord looks at the heart'" (1 Samuel 16:7). One by one, seven sons passed before Samuel—none chosen. Perplexed by the outcome, Samuel asked if there were any others. Jesse mentioned his youngest, who was tending sheep.

Once David appeared, the Lord confirmed His choice. "So Samuel took the horn of oil and anointed him in the midst of his brothers; and the Spirit of the Lord rushed upon David from that day forward. And Samuel set out and went to Ramah" (1 Samuel 16:13). David returned to shepherding his flock—unaware that soon, he would shepherd a nation.

Meanwhile, the Spirit of the Lord departed from Saul, and was replaced by a tormenting spirit (1 Samuel 16:14). To soothe Saul, his attendants suggested a musician. The choice was no accident. One servant suggested: "Behold, I have seen a son of Jesse the Bethlehemite who is a skillful musician, a valiant mighty man, a warrior, skillful in speech, and a handsome man; and the Lord is with him" (1 Samuel 16:18). In God's providence, Saul unknowingly welcomed his own replacement.

Wasting no time, Saul sent messengers to retrieve the young shepherd. David's father honored the king's request and sent him with gifts. "And Jesse took a donkey loaded with bread and a jug of wine, and he took a young goat, and sent them to Saul by his son David" (1 Samuel 16:20). Impressed by David's presence and skill, Saul quickly appointed him as his armor bearer and kept him close. "So Saul sent word to Jesse, saying, 'Let David now be my attendant for he has found favor in my sight'" (1 Samuel 16:22). When the tormenting spirit came upon Saul, David played the lyre to soothe him. "So it came about whenever the evil spirit from God came to Saul, David would take the harp and play it with his hand; and Saul would feel relieved and become well, and the evil spirit would leave him" (1 Samuel 16:23).

While Saul had once shown promise as a military leader, rescuing Jabesh in his early reign, his first true test came in the form of Goliath. Israel was again at war with the Philistines. "The Philistines were standing on the mountain on one side, while Israel was standing on the mountain on the other side, with the valley between them" (1 Samuel 17:3) With both armies positioned on high ground, an ambush was unlikely. This was to be a brutal, face-to-face confrontation, gritty as the sand beneath their feet. Goliath, the Philistine champion, emerged from Gath—home to the last of the Anakites. These were the Canaanite giants who had once struck terror into the Israelites. Though Joshua and Caleb destroyed most of them, a few enclaves remained (Joshua 11:22).

Goliath towered above all at "six cubits and a span" (1 Samuel 17:4)—approximately nine feet nine inches (3 meters) tall. He was not just massive, but also outfitted for war. "And he had a bronze helmet on his head, and he wore scale-armor which weighed five thousand shekels of bronze. He also had bronze greaves on his legs and a bronze saber slung between his shoulders" (1 Samuel 17:5–6). That coat alone weighed 125 pounds (58 kilograms). His spear was no less fearsome: "The shaft of his spear was like a weaver's beam, and the head of his spear weighed six hundred shekels of iron; and his shield-carrier walked in front of him" (1 Samuel 17:7). That's roughly 15 pounds (6.9 kilograms) at the tip alone. Covered in armor like a giant snapping turtle, Goliath presented no easy target. Only a precise strike could fell the monster of a man.

His presence sent waves of fear through Israel's ranks. "When Saul and all Israel heard these words of the Philistine, they were dismayed and very fearful" (1 Samuel 17:11). Goliath hurled insults daily, challenging any warrior brave enough to fight him in single combat. The mountainous man "stood and shouted to the ranks of Israel and said to them, 'Why do you come out to draw up in battle formation? Am I not the Philistine, and you the servants of Saul? Choose a man as your representative and have him come down to me'" (1 Samuel 17:8–9).

The taunts continued for forty days, every day and night (1 Samuel 17:16). With each day, morale eroded, and Saul's army grew more paralyzed by fear.

AN UNEXPECTED CHALLENGER

Goliath made one thing abundantly clear: this time, it was do or die. The fate of the kingdom rested on this pivotal battle. With the stakes so high, how did a young shepherd boy end up in the crossfire? The Holy Spirit guided his path.

David's three eldest brothers were fighting on the front lines for Israel, while David remained behind to help his elderly father with the work in the fields. One day, Jesse gave his youngest son a mission: deliver provisions to his older brothers and check on their welfare. He sent David with roasted grain, bread, and cheese for the commander (1 Samuel 17:12–18). David obeyed without hesitation and set off toward the battlefield.

He arrived just as tensions peaked. "So David got up early in the morning and left the flock with a keeper, and took the supplies and went as Jesse had commanded him. And he came to the entrenchment encircling the camp while the army was going out in battle formation, shouting the war cry. Israel and the Philistines drew up in battle formation, army against army" (1 Samuel 17:20–21). With haste, David delivered the goods and ran to the front lines. While he was checking in with his brothers, Goliath's ferocious taunts pierced the air once again. The psychological impact of the giant's presence cannot be overstated. "When all the men of Israel saw the man, they fled from him and were very fearful" (1 Samuel 17:24).

This fear crippled Israel's troops. As *Principles of War* states: "In general, the most important thing is confidence that the victory will be achieved, that is, the belief that the enemy will be banished from the battlefield. The guarantee of success must form the basis of the whole plan of attack" (Clausewitz 6). But Israel had no such confidence. The army of God was paralyzed by fear. It had forgotten what it meant to be strong and courageous, qualities Joshua had once embodied. Most of the army, at least.

Puzzled by the cowardice around him, David began asking questions. He learned that Saul had grown desperate. This wasn't just a battle; it was a personal challenge. Goliath was so threatening that the king had placed a bounty on his head. "And the men of Israel said, 'Have you seen this man who is coming up? Surely he is coming up to defy Israel. And it will be that the king will make the man who kills him wealthy with great riches, and will give him his daughter and make his father's house free in Israel'" (1 Samuel 17:25). It was a king's reward for a seemingly impossible task. Ambition stirred in David's heart, but his thoughts were quickly interrupted.

His oldest brother, Eliab, burned with anger when he saw David at the front. Misjudging his motives, Eliab accused him of abandoning his duties just to watch the fight. In classic older-brother fashion, the two exchanged sharp words. But even as they argued, word of David's questions reached the king. Desperate for a solution, Saul summoned the young shepherd. David wasted no time. Without skipping a beat, he volunteered for the most dangerous mission in Israel. "And David said to Saul, 'May no one's heart fail on account of him; your servant will go and fight this Philistine!'" (1 Samuel 17:32).

David's fighting spirit stunned the king. Even so, Saul denied the request. "Saul said to David, 'You are not able to go against this Philistine to fight him; for you are only a youth, while he has been a warrior since his youth'" (1 Samuel 17:33). Yet against all odds, David was uniquely suited for this mission. It is true that, "Men must be sufficiently tried before they are led against the enemy" (Vegetius 87). David may not have had battlefield experience against the Philistines, but he possessed a deadly skillset—one forged in the crucible of the wilderness.

He made his case to the king: "Your servant was tending his father's sheep. When a lion or a bear came and took a sheep from the flock, I went out after it and attacked it, and rescued the sheep from its mouth; and when it rose up against me, I grabbed it by its mane and struck it and killed it" (1 Samuel 17:34–35).

David embodied advice from the samurai, Miyamoto Musashi: "you must train day and night in order to make quick decisions. In strategy it is necessary to treat training as a part of normal life" (Musashi 4). As a shepherd, David did just that. His profession demanded both vigilance and lethal skill. Indeed, "Timing in strategy cannot be mastered without a great deal of practice" (Musashi 7). The dangers of wild predators and isolation developed not only his physical ability, but also his mental fortitude. Shepherding teaches decisive action under pressure. You cannot rely on reinforcements. It's just you, your flock, and the threats of nature. You either protect them, or you fail. David was already living in a do-or-die world.

Why is this important? Recall from earlier: the Philistines had systematically disarmed the Israelites (1 Samuel 13:16–22). This strategy seemed like a sure path to dominance. Ironically, it backfired. Denied access to swords and spears, David trained with the only weapon available to him: the sling. He didn't lose sword-fighting skills— he never had them. The sling was all he knew.

The young shepherd embodied a timeless martial truth: "the very essence of an art consists in constant practice" (Vegetius 43). And practice David did. Like many young men, he likely passed idle hours launching stones at targets—refining muscle memory, range, and timing. He didn't prepare for Goliath by accident. God's hand was in all of it.

David's weapon was not a fluke. It was perfect. He never could have defeated Goliath at close quarters. No one could. Trying to beat a specialist at his own game is folly. Goliath's size, armor, and reach made him invincible up close. However, at range, he was vulnerable—but just barely. The sling and stone were deceptively simple tools, but no less deadly. *De Re Militari* elaborates on the weapon's brutal efficiency: "Soldiers ... are often more annoyed by the round stones from the sling than by all the arrows of the enemy. Stones kill without mangling the body, and the concussion is mortal without loss of blood" (Vegetius 14).

Every boy dreams of slaying the monster. Few get the chance. David already had. Through that simple device, he had killed lions and bears. By banning swords, the Philistines became the architects of their own destruction. In the words of Sun Tzu: "Water shapes its course according to the nature of the ground over which it flows; the soldier works out his victory in relation to the foe whom he is facing" (Tzu 12). Soon, this water would wash away the Philistine champion.

Desperate times call for desperate measures. As Clausewitz observed: "if the probability of success is also against us, we still must not consider the enterprise impossible or unreasonable; it is always reasonable, since we cannot think of a better plan and if we use the scarce means, we have to achieve whatever is possible" (Clausewitz 2). Saul knew his army had scarce means, but he lacked the creativity or faith to use them.

David pressed on as he lobbied the king: "'Your servant has killed both the lion and the bear; and this uncircumcised Philistine will be like one of them, since he has defied the armies of the living God.' And David said, 'The Lord who saved me from the paw of the lion and the paw of the bear, He will save me from the hand of this Philistine.' So Saul said to David, 'Go, and may the Lord be with you'" (1 Samuel 17:36–37). David embodied another timeless principle: "Few men are born brave; many become so through care and force of discipline" (Vegetius 88).

David offers us more than an example; he gives us hope. He didn't become brave overnight. He built courage the same way one builds muscle: through repetition under resistance. His shepherding days were a crucible. When lions or bears threatened the flock, he stood alone and fought back. This wasn't just practice in aim; it was training in faith. Over time, he learned that with God's help, even modest tools in steady hands can accomplish incredible feats. We must do the work, but trust God with the outcome. As seen before, the Lord helps those who help themselves.

Saul, out of options, finally relented. He offered his blessing and invoked divine protection (1 Samuel 17:37). Then he did what worldly logic suggested: he gave David royal armor. "Then Saul clothed David with his military attire and put a bronze helmet on his head, and outfitted him with armor. And David strapped on his sword over his military attire and struggled at walking, for he had not trained with the armor. So David said to Saul, 'I cannot go with these, because I have not trained with them.' And David took them off" (1 Samuel 17:38–39).

This was the best armor in the kingdom, quite literally fit for a king. But David had the wisdom to decline. He told Saul he couldn't fight in what he hadn't tested. It was a wise choice, revealing an important lesson: *when the pressure is high, stick to what you know.*

Many warriors fall victim to the shiny object syndrome. They reach for new gear or tactics when discipline and confidence in their craft would serve better. There is a time to experiment with new techniques or equipment; a major challenge is not the time to experiment. David knew this. "Then he took his staff in his hand and chose for himself five smooth stones from the brook, and put them in the shepherd's bag which he had, that is, in his shepherd's pouch, and his sling was in his hand; and he approached the Philistine" (1 Samuel 17:40).

This wasn't just smart, it was also symbolic. Saul placed his faith in worldly things: armor, swords, appearances. David placed his faith in the Lord. God had protected him before and would do so again. For the faithful shepherd, it was just another day on duty. In the words of Sun Tzu: "If you know the enemy and know yourself, you need not fear the result of a hundred battles" (Tzu 7).

Goliath couldn't believe his eyes. For forty days, he had terrorized Saul's men. Now, some youth dared to stand up? Offended, Goliath reverted to his usual taunts. "So the Philistine said to David, 'Am I a dog, that you come to me with sticks?' And the Philistine cursed David by his gods. The Philistine also said to David, 'Come to me, and I will give your flesh to the birds of the sky and the wild animals'" (1 Samuel 17:43–44).

These cheap tactics sent shivers through Israel's army. But David remained completely unfazed. He had grown up hearing the roaring of lions and the growling of bears. The shouting of a man, even a giant, paled in comparison.

Goliath made a crucial mistake: he showed his hand. His arrogance and anger made him easier to kill. David quickly identified the opportunity. *The Art of War* warns of five dangerous faults in combat, one of which is "a hasty temper, which can be provoked by insults" (Tzu 16). Goliath was a textbook example. Despite his youth, David understood this trap and sprang it with precision. Sun Tzu would have admired his strategy. As the Chinese General advised: "If your opponent is of choleric temper, seek to irritate him. Pretend to be weak, that he may grow arrogant" (Tzu 4).

David delivered his own counter-offensive, this one verbal. He shouted: "But David said to the Philistine, 'You come to me with a sword, a spear, and a saber, but I come to you in the name of the Lord of armies, the God of the armies of Israel, whom you have defied. This day the Lord will hand you over to me, and I will strike you and remove your head from you. Then I will give the dead bodies of the army of the Philistines this day to the birds of the sky and the wild animals of the earth, so that all the earth may know that there is a God in Israel'" (1 Samuel 17:45–46).

He concluded with a truth Israel had forgotten, and Saul had never grasped: "and that this entire assembly may know that the Lord does not save by sword or by spear; for the battle is the Lord's, and He will hand you over to us!" (1 Samuel 17:47). Goliath took the bait—hook, line, and sinker. "Then it happened, when the Philistine came closer to meet David, that David ran quickly toward the battle line to meet the Philistine" (1 Samuel 17:48).

Even this maneuver showcased David's combat instincts. *Principles of War* affirms: "The third principle is not to waste time. Unless we can draw some special benefit from the delay, it is important to get to work as soon as possible" (Clausewitz 13). Likewise, *The Book of Five Rings* echoes this: "you can win quickly by taking the lead, it is one of the most important things in strategy" (Musashi 22). Once again, David's character shone. While Saul's men fled from danger, David charged toward it with resolve. Why? Because he had faith in the Lord.

Earlier, we explored the *shield of faith* (Ephesians 6:16). Though primarily a defensive tool, a shield can also serve offensively. In combat, a warrior can strike with the rim—delivering a devastating *shield bash* that stuns and staggers an enemy mid-attack. It extends range, inflicts concussive force, and breaks an opponent's momentum. In the same way, David turned his greatest defense, faith, into a weapon. He seized the initiative and turned the tide of battle before a blow was even struck.

David's whole life had prepared him for this moment. The mission was daunting. He had five stones, but in reality, just one shot. He wouldn't have time to reload. One opportunity, a single stone, to get the job done. He needed to thread the needle with absolute precision. Goliath was armored head to toe, leaving few vulnerable targets. Worse still, the giant was moving. And if David missed, Goliath would close the distance and slaughter him.

Hitting a moving target is the ultimate test of a marksman. It is not as easy as Hollywood portrays. This was truly do or die. With buttery smoothness polished by thousands of repetitions, David let the stone fly. "And David put his hand into his bag and took from it a stone and slung it, and struck the Philistine on his forehead. And the stone penetrated his forehead, and he fell on his face to the ground" (1 Samuel 17:49).

It was a perfect shot—launched with deadly precision at a small, moving target. The projectile hit with such force that it sank into the giant's skull. This is particularly striking given that the human forehead is one of the hardest bones in the body.

Whether he knew it or not, David embodied the core principle of jiu jitsu: use leverage to generate maximum output with minimal input. His sling gave him exponentially greater speed and force than a thrown stone could achieve alone. This is the martial mindset at its finest: a trained, humble shepherd defeating a heavily armed titan through skill, discipline, and faith.

As Sun Tzu put it, "a clever fighter is one who not only wins, but excels in winning with ease" (Tzu 8). David also exemplified Clausewitz's principle of bold initiative: "One of the most important principles of offensive war is to surprise the enemy with a fast attack. The more unexpected the attack, the more successful it will be" (Clausewitz 8). "So David prevailed over the Philistine with the sling and the stone: he struck the Philistine and killed him, and there was no sword in David's hand" (1 Samuel 17:50).

But the battle was not quite finished.

David claimed victory with a bold, symbolic act: "Then David ran and stood over the Philistine, and took his sword and drew it out of its sheath and finished him, and cut off his head with it. When the Philistines saw that their champion was dead, they fled" (1 Samuel 17:51).

This act sent a powerful message. The Philistines had long oppressed Israel, even outlawing blacksmiths to deny them weapons (1 Samuel 13:19). In God's infinite wisdom, He turned the enemy's own strategy against it. David decapitated Goliath with Goliath's own sword—symbolizing how the Lord used the tools of Philistine oppression against them.

The beheading had a devastating psychological effect. As with Jonathan and Gideon before him, one bold act sparked a chain reaction. The Israelites surged forward and chased the Philistines back to their territory. Indeed, "In large-scale strategy, when the enemy starts to collapse you must pursue him without letting the chance go. If you fail to take advantage of your enemies' collapse, they may recover" (Musashi 24). Likewise, "Opportunity in war is often more to be depended on than courage" (Vegetius 87).

Israel wasn't letting this opportunity slip from its fingers as it forced a retreat. The Israelites surged forward, slaughtering the Philistines until they retreated to their territory (1 Samuel 17:52).

DAVID'S NEW ROLE AND SAUL'S GROWING PARANOIA

David's epic feat solidified his role in Saul's court. The young shepherd had already impressed the king, but his latest feat ensured a permanent appointment (1 Samuel 18:2). Saul saw the utility of keeping David around, but Jonathan appreciated him on a deeper level. The two young warriors were kindred spirits. This is no surprise. Both of them displayed courage and initiative in battle. The two were more like brothers than friends. After David slayed the giant, Jonathan swore his loyalty to him (1 Samuel 18:3–4). Despite their different social statuses, the Prince of Israel viewed the shepherd as an equal. This highlights one of the most beautiful aspects of combat. Combat reveals the truth. It doesn't matter who you are or where you came from, only the results matter. Even adversarial warriors can appreciate the other's skillset, because few people realize what it takes to achieve such feats.

David's future looked incredibly bright. Saul continued to send David off to battle. His performance was nothing short of exceptional. "And David went into battle wherever Saul sent him, and always achieved success; so Saul put him in charge of the men of war. And it was pleasing in the sight of all the people, and also in the sight of Saul's servants" (1 Samuel 18:5). This success was due to his faith in the Lord and his lead-from-the-front leadership style. In fact, David performed a little bit *too* well for the insecure king.

In David's youthful naivete, he unknowingly violated the first law of power: *never outshine the master* (Greene 1). David didn't seek this attention and fanfare from the local community. But his accomplishments were too magnificent to ignore. He wasn't king yet, but he was certainly the people's champion. "Now it happened as they were coming, when David returned from killing the Philistine, that the women came out of all the cities of Israel, singing and dancing, to meet King Saul, with tambourines, with joy and with other musical instruments. The women sang as they played, and said, 'Saul has slain his thousands, And David his ten thousands'" (1 Samuel 18:6–7). The songs of praise enraged the king! "Then Saul became very angry, for this lyric displeased him; and he said, 'They have given David credit for ten thousands, but to me they have given credit for only thousands! Now what more can he have but the kingdom?'" (1 Samuel 18:8).

While the locals praised the up-and-coming warrior, they unwittingly placed a target on his back! Given Saul's long history with insecurity and paranoia, this was one of the worst outcomes for David. David had the skillset to slay the giant. But how could he protect himself from a mad king? Ironically, what made him so effective against Goliath was now a weakness. Oftentimes our greatest strength can be our biggest weakness when circumstances inevitably change. Saul wanted David to slay the giant, but this act of heroism sparked jealousy in Saul. Now Saul had a different giant to deal with. Not in stature or brute strength, but a larger-than-life reputation that overshadowed the king's gravitas. Indeed, this rising star made King Saul resentful. After that Saul watched David like a hawk (1 Samuel 18:8–9).

Jealousy is one of the most insidious emotions we have. It is a nasty thing, akin to a festering wound. The irony is that while jealous people harbor hate in their hearts *they* are the ones most damaged by it. If kept at a low level, it only harms the person feeling this damaging emotion. However, once things reach a boiling point, these feelings spill out and harm others as well. Jealousy is a corrosive emotion. It weakens and erodes loving bonds and social cohesion. Indeed, "A tranquil heart is life to the body, But jealousy is rottenness to the bones" (Proverbs 14:30).

To elucidate this point, let's examine the bucket of crabs analogy. As the story goes, if you place a solitary crab in a bucket, it will climb and scamper out in no time. However, should you place two or more crabs in the same bucket, things change dramatically. Following its instincts, one ambitious crab attempts to climb out to freedom. Yet, no matter what its strength and willpower, it remains trapped. This is because the crabs in the bottom of the bucket pull it back down into their self-made prison! They would rather condemn their fellow crustacean than watch another succeed without them. The Apostle Paul warned us of this mentality in his letter to the Galatians. "But if you bite and devour one another, take care that you are not consumed by one another" (Galatians 5:15).

There is a reason the last of the Ten Commandments is "You shall not covet ..." (Exodus 20:17). Society cannot exist with too much envy. This is because envious people are more content to go scorched earth than to accept the success of others. Misery *loves* company! The envious prefer to poison their own wells if it means denying you a drink. Despite the world being filled with abundance, they simply can't tolerate the thought of someone else having more than them in some way.

This myopic reasoning is what makes jealous people so dangerous and why many are content to destroy themselves. If you and the jealous person are both dead, you are equals in a morbid sense. This is the classic if I can't have it, no one will! mentality. Yet, envy isn't the root of the issue. Most envy comes from narcissism. It's the belief that you deserve special attention, praise, or unearned rewards. Once again, it all stems back to pride.

There is a reason pride is often referred to as the mother of all sins. C. S. Lewis does an excellent job of explaining this important concept in his book *Mere Christianity*: "According to Christian teachers, the essential vice, the utmost evil, is Pride. Unchastity, anger, greed, drunkenness, and all that, are mere fleabites in comparison: it was through Pride that the devil became the devil: Pride leads to every other vice: it is the complete anti-God state of mind" (Lewis 121).

Additionally, Lewis elaborates on the unique competitiveness of a prideful heart. It is why coveting is so harmful. "Now what you want to get clear is that Pride is essentially competitive.... Pride gets no pleasure out of having something, only out of having more of it than the next man. We say that people are proud of being rich, or clever, or good-looking, but they are not. They are proud of being richer, or cleverer, or better-looking than others. If everyone else became equally rich, or clever, or good-looking there would be nothing to be proud about. It is the comparison that makes you proud: the pleasure of being above the rest" (Lewis 121–122).

By utilizing this framework, we can make better sense of Saul's hostility toward David. On paper, Saul had it all! He was blessed with handsome looks, great height, and a wealthy family. After his anointing, he had virtually unlimited wealth and power at his disposal. He had many challenges as King of Israel. Yet the people still obeyed him. Goliath presented an existential threat to his kingdom. David quickly solved this problem. Yet instead of celebrating this miracle victory, Saul made it all about himself. David was a humble farm boy. The thought of a peasant getting more praise than the king drove him mad! David was a blessing to Israel, but Saul viewed him more as a curse.

The King of Israel could have had it all! But this perceived slight led to an ever-growing rift between the pair. This is yet another example of how Saul embraced the ways of the world. The early church leader, James, remarked on the dangers of these unchecked emotions: "But if you have bitter jealousy and selfish ambition in your heart, do not be arrogant and so lie against the truth. This wisdom is not that which comes down from above, but is earthly, natural, demonic. For where jealousy and selfish ambition exist, there is disorder and every evil thing" (James 3:14–16).

Jealousy mixed with power is extraordinarily dangerous. Saul demonstrated this well. Shortly after he heard the songs of praise towards David, he snapped. David played the lyre to soothe the king, as usual. Yet this time was different. The evil Spirit that was tormenting Saul pushed him over the edge. "Then Saul hurled the spear, for he thought, 'I will pin David to the wall.' But David escaped from his presence, twice" (1 Samuel 18:11). Fortunately, the Lord was watching over the young warrior. He eluded the mad king, unscathed. After this near-fatal incident, Saul sent David away.

Slowly, Saul's faculties of reason returned to him. On some level, Saul knew it unwise blatantly to kill off the people's champion, and he relented. "Now Saul was afraid of David, because the Lord was with him but had left Saul. So Saul removed him from his presence and appointed him as his commander of a thousand; and he went out and came in before the people" (1 Samuel 18:12–13). While this provided some relief to the king's ego, it made his problem worse in the long run. Saul took out his frustrations on David, but really he was rebelling against the Lord himself. This is why Saul's attempts to subvert David were so futile. David couldn't fail because he had a secret weapon against dark forces. Indeed, "David was successful in all his ways, for the Lord was with him" (1 Samuel 18:14).

SAUL'S PERSECUTION OF DAVID

Saul's attempt to isolate David backfired spectacularly! David was an excellent military leader who only grew more successful and famous. David was doing the Lord's work, and the people of Israel took notice. A lead-from-the-front style tempered with humility is a certain path to victory. The more the public saw of David, the more they liked the up-and-coming leader (1 Samuel 18:16).

Of course, this only widened the growing hostility between David and Saul. David's popularity exacerbated Saul's insecurities and paranoia (1 Samuel 18:15). According to Carl von Clausewitz, the main tasks of war are threefold. First, gain victory and destroy the armed forces of the enemy. Second, obtain the material means of combat and other resources of the hostile army. Last, and where Saul struggled most, capture public opinion (Clausewitz 11).

Saul knew that the people were with David. Should he go too hard on David, the people might rebel and form a mutiny. His soldiers had already stood up to him once when he attempted to kill his son, Jonathan. Why wouldn't they do the same with their respected military commander? Realizing how few his options were, he had a moment of dark inspiration. Saul decided to use his own daughter as bait in an elaborate trap! Recall from earlier the generous bounty for Goliath's head, which included his daughter's hand in marriage (1 Samuel 17:25)?

In a moment of political savvy, Saul knew he had to be more discreet. Embodying the Serpent from the Garden of Eden, Saul offered David the bait. "Then Saul said to David, 'Here is my older daughter Merab; I will give her to you as a wife, only be a valiant man for me and

217

fight the Lord's battles'" (1 Samuel 18:17). At first glance it appeared Saul was merely honoring the bounty. Plausible deniability is the currency of the dirty world of politics. Unfortunately, this proposition was fueled by sinister motivations. While lobbying David, Saul thought to himself, "My hand shall not be against him, but let the hand of the Philistines be against him" (1 Samuel 18:17).

Here, we see a deep parallel to the New Testament. Earlier, we discussed the Roman Governor, Pontius Pilate. According to the Gospel of Matthew: "Now when Pilate saw that he was accomplishing nothing, but rather that a riot was starting, he took water and washed his hands in front of the crowd, saying, 'I am innocent of this Man's blood; you yourselves shall see'" (Matthew 27:24). Some things in life never change. We will always have death, taxes, and scummy politicians.

Once again, the shepherd boy outmaneuvered the king. With a humble heart, David thought he was unworthy of becoming royalty. "But David said to Saul, 'Who am I, and who is my family, or my father's family in Israel, that I should be the king's son-in-law?'" (1 Samuel 18:18). Stunned at David's response, he married her off to another man (1 Samuel 18:19). Sun Tzu would be pleased with David's selflessness and dedication to service. According to *The Art of War*, "The general who advances without coveting fame and retreats without fearing disgrace, whose only thought is to protect his country and do good service for his sovereign, is the jewel of the kingdom" (Tzu 21). David was the crowning jewel of Israel, no question. This unexpected roadblock sent Saul's plan to the back burner for some time. However, later on he saw another opportunity to dispatch David. Saul learned that his other daughter, Michal, was in love with David. The king was delighted he had a second chance to subvert his foe. "For Saul thought, 'I will give her to him so that she may become a trap for him, and that the hand of the Philistines may be against him.' Therefore Saul said to David, 'For a second time you may become my son-in-law, today'" (1 Samuel 18:21).

This time, however, Saul left little to chance. Clever as a fox, he orchestrated a conspiracy against David. Saul told David that he would have a second chance to become his son-in-law. What he *didn't* tell David is that he was stacking the deck against him. For this mission, Saul recruited help. Covertly, Saul ordered his attendants to whisper falsehoods in David's ear. "Then Saul commanded his servants, 'Speak to David in secret,' saying, 'Behold, the king delights in you, and all his servants love you; now then, become the king's son-in-law'" (1 Samuel 18:22).

David heard their advice, but once again appealed to his lowly status. He truly thought himself unworthy of becoming royalty. Yet this wasn't going to stop Saul's plan. He had already tipped his hand once by losing his temper and throwing his spear at the young shepherd. He needed to keep his hands clean of this matter. It was crucial to have the Philistines kill David. He gave his servants more instructions. He ordered them, "Saul then said, 'This is what you shall say to David: The king does not desire any dowry except a hundred foreskins of the Philistines, to take vengeance on the king's enemies.' But Saul plotted to have David fall by the hand of the Philistines" (1 Samuel 18:25). Given his longstanding feud with the Philistines, this was a good cover story for his ulterior motives.

David took the bait! This offer changed everything. Here, we see more of David's character on display. Despite already proving himself with Goliath, he didn't want any handouts. David wanted to earn his station. Filled with a fighting spirit, David went to battle enthusiastically. The young warrior went above and beyond the call of duty. "David set out and went, he and his men, and fatally struck two hundred men among the Philistines. Then David brought their foreskins, and they presented all two hundred of them to the king, so that he might become the king's son-in-law. And Saul gave him his daughter Michal as a wife" (1 Samuel 18:27). There are many parallels to David and Jesus Christ. This event may serve as foreshadowing to part of Christ's Sermon on the Mount. "Whoever forces you to go one mile, go with him two" (Matthew 5:41).

Presented with another monumental challenge, David not only survived, but also thrived! After this stellar performance, Saul had no choice but to give Michal's hand in marriage. With Saul's plot failing twice in a row, he became even more desperate and fearful. "When Saul saw and realized that the Lord was with David, and that Michal, Saul's daughter, loved him, then Saul was even more afraid of David. So Saul was David's enemy continually" (1 Samuel 18:28–29).

Perhaps David was thinking about this tumultuous time in his life when he later wrote the following: "The Lord is for me; I will not fear; What can man do to me?" (Psalm 118:6). The young warrior continued to grow in his skill and fame. With the Lord firmly on his side, this is no surprise. "Then the commanders of the Philistines went to battle, and it happened as often as they went out, that David was more successful than all the servants of Saul. So his name was held in high esteem" (1 Samuel 18:30).

With every failed plot to dispatch David, Saul became a little more desperate. Saul shifted gears and became more direct in his hostility. He even tried to recruit his son, David's closest friend, to do the dirty work for him. "Now Saul told his son Jonathan and all his servants to put David to death. But Jonathan, Saul's son, greatly delighted in David" (1 Samuel 19:1). In terms of strategy, this is one of the worst approaches Saul could have taken.

While strategically poor, Saul's methodology highlights an important aspect of sin. The more we embrace sin, the further we depart from our faculties of reason. Why? Sin tempts us on our most base and animal instincts. In effect, *sin makes us behave stupidly*. Even intelligent and wise people act foolishly when they embrace sinful ways.

We see examples of this constantly in daily life. Think of the wealthy executive who gets busted having an affair with his secretary. His influence and wealth suffer as he's fired and his assets liquidated in divorce proceedings. Picture a superstar athlete who is skilled and well respected. After a few drinks, he sucker punches some guy at the bar only to get arrested for drunk driving while fleeing the scene. He loses valuable sponsorships and then gets further drained in court settlements and legal fees. Of course all this means less time on the practice field. Without a serious course correction, this leads to bad performances and less negotiating power with future contracts and deals. Envision the politician who gets caught red handed in a bribery scandal. He summarily gets voted out of office or perhaps arrested. He loses all his influence and gravitas, along with his earning potential. Not even criminals wish to do business with this man. As discussed previously, plausible deniability is one of the principal currencies of the political world. With his main leverage and assets forever destroyed, the once prominent figure fades away into irrelevancy. Lust, wrath and greed are some of the worst offenders for interpersonal relationships. Yet, it's worth repeating: *all sin stems from a prideful heart.*

Fortunately for David's sake, cooler heads prevailed. For now, that is. Jonathan warned his friend of this fiendish plot. "So Jonathan informed David, saying, 'My father Saul is seeking to put you to death. Now then, please be on your guard in the morning, and stay in a hiding place and conceal yourself'" (1 Samuel 19:2). Knowing Saul's bad side all too well, David concealed himself as Jonathan lobbied on his behalf. "Jonathan spoke well of David to Saul his father and said to him, 'Let not the king do wrong to his servant David; he has not wronged you, and what he has done has benefited you greatly. He took his life in his hands when he killed the Philistine. The Lord won a great victory for all Israel, and you saw it and were glad. Why then would you do wrong to an innocent man like David by killing him for no reason?'" (1 Samuel 19:4–5).

Jonathan's appeal was quite reasonable. However, the funny thing about reason is that it only works on *reasonable* people! When dealing with petty people we must appeal to *their* self-interests. While noble, appealing to grand causes like justice, honor, truth, God, et cetera fall on deaf ears with these types. This is also true when dealing with superiors. It's unfortunate, but many people are only motivated to act by what serves them personally.

With a forked tongue, Saul concealed his sinister intentions once more. "Saul listened to the voice of Jonathan, and Saul vowed, 'As the Lord lives, David shall not be put to death'" (1 Samuel 19:6). David returned to his royal duties right after Saul buried the hatchet. Things briefly returned to normal in the king's court. To no one's surprise, David continued his successful military campaign. "When there was war again, David went out and fought the Philistines and defeated them with great slaughter, so that they fled from him" (1 Samuel 19:8).

Everything was well in the kingdom of Israel. But in our fallen world, peace is as fleeting as the wind. Most kings would be jumping for joy with such effective victories over their main rival. Yet this once again cast David's larger-than-life shadow over the King of Israel. Saul simply couldn't let it go. "Now there was an evil spirit from the Lord on Saul as he was sitting in his house with his spear in his hand, and David was playing the harp with his hand. And Saul tried to pin David to the wall with the spear, but he escaped from Saul's presence, so that he stuck the spear into the wall. And David fled and escaped that night" (1 Samuel 19:9–10).

David could deny it no longer; the king wanted him dead. The man who had once faced a single giant now faced a far greater threat: the full resources of the kingdom. But David wasn't alone. The Holy Spirit was with him, and so was his wife, Michal. "Then Saul sent messengers to David's house to watch him, in order to put him to death in the morning. But Michal, David's wife, informed him, saying, 'If you do not save your life tonight, tomorrow you will be put to death!'" (1 Samuel 19:11).

Saul underestimated his daughter's loyalty. Like Rahab before her, Michal risked everything to protect a servant of God. "So Michal let David down through a window, and he went and fled, and escaped" (1 Samuel 19:12). She then employed a classic principle of warfare: deception. She made a decoy in David's bed and claimed he was ill. When Saul's men returned to verify, the trap was exposed, but too late. David was long gone.

Michal's love had bought David a critical head start. Saul's plan backfired again. His schemes were like a boomerang: the harder he threw them, the faster they returned to strike him. When interrogated, Michal maintained her story. So Saul said to Michal, 'Why have you betrayed me like this and let my enemy go, so that he has escaped?' And Michal said to Saul, 'He said to me, "Let me go! Why should I put you to death?"'" (1 Samuel 19:17). Her courage and quick thinking proved pivotal in the life of Israel's future king.

Not squandering his head start, David fled to someone he could trust, Samuel. He went to Ramah and reported everything. The wise prophet took him in and helped to hide him, but Saul quickly learned of their location. Once again, he sent men to capture David. But this time, they encountered a power no soldier could overcome. "Then Saul sent messengers to take David, but when they saw the company of prophets prophesying, with Samuel standing and presiding over them, the Spirit of God came upon the messengers of Saul; and they also prophesied" (1 Samuel 19:20). Saul sent a second group—and then a third—but all of them were overtaken by the Spirit and began to prophesy too.

Saul should have taken this as a warning. Instead, blinded by rage, he went to Ramah himself. But the same fate befell him (1 Samuel 19:23–24). The Lord had made His point: no king—not even Saul— could lay a hand on David without divine permission.

JONATHAN'S AID

With the Holy Spirit distracting Saul, David once again fled. He utilized this head start by again seeking counsel from Jonathan. David pled earnestly, "What have I done? What is my guilt? And what is my sin before your father, that he is seeking my life?" (1 Samuel 20:1). Jonathan's emotions towards his father clouded his judgement. In spite of Saul's long history of erratic behavior, Jonathan took his father's words at face value. He truly believed his father had turned over a new leaf and explained this to David. "He said to him, 'Far from it, you shall not die! Behold, my father does nothing either great or small without informing me. So why would my father hide this thing from me? It is not so!'" (1 Samuel 20:2).

David wholeheartedly disagreed with Jonathan's read about the situation. Being on the receiving end of a spearpoint has a funny way of sharpening the senses; even more so the second time around! David appealed to reason as he responded to his friend: "Your father is well aware that I have found favor in your sight, and he has said, 'Jonathan is not to know this, otherwise he will be worried.' But indeed as the Lord lives and as your soul lives, there is just a step between me and death" (1 Samuel 20:3). Finally, Jonathan started to come around. He agreed to help David however possible.

The pair decided to gather more intel before planning their next move. David came up with a cover story to gauge Saul's intentions. This was a wise approach. Indeed, "in every kind of warfare it is deemed of the highest importance to spy out and get to know thoroughly the habits of the enemy" (Vegetius 107). Gathering intel straight from the source is more valuable than gold. Typically, these types of spies are handsomely rewarded. Yet, this wasn't about money or winning favors. Jonathan was willing to risk it all for his friend. He embodied the strong and courageous fighting spirit discussed in the Book of Joshua.

David told the plan to Jonathan. "So David said to Jonathan, 'Behold, tomorrow is the new moon, and I am obligated to sit down to eat with the king. But let me go so that I may hide myself in the field until the third evening. If your father misses me at all, then say, "David earnestly requested leave of me to run to Bethlehem, his city, because it is the yearly sacrifice there for the whole family"'" (1 Samuel 20:5–6). David then explained that if Saul lost his temper at this news, he meant ill towards him.

Jonathan agreed with the plan and reaffirmed his loyalty to David. "So Jonathan made a covenant with the house of David, saying, 'May the Lord demand it from the hands of David's enemies'" (1 Samuel 20:16). Leaving no room for chance, Jonathan devised a secret communication plan to convey Saul's intentions. He would shoot arrows in the field where David was hiding. Different locations in the field signified either to flee or to return to safety (1 Samuel 20:19–22). With the logistics arranged, the two carried out the plan. On the second day of the feast, Saul questioned Jonathan about David's whereabouts. It was no small thing to deny an invitation to the king's table. Sticking to the plan, he told Saul the cover story. The mad king exploded at his son as he confirmed David's worst fear. The king's anger raged on as he said: "For, as long as the son of Jesse lives on the earth, neither you nor your kingdom will be established. Now then, send men and bring him to me, for he is doomed to die!" (1 Samuel 20:31). Yet again, Jonathan asked what wrong David had done to Saul. The king answered his protest with extreme violence. Enraged, Saul hurled his spear at his son. Jonathan knew David had made the right choice (1 Samuel 20:33).

Saul's actions were as tragic as they were predictable. Saul's self-destructive nature, paired with Jonathan's naivete echoes a wise folk tale. The tale of the *Scorpion and the Frog*. The story goes something like this. One day, a scorpion met a frog at the bank of a river. Unable to traverse the current, he asked the frog for a lift across the river. Trusting his instincts, the frog hesitated, fearing the lethal venom in the scorpion's tail. The frog came to his senses and denied the scorpion's request. The scorpion didn't skip a beat as he appealed to reason. With iron-clad logic he told the frog, "If I sting you, we'll both drown." It's hard to argue with such a straightforward cause-and-effect relationship. Fearing he may have misjudged the armored arthropod, the frog allowed the scorpion to climb on his back. Midway across the river, the scorpion plunged his venomous tail deep into the frog's back! Paralysis quickly set in. Both the frog and the scorpion began to sink into the turbulent waters. With his dying breath, the frog croaked out his final words. "Why did you do that? Now we'll both die!" With a cold-blooded heart the scorpion replied, "I couldn't help it. It's in my nature." The back-stabbing scorpion drowned still clinging to his good-hearted helper.

God gave us the gift of discernment for a reason. Simply appealing to scripture or logic is not enough. In his wicked ways, Satan can warp these divine gifts against us. By twisting contexts, anyone can abuse scripture to fit personal agendas or political goals. Likewise, logic and reason only appeal to reasonable people. Fortunately, Christ shows us how to deal with such underhanded tactics and bad faith actors. One of the best examples comes from the Gospel of Matthew.

The Holy Spirit had led Jesus into the wilderness. He had been fasting for forty days. Tired and hungry, Satan began to tempt Him. Like a patrolling shark, Satan can smell the blood in the water. He often tempts us when we are weak, isolated, or otherwise vulnerable. Satan's opening shot targeted Christ's human side. "If You are the Son of God, command that these stones become bread" (Matthew 4:3). Referencing Moses, Jesus answered with the Word of God. "It is written: 'Man shall not live on bread alone, but on every word that comes out of the mouth of God'" (Matthew 4:4/Deuteronomy 8:3).

Realizing he needed to alter his tactics, Satan changed his attack vector. The fallen angel appealed to Christ's ego (pride) as he challenged His holy status. Satan took Him to the highest point of the temple and twisted scripture yet again. "If You are the Son of God, throw Yourself down; for it is written: 'He will give His angels orders concerning You'; and 'On their hands they will lift You up, So that You do not strike Your foot against a stone'" (Matthew 4:6/Psalm 91:11–12). Again, Jesus fought back with an honest interpretation of scripture. HE countered Satan succinctly. "On the other hand, it is written: 'You shall not put the Lord your God to the test'" (Matthew 4:7/Deuteronomy 6:16).

Scrambling for a comeback, Satan desperately appealed to greed. Satan took Him atop a high mountain. They could see all the kingdoms of the world from this magnificent spot. Surely it was a stunning view; however, earthly kingdoms pale in comparison to the Kingdom of God. Satan tried once more to sway our one true King. "Then Jesus said to him, 'Go away, Satan! For it is written: "You shall worship the Lord your God, and serve Him only"'" (Matthew 4:8–9). Jesus promptly told him to go pound sand! "Away from me, Satan! For it is written: 'Worship the Lord your God, and serve him only'" (Matthew 4:10/Deuteronomy 6:13). Realizing his defeat, Satan tucked tail and fled before the angels of God arrived (Matthew 4:11).

Furthermore, Christ again warns us of these tactics in his Sermon on the Mount: "Do not give what is holy to dogs, and do not throw your pearls before pigs, or they will trample them under their feet, and turn and tear you to pieces" (Matthew 7:6). Christ did not say this to be callous or coldhearted. Rather, it is a call to marshal our resources and focus where we can do the most good at the time. A myriad of scriptures support this approach, in both the Old and the New Testament. For example, the First Letter of John warns us about the danger of wolves in sheep's clothing. "Beloved, do not believe every spirit, but test the spirits to see whether they are from God, because many false prophets have gone out into the world" (1 John 4:1).

In this same way, the author of Hebrews discusses discernment as a pivotal component of maturity. "For everyone who partakes only of milk is unacquainted with the word of righteousness, for he is an infant. But solid food is for the mature, who because of practice have their senses trained to distinguish between good and evil" (Hebrews 5:13–14). This passage provides great insight to discernment. Hebrews teaches us that discernment is a *skill* as well as a virtue. This is encouraging because we can improve any skill with practice. Additionally, discernment is praised in the Old Testament. "The Lord founded the earth by wisdom, He established the heavens by understanding. By His knowledge the ocean depths were burst open, And the clouds drip with dew. My son, see that they do not escape from your sight; Comply with sound wisdom and discretion" (Proverbs 3:19–21).

Now, back to Jonathan.

All of the above is to say that Jonathan did *not* have access to these wise teachings. He had to learn these painful lessons the hard way. The average man learns from his own mistakes. The wise man learns from the mistakes of others. Yet, the fool learns from nothing. We can learn from Jonathan's experiences, so we are not doomed to repeat them. After Saul's spearpoint nearly impaled him, Jonathan realized that no amount of negotiation could save him or his best friend.

He promptly stormed out to warn his fugitive friend. "Then Jonathan got up from the table in the heat of anger, and did not eat food on the second day of the new moon, because he was worried about David since his father had insulted him" (1 Samuel 20:34). Jonathan was distraught, but he still had a job to do. Utilizing their prearranged signal with the arrows, he retrieved David from his hiding spot. Emotions ran high as the two greeted each other. With teary eyes, Jonathan reaffirmed his loyalty to David. This was deeper than two friends saying goodbye. By affirming his loyalty to David, Jonathan was siding with the Lord. Jonathan delivered this parting message to his friend and closest ally. "Go in safety, since we have sworn to each other in the name of the Lord, saying, 'The Lord will be between me and you, and between my descendants and your descendants forever.' So David set out and went on his way, while Jonathan went into the city" (1 Samuel 20:42). The two friends then went their separate ways, before Saul caught onto them.

DAVID ON THE RUN

Things had changed forever. In the eyes of Saul, David was officially an outlaw! Realizing Saul wouldn't relent until one of them was dead, David fled. With few options, he fled to the town of Nob to seek aid and provisions. Being a man of God, he went to the tabernacle for help in this daunting mission. The mighty warrior's unexpected visit startled the priest, Ahimelech. Trembling with apprehension, he asked David why he was alone. David gave the priest a plausible cover story. He likely did this to protect Ahimelech should Saul interrogate him. In the eyes of royalty, aiding and abetting a fugitive of the law was no small matter. "David said to Ahimelech the priest, 'The king has commissioned me with a matter and has said to me, "No one is to know anything about the matter on which I am sending you and with which I have commissioned you; and I have directed the young men to a certain place"'" (1 Samuel 21:2). After that, David asked the priest for food and supplies for this secret mission.

The priest told the young warrior he only had bread that was consecrated for religious ceremonies. He offered it to David, provided he and his men were ceremonially clean. David happily accepted this gift from Ahimelech. Perhaps the priest could sense the Holy Spirit was with David. At first glance, this interaction may appear sacrilege. The consecrated bread was *only* for priests. Moses made this point abundantly clear (Leviticus 24:8–9). Thus, by aiding David in this way, the priest actually violated ceremonial law. As discussed previously, if there is a dilemma between God and the law, choose God every time. Likely, Ahimelech could tell that the Lord was with David. By allowing

David to eat this blessed food, he essentially acknowledged David's divine appointment.

MERCY VERSUS LEGALISM

Interestingly, Christ referenced this very account in one of his many altercations with the Pharisees. The Gospel of Matthew provides clarity on the tension between rules and faith. Both are necessary pillars of Christianity. True wisdom is knowing *when* to lean more on one than the other. Fortunately, Christ guides us. One day, Jesus and his disciples were passing through a grainfield. Famished and tired, they picked a modest amount of grain to nourish themselves for the day. Never wasting an opportunity to condemn another, the Pharisees pounced on Jesus and his followers. They promptly chimed in to criticize Jesus. Ever wise, Christ responded to them with scripture. "But He said to them, 'Have you not read what David did when he became hungry, he and his companions— how he entered the house of God, and they ate the consecrated bread, which was not lawful for him to eat nor for those with him, but for the priests alone?'" (Matthew 12:3–4). Jesus didn't stop there! He immediately threw the hypocrisy of the legalists back at them. "Or have you not read in the Law that on the Sabbath the priests in the temple violate the Sabbath, and yet are innocent?" (Matthew 12:5).

This point was absolutely crucial. The Pharisees essentially accused Jesus and his disciples of working on the Sabbath, simply because they picked a few pieces of grain to sustain themselves; thus continuing God's work. In this same way, the Pharisees were required to perform ceremonies and duties on the Sabbath. This may have technically been work, but it was in service of the Lord's work and thus, permissible. That is what's important in Christ's interpretation. Christ was a fierce opponent of the strict legalistic framework used by the Pharisees. As shown above, making an idol of legalism quickly gets in the way of the Lord's true work. The Messiah then closed his rebuttal by referencing scripture once more. "But I say to you that something greater than the temple is here. But if you had known what this means: 'I desire compassion, rather than sacrifice,' you would not have condemned the innocent" (Matthew 12:6–7/Hosea 6:6).

Bureaucracy isn't virtuous; mercy is.

Now, back to David's story.

THE SWORD OF GOLIATH

During their hasty meeting, David and the priest were under watchful eyes. "Now one of the servants of Saul was there that day, detained before the Lord; and his name was Doeg the Edomite, the chief of Saul's shepherds" (1 Samuel 21:7). One of the greatest strengths of monarchy is the presence of eyes and ears throughout the kingdom. In effect, every citizen or guest is an intelligence asset. Doeg was no exception to this. A fly on the wall, he took note of the pair. With rations secured, David moved on to weaponry. "David said to Ahimelech, 'Now is there no spear or sword on hand? For I brought neither my sword nor my weapons with me, because the king's matter was urgent'" (1 Samuel 21:8). This was the house of the Lord, not an armory. There was only one proper weapon in the building. Preserved as a testament to God's power, this blade was truly one of a kind. The priest aided David once more as he responded to the young warrior. "Then the priest said, 'The sword of Goliath the Philistine, whom you killed in the Valley of Elah, behold, it is wrapped in a cloth behind the ephod; if you would take it for yourself, take it. For there is no other except it here.' And David said, 'There is none like it; give it to me'" (1 Samuel 21:9). David departed hastily, with the oversized sword in hand!

Aside from this quick pitstop, there was little rest for David. In David's flight, he found himself deep in Philistine territory. The fugitive leapt out of the frying pan only to land in the fire! Ironically, he wound up in Goliath's hometown of Gath. The servants of the local king, Achish, realized who David was (1 Samuel 21:11). Like a turtle stripped of his shell, he had nowhere to hide. The reality of the situation came

233

crashing down on him like a lead-filled balloon. The vulnerability filled him with terror and dread. Rightly so: his current situation was a far cry from the usual military assets and army at his disposal. Desperate times call for desperate measures. David played the only move he could. Much like his single shot at Goliath, there was only one opportunity for the following strategy to work.

After David heard the king's servants talking about him, he knew his cover was blown. Taking notes from a rabid dog, David vigorously pursued a strategy of deterrence. "David took these words to heart and greatly feared Achish king of Gath. So he disguised his sanity while in their sight and acted insanely in their custody, and he scribbled on the doors of the gate, and drooled on his beard" (1 Samuel 21:12–13). Imagine the chaos of the scene. Out of nowhere, a man responsible for slaying tens of thousands shows up at your front gate. This would be formidable enough, but things continued to escalate. Additionally, this seasoned warrior bested your gigantic champion singlehandedly. The shocking spectacle caused the entire army to flee the battlefield as a result. To add insult to injury, the mad man is wielding a giant sword, which he took as a trophy from the fallen champion. Before any potential negotiations, this warrior lashed out with animalistic rage. With no provocation, he foamed at the mouth and clawed the doors! You can't tell where the beast ends and the man begins! Any one of the above factors is cause for concern. All of them together were enough to scare off the King of Gath, post haste.

David understood one of the greatest leverage points of powerful people; they have *much* to lose! Cunning combatants know how to utilize this advantage to their favor. If performed correctly, this is *by far* one of the most effective strategies in warfare. It saves valuable time, money, resources, and equipment. Most importantly, no lives are lost. Every foe who backs down is a flawless victory for you. By utilizing deterrence, the wise commander can build an even greater army to save for the major threats.

The unpredictability of this situation shook King Achish to his core. *No one* wants to fight a legitimately insane person. Even skilled fighters avoid this. The costs are simply too high for too little reward. Unless it's truly a do-or-die situation, these types are best left to their own devices. David had already dispatched King Achish's strongest warrior without breaking a sweat. He knew what the young warrior was capable of. Goliath wrote a check that King Achish couldn't cash. With an arrogance matched only by his stature, the giant wagered kingdom for kingdom. Recall that Goliath had offered Saul representative combat—whatever side lost would be subject to the victor (1 Samuel 17:8–9).

From his perspective, Achish likely thought David had come to collect the debt. In a desperate bid to save his kingdom, he dismissed David before he got too close. "Then Achish said to his servants, 'Look, you see the man is behaving like an insane person. Why do you bring him to me? Do I lack insane people, that you have brought this one to behave like an insane person in my presence? Shall this one come into my house?'" (1 Samuel 21:14–15). The plan worked! David allowed King Achish to save face as he made his departure.

It's best not to gloat in victory. We see yet another parallel between David and the teaching of Jesus. In the words of Christ, "everyone who exalts himself will be humbled, and the one who humbles himself will be exalted" (Luke 14:11). In his heart, David knew he had won decisively. Sun Tzu would be proud of David's performance! According to him, "to fight and conquer in all your battles is not supreme excellence; supreme excellence consists in breaking the enemy's resistance without fighting" (Tzu 6). David had a firm understanding of this timeless principle. While explaining the application of warfare principles, Clausewitz wrote the following. "Therefore, we do not need extensive education, a scientific mentality, or even particularly outstanding qualities of mind. If there were anything apart from good judgment that would be required at all, it would be cunning or resourcefulness" (Clausewitz 27). Fortunately for David's sake, he possessed good judgement, cunning, and resourcefulness in spades!

After achieving supreme excellence in Gath, he moved on to safer territory. David took refuge in the cave of Adullam. His family, fearing for its own safety, joined him as well. It was here that David cemented his outlaw status in the eyes of Saul's regime. In fact, David became a leader of outlaws as he started a new army. "Then everyone who was in distress, and everyone who was in debt, and everyone who was discontented gathered to him; and he became captain over them. Now there were about four hundred men with him" (1 Samuel 22:2). This ragtag band of misfits and outlaws were foundational to David's band of loyal warriors. Later on, they earned the name of David's Mighty Men, with good reason (2 Samuel 23).

We see yet another parallel to Jesus Christ. A notable example is His calling of the disciple Matthew (also known as Levi). To understand the significance, we must first consider the cultural context. Matthew was a Jewish tax collector—an agent of the Roman Empire. But tax collectors in ancient Judea were nothing like today's bureaucrats. They were more like mafia bosses, shaking down their own people to fund a foreign oppressor.

Rome ruled through centralized power but outsourced tax collection to local contractors. Once a tax quota was set, Roman authorities gave collectors broad freedom to extract it however they wished. As long as the treasury got its due, no questions were asked. But when the official amount was met, many continued collecting for their own gain. This practice, often called tax farming, was highly lucrative—and deeply corrupt. Some shed blood to meet their goals; others relied on extortion and threats, backed by Roman soldiers. No wonder the public viewed them as traitors. Their greed and violence made them social outcasts.

That's what made Jesus's invitation to Matthew so shocking. Christ approached the tax booth and invited Matthew to follow. And Matthew did. He left his post and hosted Jesus in his home, where many others, also considered outcasts, joined them for a meal. The Pharisees were quick to criticize. "And when the Pharisees saw this, they said to His disciples, 'Why is your Teacher eating with the tax collectors and sinners?' But when Jesus heard this, He said, 'It is not those who are healthy who need a physician, but those who are sick'" (Matthew 9:11–12). Then He cited the prophet Hosea to drive the point home: "Now go and learn what this means: 'I desire compassion, rather than sacrifice,' for I did not come to call the righteous, but sinners" (Matthew 9:13/Hosea 6:6).

Now, back to David and his forming coalition.

David now had some fighting men on his side. Yet he still had to protect his family. When multiple parties are in conflict, always remember: *the enemy of my enemy is my friend*. The young warrior understood this enduring principle. Luckily for the refugee, Saul had made many enemies in his career (1 Samuel 14:47). With this in mind, David reached out to the King of Moab for protection.

Defectors make excellent allies. They bring critical intel and usually have good reason to share it with their new friends. The king welcomed them with open arms as he ushered them to a stronghold (1 Samuel 22:3–4). However there is little rest for outlaws. With his family in safe keeping, the prophet Gad offered David sage advice. "Do not stay in the stronghold; leave, and go into the land of Judah" (1 Samuel 22:5). Additionally, this advice correlates to *Principles of War*. "The most important principle for the defender, one that should be regarded as a cornerstone of the theory of defense—never rely entirely on the favorable conditions of the terrain; therefore, never succumb to the temptation of passive defense" (Clausewitz 9–10).

Oftentimes, one's greatest strength becomes a weakness when circumstances change. Fortresses are a prime example. Their towering walls are excellent for repelling invaders. But that same strength can become a deadly liability. If an enemy cuts off your supply lines, those impenetrable walls turn into a cage. Trapped within, you grow weak, isolated, and desperate, with no escape.

The prophet Gad ensured that David would avoid such a fate. Obedient to the Lord's will, David didn't entrench himself behind walls. Instead, he safeguarded his family and returned to hostile territory. With a small band of fellow outlaws, he embraced the wilderness, choosing movement over entrapment—faith over fear.

SAUL'S WRATH

S aul's paranoia and rage continued to erode his faculties of reason. He embraced these dark emotions in spite of the damage they caused, to himself, his reputation, and his nation. Once again, the mad king caught wind of David's whereabouts. Though his information was vague, he pressed his officials for answers. Spear in hand, he returned to the ways of the world through bribes and threats.

Saul appealed to greed and ambition: "Hear now, you Benjaminites! Will the son of Jesse really give all of you fields and vineyards? Will he make you all commanders of thousands and commanders of hundreds?" (1 Samuel 22:7). He spoke to his officials not as fellow Israelites with a divine mission, but as mercenaries for hire. It was a crude transaction, power and wealth in exchange for loyalty to his personal vendetta. "For all of you have conspired against me so that there is no one who informs me when my son makes a covenant with the son of Jesse, and there is none of you who cares about me or informs me that my son has stirred up my servant against me to lie in ambush, as it is this day" (1 Samuel 22:8).

But the truth was, his men were just as in the dark as he was. David and Jonathan had covered their tracks well. And the officials had good reason to keep their mouths shut; Saul was holding the very same spear he had hurled at both David and Jonathan! If you repeatedly shoot the messenger, don't act surprised when no one delivers the message.

David had largely succeeded in concealing his movements. But in his haste, he left one loose end. Likely motivated by royal rewards, Doeg the Edomite chimed in: "I saw the son of Jesse coming to Nob, to Ahimelech the son of Ahitub. And he inquired of the Lord for him, gave

238

him provisions, and gave him the sword of Goliath the Philistine" (1 Samuel 22:9–10).

With a viable lead, Saul acted quickly. He summoned Ahimelech and all the male members of his family, priests of the town of Nob. Without hesitation, he launched into an interrogation, having already decided their guilt: "Why have you and the son of Jesse conspired against me, in that you have given him bread and a sword, and have inquired of God for him, so that he would rise up against me by lying in ambush as it is this day?" (1 Samuel 22:13).

This exchange reveals how far Saul had strayed. His kingship was never absolute; it was contingent on his faithfulness to the Lord. But now, consumed by ego and vengeance, Saul treated the priesthood as if it answered to him. Ahimelech, ever faithful to God's calling, was stunned by the accusation. David had shown nothing but loyalty. The priest's response echoed Jonathan's own appeal—spoken at the New Moon feast, just before Saul hurled a spear at his own son.

The priest spoke up to defend himself: "And who among all your servants is as faithful as David, the king's own son-in-law, who is commander over your bodyguard, and is honored in your house? Did I just begin to inquire of God for him today? Far be it from me! Do not let the king impute anything against his servant or against any of the household of my father, because your servant knows nothing at all of this whole affair" (1 Samuel 22:14–15). Ahimelech's response was clear, honest, and restrained. His simple retort was truthful and well-reasoned. Unfortunately, *tyrants only care about truth when it serves their interests.*

In a chilling echo of the rash curse on Jonathan, Saul condemned Ahimelech and his entire family to death! "You shall certainly die, Ahimelech, you and all your father's household!" (1 Samuel 22:16). The King of Israel completely turned his back to the Lord. "And the king said to the guards who were attending him, 'Turn around and put the priests of the Lord to death, because their hand also is with David and because they knew that he was fleeing and did not inform me.' But the servants of the king were unwilling to reach out with their hands to attack the priests of the Lord" (1 Samuel 22:17).

Something incredible happened. Courageous as lions, all his officials rejected their king's order! Just as they had interceded on Jonathan's behalf, they were unwilling to kill the Lord's servants. (1 Samuel 18:45). Yet Saul refused to relent until the ground was soaked with blood. With few options, he ordered Doeg the Edomite to fulfil the dirty job. It's worth noting that despite working for the King of Israel, Doeg was *not* an Israelite.

Doeg jumped at the opportunity to prove himself. "And Doeg the Edomite turned around and attacked the priests, and he killed on that day eighty-five men who wore the linen ephod" (1 Samuel 22:18). While tragic, this atrocity partially fulfilled an earlier prophecy against the house of Eli (1 Samuel 2:31–33) This prophecy was later completed in the time of Solomon (1 Kings 2:26–27).

Not even this slaughter was enough to satisfy Saul's bloodlust. In a rampage that surely made Satan proud, Saul ordered Doeg to eradicate the *entire* town of Nob! Earlier in the book we discussed the slippery nature of sin. The gruesome nature of this attack is another example. With all the men in town dead, the women and children were truly helpless now. "He also struck Nob the city of the priests with the edge of the sword, both men and women, children and infants; he also struck oxen, donkeys, and sheep with the edge of the sword" (1 Samuel 22:19). What began as insecurity ended in blood-soaked madness. Here we see, yet again, the spiraling effect of unrepented sin.

Previously we discussed the important distinction between peacefulness and harmlessness. *Pacifism is a luxury good; cowards buy it with the blood of others.* Fools think it works great—until the barbarians storm the gate. The town of Nob had over eighty grown men. One wonders whether they might have resisted, had they been more like David in spirit. Eighty five men versus one are excellent odds. But lacking either the will or the training, they were left defenseless before Doeg's blade. To be genuinely peaceful, one must be capable of effective violence. When David worked the fields, he always carried his trusty sling and staff. With these simple tools, he protected his flock from lions and bears. The priests of Nob would have been wise to take note of his tactics.

Additionally, Christ later spoke on this critical subject in the Gospel of John. "I am the good shepherd; the good shepherd lays down His life for the sheep. He who is a hired hand, and not a shepherd, who is not the owner of the sheep, sees the wolf coming, and leaves the sheep and flees; and the wolf snatches them and scatters the flock. He flees because he is a hired hand and does not care about the sheep" (John 10:11–13). This again highlights the distinction between mercenaries and warriors. As Christians, we are called to follow the ways of Christ. The good shepherd doesn't throw away his staff and flee when the wolf comes. Rather, he stands firm and confronts the beast with resolve. The legendary warrior–poet Musashi spoke to this same principle: "Generally speaking, the way of the warrior is resolute acceptance of death.... This is the truth: when you sacrifice your life, you must make fullest use of your weaponry. It is false not to do so, and to die with a weapon yet undrawn" (Musashi 1, 5).

Only one man escaped Saul's gruesome wrath, Ahimelech's son, Abiathar. He fled the town of Nob and sought refuge with David and his band of outlaws. When he told David all the horrors he had seen, David responded not with excuses but with responsibility. He welcomed Abiathar with open arms and took ownership of the situation and apologized (1 Samuel 22:22). Abiathar would go on to become one of David's spiritual advisors. For now, David offered protection to the weary priest. "Stay with me; do not be afraid, even though he who is seeking my life is seeking your life. For you are safe with me" (1 Samuel 22:23). Their common enemy forged an unbreakable alliance.

As David gathered the broken and faithful into his camp, a new threat emerged. The Philistines again raided Israelite territory, this time plundering grain from the town of Keilah. Despite being an outlaw to Saul's regime, David remained faithful to the Lord and sought guidance in prayer. God answered clearly: attack the Philistines and save the town (1 Samuel 23:2).

While David prepared the mission, his inexperienced troops voiced concern: "But David's men said to him, 'Behold, we are fearful here in Judah. How much more then if we go to Keilah against the ranks of the Philistines?'" (1 Samuel 23:3). Their fear was not unfounded. The Philistines were seasoned warriors. David had proved himself in combat, but his new followers had little experience. As Vegetius warns, "A handful of men, accustomed to war, proceed to certain victory, while on the contrary numerous armies of raw and undisciplined troops are but multitudes of men dragged to a slaughter" (Vegetius 4). Could

David's leadership and example shape this unlikely band into a fighting force?

Thankfully, they were not alone. With the Holy Spirit guiding their cause, even a small and battered company could become instruments of divine justice.

David again sought the Lord's counsel to calm his men's fears. In His mercy, God gave an order that echoed His words to Joshua and Gideon: "So David inquired of the Lord once more. And the Lord answered him and said, 'Arise, go down to Keilah, for I am going to hand the Philistines over to you'" (1 Samuel 23:4). With this divine assurance, David struck swiftly. The battle was decisively one-sided. David's men dealt heavy losses to the Philistines and reclaimed livestock and provisions from the enemy (1 Samuel 23:5).

This was a tactical victory in more ways than one. Smaller forces move more rapidly, hide more easily, and can live better off enemy resources. As Sun Tzu put it, "One cartload of the enemy's provisions is equivalent to twenty of one's own" (Tzu 5).

But they couldn't rest long. Saul caught word of David's position and believed he had him trapped. "When it was reported to Saul that David had come to Keilah, Saul said, 'God has handed him over to me, for he shut himself in by entering a city with double gates and bars'" (1 Samuel 23:7). Now we see the wisdom behind the prophet Gad's earlier warning to leave the stronghold.

Preparing for a siege, Saul set his army in motion. Realizing the threat, David called on Abiathar to seek God's counsel. The answer was grim: Saul was coming, and the very people David had saved would surrender him (1 Samuel 23:9–12). Once again, David used speed and agility to escape. "Then David and his men, about six hundred, rose up and departed from Keilah, and they went wherever they could go. When it was reported to Saul that David had escaped from Keilah, he gave up the pursuit" (1 Samuel 23:13).

The term "guerrilla" translates roughly to "little war." Appropriately, the guerrilla fighter is often likened to a flea, not as an insult, but a tribute to pound-for-pound efficacy. The flea is nearly invisible until it strikes. By the time the big dog reacts, the flea has vanished. You cannot kill what you cannot find. The constant, unseen bites frustrate the larger foe, wearing him down, causing him to lash out blindly, harming himself in the process. In time, festering wounds appear. Other dogs sense the weakness and take their shot. The dog forgets about the flea entirely until it's too late.

David later used this very analogy in his confrontation with Saul: "Now then, do not let my blood fall to the ground far from the presence of the Lord; for the king of Israel has come out to search for a single flea, just as one hunts a partridge in the mountains" (1 Samuel 26:20). He understood the art of evasion as he slipped through wilderness strongholds and the hills of Ziph. "David stayed in the wilderness in the strongholds, and remained in the hill country in the wilderness of Ziph. And Saul searched for him every day, but God did not hand him over to him" (1 Samuel 23:14). At one point, Jonathan found David and offered heartfelt encouragement. "Do not be afraid, because the hand of Saul my father will not find you, and you will be king over Israel, and I will be second in command to you; and Saul my father knows that as well" (1 Samuel 23:17). The two renewed their covenant before God, and Jonathan departed. But the monarchy's spies were everywhere.

The Ziphites informed Saul of David's approximate location. Weary and discouraged from repeated failure, Saul asked for more detail: "So look, and learn about all the hiding places where he keeps himself hidden, and return to me with certainty, and I will go with you; and if he is in the land, I will search him out among all the thousands of Judah" (1 Samuel 23:23).

David's evasive strategy was working. While Saul wasted time and resources, morale eroded. Still, the Ziphites' tip narrowed the search. The chase continued—deserts, ravines, mountains. Then, a critical moment: David and his men were boxed in by rough terrain. Saul's troops closed in. It was the closest he had come.

Just as the trap was about to spring, a miracle intervened. "Saul went on one side of the mountain, and David and his men on the other side of the mountain; and David was hurrying to get away from Saul, while Saul and his men were surrounding David and his men to apprehend them. But a messenger came to Saul, saying, 'Hurry and come, for the Philistines have launched an attack against the land!'" (1 Samuel 23:26–27).

The threat of invasion forced Saul to retreat. The strategy of the flea had worked again. One vicious dog distracted another. It is often said that the Lord has perfect timing. This is a textbook example. David's mortal enemy, the Philistines, rescued him from a more aggressive enemy. To this day, this miraculous location is referred to as the *rock of parting* (1 Samuel 23:28). David seized the opportunity and vanished once more.

DAVID'S MERCY

David and his men fled to the desert of En Gedi. After responding to the Philistine raids, Saul resumed his pursuit. His hunt was as relentless as it was bloodthirsty, and this time he brought three thousand fighting men (1 Samuel 24:2). Much like Abraham, David now faced a series of tests from the Lord. Seeking shelter from the desert heat, David and his men took refuge in a cave. Unbeknownst to them, Saul entered the very same cave. "And he came to the sheepfolds on the way, where there was a cave; and Saul went in to relieve himself. Now David and his men were sitting in the inner recesses of the cave" (1 Samuel 24:3).

David's men couldn't believe their luck. They encouraged him to strike the mad king while he was vulnerable. Yet David chose a different path. With quick wits and light footing, he crept forward. Instead of killing Saul, he cut off a corner of his robe.

Perhaps David's men thought of how Ehud had assassinated King Eglon in the book of Judges. From that perspective, this was the perfect opportunity. But David understood this was not the same. Ehud had been called by the Lord to strike down His enemies. That was not the case here. Despite all Saul's madness, David still respected the Lord's anointed. He saw an opportunity for restraint and mercy: "So he said to his men, 'Far be it from me because of the Lord that I would do this thing to my lord, the Lord's anointed, to reach out with my hand against him, since he is the Lord's anointed.' And David rebuked his men with these words and did not allow them to rise up against Saul. And Saul got up, left the cave, and went on his way" (1 Samuel 24:6–7).

Before Saul got too far, David emerged from the cave and called out to him. This took immense courage. David stepped into the open, away from cover, with three thousand enemy soldiers close by. He bowed low and addressed the king. "And David said to Saul, 'Why do you listen to the words of men who say, "Behold, David is seeking to harm you?"' Behold, this day your eyes have seen that the Lord had handed you over to me today in the cave, and someone said to kill you, but I spared you; and I said, 'I will not reach out with my hand against my lord, because he is the Lord's anointed'" (1 Samuel 24:9–10).

Then David revealed the evidence. "So, my father, look! Indeed, look at the edge of your robe in my hand! For by the fact that I cut off the edge of your robe but did not kill you, know and understand that there is no evil or rebellion in my hands, and I have not sinned against you, though you are lying in wait for my life, to take it. May the Lord judge between you and me, and may the Lord take vengeance on you for me; but my hand shall not be against you" (1 Samuel 24:11–12).

David continued this appeal to Saul's conscience, hoping to bury the hatchet. Saul, stunned by the confrontation, could only listen. Then, he praised David for this act of mercy: "You have declared today that you have done good to me, that the Lord handed me over to you and yet you did not kill me. Though if a man finds his enemy, will he let him go away unharmed? May the Lord therefore reward you with good in return for what you have done to me this day" (1 Samuel 24:18–19). Then Saul acknowledged the inevitable: "Now, behold, I know that you will certainly be king, and that the kingdom of Israel will be established in your hand" (1 Samuel 24:20).

Now humbled, Saul pleaded for mercy: "So now swear to me by the Lord that you will not cut off my descendants after me, and that you will not eliminate my name from my father's household" (1 Samuel 24:21). Likely thinking of Jonathan, David gave his word.

After this oath, Saul returned home. But David led his men to another stronghold. Despite the outward diplomacy, David still harbored serious doubts. A quote from the *Dune* series sums up this tension perfectly: "When I am weaker than you, I ask for freedom because that is according to your principles. When I am stronger than you, I take away your freedom because that is according to my principles" (Herbert 235). Was Saul sincere? Or had he simply been caught with his pants down? Saul's record of duplicity suggested the latter. For now, they parted. The Lord would test David again very soon.

DAVID AND NABAL

S amuel, the prophet and judge, died of old age. Israel mourned the loss of this spiritual leader and wise elder. His death marked the dawn of a new era. While prophets would remain, Samuel's passing marked the end of the judges. For better or worse, the monarchy was here to stay. With the loss of his mentor, David entered uncharted territory. After mourning Samuel, David and his men went to the Desert of Paran (1 Samuel 25:1). He was soon tested by a figure who, like Saul, would respond to goodwill with contempt.

His name was Nabal, which roughly translates to fool (1 Samuel 25:25). Despite his foolishness, he was quite wealthy with his business in Carmel (1 Samuel 25:2). Carmel was the same town where Saul had once built a monument to himself after disobeying the Lord (1 Samuel 15:12). But not everyone in Carmel was so vain. One woman in particular stood out—Nabal's wife, Abigail. She was intelligent, kind, and beautiful.

For some time, David had protected Nabal's shepherds and livestock. He had risked himself to guard Nabal's investments, serving as a faithful shepherd. Drawing from his experience in the wilderness, David ensured nothing was taken or harmed. When sheep-shearing season came, a festive time of generosity, David politely asked for provisions. His men approached with humility (1 Samuel 25:7–8).

But Nabal dismissed them with arrogance. "Who is David? And who is the son of Jesse? There are many servants today who are each breaking away from his master. Shall I then take my bread and my water and my meat that I have slaughtered for my shearers, and give it to men whose origin I do not know?" (1 Samuel 25:10–11). Much like

Saul, Nabal repaid good with evil. Infuriated, David snapped. "Then David said to his men, 'Each of you strap on his sword.' So each man strapped on his sword. And David also strapped on his sword, and about four hundred men went up behind David, while two hundred stayed with the baggage" (1 Samuel 25:13).

Meanwhile, Nabal's servants were alarmed: "Now one of the young men told Abigail, Nabal's wife, saying, 'Behold, David sent messengers from the wilderness to greet our master, and he spoke to them in anger. Yet the men were very good to us, and we were not harmed, nor did anything go missing as long as we went with them, while we were in the fields. They were a wall to us both by night and by day, all the time we were with them tending the sheep'" (1 Samuel 25:14–16). They pleaded for her to act quickly. "Now then, be aware and consider what you should do, because harm is plotted against our master and against all his household; and he is such a worthless man that no one can speak to him" (1 Samuel 25:17).

Abigail understood the mortal danger: "Then Abigail hurried and took two hundred loaves of bread and two jugs of wine, and five sheep already prepared and five measures of roasted grain, and a hundred cakes of raisins and two hundred cakes of figs, and she loaded them on donkeys" (1 Samuel 25:18). She sent the servants ahead and followed close behind, hoping to intercept David before blood was spilled.

As David and his men moved through a mountain ravine, he vented his anger: "Now David had said, 'It is certainly for nothing that I have guarded everything that this man has in the wilderness, so that nothing has gone missing of all that belonged to him! For he has returned me evil for good'" (1 Samuel 25:21). David was fully prepared to wipe out Nabal. Then he saw Abigail. "When Abigail saw David, she hurried and dismounted from her donkey, and fell on her face in front of David and bowed herself to the ground" (1 Samuel 25:23). She apologized on Nabal's behalf and pleaded for mercy: "Please do not let my lord pay attention to this worthless man, Nabal, for as his name is, so is he. Nabal is his name, and stupidity is with him; but I your slave did not see the young men of my lord whom you sent" (1 Samuel 25:25).

Abigail offered David the generous provisions and called on the Lord to protect him from enemies like Saul. Moved by her humility and wisdom, David relented. "Then David said to Abigail, 'Blessed be the

Lord God of Israel, who sent you this day to meet me, and blessed be your discernment, and blessed be you, who have kept me this day from bloodshed and from avenging myself by my own hand'" (1 Samuel 25:32–33). Peace was restored, and Abigail returned home.

That night, Nabal held a lavish feast. He was drunk and oblivious. "Then Abigail came to Nabal, and behold, he was having a feast in his house, like the feast of a king. And Nabal's heart was cheerful within him, for he was very drunk; so she did not tell him anything at all until the morning light. But in the morning, when the wine had gone out of Nabal, his wife told him these things, and his heart died within him so that he became like a stone. About ten days later, the Lord struck Nabal and he died" (1 Samuel 25:36–38). David saw the Lord's justice. "When David heard that Nabal was dead, he said, 'Blessed be the Lord, who has pleaded the cause of the shame inflicted on me by the hand of Nabal, and has kept back His servant from evil. The Lord has also returned the evildoing of Nabal on his own head.' Then David sent a proposal to Abigail, to take her as his wife" (1 Samuel 25:39).

Perhaps reflecting on this moment, David later wrote, "Rest in the Lord and wait patiently for Him; Do not get upset because of one who is successful in his way, Because of the person who carries out wicked schemes. Cease from anger and abandon wrath; Do not get upset; it leads only to evildoing. For evildoers will be eliminated, But those who wait for the Lord, they will inherit the land" (Psalm 37:7–9). His son Solomon would echo this wisdom: "Do not rejoice when your enemy falls, And do not let your heart rejoice when he stumbles, Otherwise, the Lord will see and be displeased, And turn His anger away from him. Do not get upset because of evildoers Or be envious of the wicked; For there will be no future for the evil person; The lamp of the wicked will be put out" (Proverbs 24:17–20).

Recognizing Abigail's courage and discernment, David asked her to become his wife; she gladly accepted (1 Samuel 25:42). But Abigail wasn't his only bride. "David had also taken Ahinoam of Jezreel, and they both became his wives. But Saul had given his daughter Michal, David's wife, to Palti the son of Laish, who was from Gallim" (1 Samuel 25:43–44). This would have ripple effects in the future. As with Gideon's son Abimelech, seeds of conflict are often sown in a divided household. But for now, David's test of wrath had passed—thanks in large part to a wise and fearless woman.

DAVID'S CONTINUED MERCY

There was little rest for David. Whether greed or fear motivated the Ziphites, we will never know. What we do know is they yet again attempted to court Saul's favor. They once more reported David's location to the king. Tempted yet again, Saul's true nature showed its ugly head. He simply couldn't let go of his perceived rival. Not playing around, he brought a small army of hand-picked troops. For this high-stakes mission, only the best would suffice. "So Saul set out and went down to the wilderness of Ziph, taking with him three thousand chosen men of Israel, to search for David in the wilderness of Ziph" (1 Samuel 26:2). Saul's elite fighting force made camp nearby. David stayed in the shadows to collect more intel on his pursuer. He sent out scouts to confirm Saul's arrival (1 Samuel 26:4). A wise strategy, for, "in every kind of warfare it is deemed of the highest importance to spy out and get to know thoroughly the habits of the enemy" (Vegetius 107). Even more so when you are severely outnumbered.

David crept nearby for a closer look. What he saw was nothing short of daunting. Like a python wrapped around her clutch of eggs, Saul's warriors enveloped him. "David then set out and came to the place where Saul had camped. And David saw the place where Saul lay, and Abner the son of Ner, the commander of his army; and Saul was lying in the circle of the camp, and the people were camped around him" (1 Samuel 26:5). Gears turned in David's mind as he thought of a way to outmaneuver his opponent. He refused to harm the Lord's anointed, despite King Saul abandoning his divine duties and the Lord abandoning him as a result. He needed a new tactic; something no one would expect. Like Gideon, David employed a strategy of psychological warfare (PSYOP). When one is greatly outnumbered, PSYOPs are an excellent tool. They often cost little and can have widespread effects. Sun Tzu would approve of David's tactical reasoning. According to the Chinese general, "Do not repeat the tactics which have gained you one victory, but let your methods be regulated by the infinite variety of circumstances" (Tzu 12). David asked for some volunteers for this clandestine mission. Only one man answered the call, Abishai.

With the hearts of lions, David and his companion entered the dragon's den. His plan was so audacious that no one could have reasonably expected it. "So David and Abishai came to the people by night, and behold, Saul lay sleeping inside the circle of the camp with his spear stuck in the ground at his head; and Abner and the people were lying around him" (1 Samuel 26:7). There was no room for error. One slip up and the entire camp would surely pounce. Abishai, while fearless, misunderstood the mission. He was itching for a fight. Adrenaline pumped through his veins as he told David the following: "Today God has handed your enemy over to you; now then, please let me pin him with the spear to the ground with one thrust, and I will not do it to him a second time" (1 Samuel 26:8). David reprimanded his subordinate and reminded him of Saul's anointed status.

Moreover, even if he impaled Saul, they had three thousand loyal troops to contend with. There is a difference between bravery and recklessness. "David also said, 'As the Lord lives, the Lord certainly will strike him, or his day will come that he dies, or he will go down in battle and perish'" (1 Samuel 26:10). Before anyone discovered them, David grabbed some mission-critical items and ordered Abishai to follow. Quiet as mice, they left the enemy camp. David passed the test! "So David took the spear and the jug of water that were at Saul's head, and they left; and no one saw or knew about it, nor did anyone awaken, for they were all asleep, because a deep sleep from the Lord had fallen on them" (1 Samuel 26:12).

At the time, David was unaware the Lord had aided him in this way. He knew he needed to send a message to Saul without bloodshed. As we have seen in previous chapters, the Lord helps those who help themselves. David and Abisahi continued their stealthy egress until they reached a safe distance from the enemy camp on a hill top. With distance between them, David made the point to embarrass the commander of Saul's army, Abner.

With a roaring shout, David called out to the enemy camp. The more ears the better. After he captured Abner's attention, a verbal beat down ensued: "So David said to Abner, 'Are you not a man? And who is like you in Israel? Why then have you not guarded your lord the king? For one of the people came to kill the king your lord! This thing that you have done is not good. As the Lord lives, all of you undoubtedly must die, because you did not guard your lord, the Lord's anointed. And now, see where the king's spear is and the jug of water that was at his head!'" (1 Samuel 26:15–16).

All the commotion roused the King of Israel from his slumber. He called out, asking if the voice belonged to David. David responded earnestly: "Why then is my lord pursuing his servant? For what have I done? Or what evil is in my hand?" (1 Samuel 26:18). David then reiterated how people kept leading Saul astray. David concluded his criticism by pointing out how ridiculous the king's pursuit was. Saul put such an unreasonable effort into killing an innocent man. Realizing the absurdity of the situation, Saul pleaded for David to return. "Then Saul said, 'I have sinned. Return, my son David, for I will not harm you again since my life was precious in your sight this day. Behold, I have played the fool and have made a very great mistake'" (1 Samuel 26:21). This was not the first time Saul had shed crocodile tears. Saul's actions proved that his words were scarcely more than hot air. David was finally coming around to this unfortunate reality.

Twice Saul had tried to spear him. Twice now, David had spared his life. David left Saul's spear on the hilltop and called for one of his men to fetch it. He wasn't going to harm Saul. But he wasn't pretending to be his friend either. He knew he must move on. Before departing, David shared some words that we should all remember: "And the Lord will repay each man for his righteousness and his faithfulness; for the Lord handed you over to me today, but I refused to reach out with my hand against the Lord's anointed. Therefore behold, just as your life was highly valued in my sight this day, so may my life be highly valued in the sight of the Lord, and may He rescue me from all distress" (1 Samuel 26:23–24). Stunned at David's continued mercy, Saul thanked him for sparing his life once more. He likely knew it was more gracious than he deserved. "Then Saul said to David, 'Blessed are you, my son David; you will both accomplish much and assuredly prevail.' So David went on his way, and Saul returned to his place" (1 Samuel 26:25).

The pair went their separate ways yet again. David's performance on this mission was nothing short of phenomenal. He demoralized the opposing forces without spilling a single drop of blood. Sun Tzu surely would approve of such a strategy. According to his framework, David's PSYOP was the ultimate expression of warfare. "Hence to fight and conquer in all your battles is not supreme excellence; supreme excellence consists in breaking the enemy's resistance without fighting…. Thus the highest form of generalship is to balk the enemy's plans" (Tzu 6). Any random street thug can stab someone in their sleep, as Abishai suggested. While that might be effective in certain scenarios, it doesn't take a tactical genius to figure that out.

Contrast that to instilling paralyzing fear into your enemy, alongside his three thousand elite warriors, no less. Now *that* takes skill! None of their experience or technology protected them in the moment, and they knew it. A complete slaughter is less embarrassing than being utterly duped. At least then they could fight and die with valor. However, to be completely circumvented leaves a lingering feeling of doubt and helplessness. Using his wits alone, David outmaneuvered the king and his men. It's hard to be more efficient or cost effective than that. Once more, we see a familiar Biblical pattern. When fighting the Lord's battles, we are capable of incredible feats. All we need is a fighting spirit and faith. God helps those who help themselves. General Carl von Clausewitz concurs: "remember that no military leader can ever become great without a dash of boldness in him" (Clausewitz 3). Faith is our secret weapon. Faith fuels boldness, and boldness begets action.

DAVID AND THE PHILISTINE KING

David had never heard the story of *The Scorpion and the Frog*, yet he had finally internalized its message. He needed to avoid Saul at all costs. "Then David said to himself, 'Now I will perish one day by the hand of Saul. There is nothing better for me than to safely escape into the land of the Philistines. Then Saul will despair of searching for me anymore in all the territory of Israel, and I will escape from his hand'" (1 Samuel 27:1). With few options, David and his mighty men fled deep into enemy territory. Despite Saul's paranoia and bloodthirst, even he knew better than to follow David this time. David returned to Achish, King of Gath: the same location where he had pretended to be insane! Yet as the classic axiom goes: *the enemy of my enemy is my friend*. Both Achish and David understood this practical strategy. As mentioned earlier, defectors can prove themselves immensely valuable. They can provide intel not even the best spies can dream of. Additionally, they very often have an axe to grind. Thus, they are greatly motivated to work with their new allies.

Actions speak louder than words. David and his mighty men brought their wives and children along; they were in it for the long haul (1 Samuel 27:2–3). This almost certainly helped to sway King Achish's decision to let them stay. One maniac might risk his family for a tactical advantage. But would six hundred men do that? Highly unlikely. This bold action was enough for Saul to relent (1 Samuel 27:4). True to his humble nature, David asked for modest accommodations. He asked for a place out in the countryside. Much like when he turned down a wife from Saul, he considered himself unworthy of the king's presence.

Achish happily obliged. "So Achish gave him Ziklag that day; therefore Ziklag has belonged to the kings of Judah to this day" (1 Samuel 27:6). Ziklag was David's new base of operations. It remained under the control of Judah for generations to come.

King Achish knew the power of David's skillset. He gave David assignments early and often. David raided the nearby tribes and plundered the remaining supplies (1 Samuel 27:9). Yet there was more to this. David continued fighting the enemies of the Israelites. In reality, he continued waging war against the Lord's enemies, who had slipped from Joshua's grasp. Even Joshua knew this fight wouldn't be over in his lifetime. David picked up where he left off.

To maintain the illusion of loyalty, David lied to Achish. To preserve his cover story to the king, David left no loose ends untied. Here we see the brutal logic of warfare. "And David did not leave a man or a woman alive to bring to Gath, saying, 'Otherwise they will tell about us, saying, "This is what David has done, and this has been his practice all the time that he has lived in the country of the Philistines"'" (1 Samuel 27:11), As the old adage goes: *dead men tell no tales*. An unfortunate reality. While modern audiences might scoff at this level of brutality, no one questions the results. David's plan worked both spiritually and pragmatically. First, David continued to fulfil Joshua's conquest of Canaan, thus continuing the Lord's work (Joshua 13:1). In addition, Saul abandoned this divine calling early in his career, thus invalidating his conditional kingship (1 Samuel 15:20−29).

Second, this was highly practical. Should only one survivor flee, it would spell the end for David and his mighty men in addition to their families, no less. Achish took the bait. In fact, he was delighted to have such a prodigious asset on his team. "So Achish believed David, saying, 'He has undoubtedly made himself repulsive among his people Israel; therefore he will become my servant forever'" (1 Samuel 27:12).

King Achish made it clear to David that they were more than simple mercenaries. Things reached a boiling point with Saul's army. Achish prepared a full-scale attack on his longtime rival and prepared David accordingly. "Now it came about in those days that the Philistines gathered their armed camps for war, to fight against Israel. And Achish said to David, 'Know for certain that you will go out with me in the camp, you and your men'" (1 Samuel 28:1). Never one to back down from a challenge, David responded with the following. "David said to Achish, 'Very well, you will learn what your servant can do.' So Achish said to David, 'Then I will assuredly make you my bodyguard for life!'" (1 Samuel 28:2).

Saul's worst nightmare slowly became reality. When he realized the danger of his two rivals pairing up, Saul panicked. The Philistine army made its encampment as it prepared to invade, the sight of which took the wind right out of his sails. "When Saul saw the camp of the Philistines, he was afraid and his heart trembled greatly. So Saul inquired of the Lord, but the Lord did not answer him, either in dreams, or by the Urim, or by the prophets" (1 Samuel 28:5–6). Saul failed to realize that God's apparent lack of an answer *was* His answer! Saul had violated the commandments of his theocratic position, and thus God's grace had departed. This scenario echoes back to Eli's rebuke of his two wicked sons. "If one person sins against another, God will mediate for him; but if a person sins against the Lord, who can intercede for him?" (1 Samuel 2:25). Saul had dug his own grave some time ago. The realization was only now beginning to dawn on the first King of Israel.

Without Samuel for guidance, Saul defaulted to his cowardly ways. His desperation led to even more hypocrisy and apostasy. In one of his few good feats, Saul had expelled the mediums and spiritists from Israel (1 Samuel 28:3). Saul was wise to do this, as God made this point abundantly clear to Moses early on. "Do not turn to mediums or spiritists; do not seek them out to be defiled by them. I am the Lord your God" (Leviticus 19:31). This advice was later elaborated in Deuteronomy. "There shall not be found among you anyone who makes his son or his daughter pass through the fire, one who uses divination, a soothsayer, one who interprets omens, or a sorcerer, or one who casts a spell, or a medium, or a spiritist, or one who consults the dead. For whoever does these things is detestable to the Lord; and because of these detestable things the Lord your God is going to drive them out before you" (Deuteronomy 18:10–12).

The same underlying theme continued throughout the New Testament as well. The Apostle Paul remarked on this in his second letter to the Corinthians. While referring to false prophets, he wrote: "No wonder, for even Satan disguises himself as an angel of light. Therefore it is not surprising if his servants also disguise themselves as servants of righteousness, whose end will be according to their deeds" (2 Corinthians 11:14–15). Additionally, Jesus condemned these practices. In the epilogue of Revelation, He left us this message: "I am the Alpha and the Omega, the first and the last, the beginning and the end. Blessed are those who wash their robes, so that they will have the right to the tree of life, and may enter the city by the gates. Outside are the dogs, the sorcerers, the sexually immoral persons, the murderers, the idolaters, and everyone who loves and practices lying" (Revelation 22:13–15).

Yet Saul was a far cry from the strong and courageous Biblical warriors. He had a habit of seizing defeat from the jaws of victory. His lack of faith led to cowardice, and cowardice never leads to anything good. Desperate for guidance, he backtracked on one of his few good feats, as he ordered his servants to locate a medium (1 Samuel 28:7). The servants located an occult practitioner in short order. Saul disguised himself and inquired about her demonic services. Ever so cautious, she deflected the undercover king. "But the woman said to him, 'Behold, you know what Saul has done, that he has eliminated the mediums and spiritists from the land. Why are you then setting a trap for my life, to bring about my death?'" (1 Samuel 28:9). After some negotiation, she complied. Saul requested to speak with the spirit of Samuel. "When the woman saw Samuel, she cried out with a loud voice; and the woman spoke to Saul, saying, 'Why have you deceived me? For you are Saul!'" (1 Samuel 28:12). The necromancer feared for her life, but Saul assured her she was in no danger. The disgraced king implored her to keep going.

SAUL'S DOWNFALL FORETOLD

When Samuel's spirit appeared, Saul threw himself down and bowed to the deceased prophet. Samuel asked why Saul had disturbed him. Wasting no time on formalities, Saul told him that he was in great distress. As he rambled on, desperation poured out of him like a foul sweat as he said the following. "And Samuel said to Saul, 'Why have you disturbed me by bringing me up?' Saul replied, 'I am very distressed, for the Philistines are waging war against me, and God has abandoned me and no longer answers me, either through prophets or in dreams; therefore I have called you, so that you may let me know what I should do'" (1 Samuel 28:15).

Samuel responded with criticism to Saul's foolishness. He then issued a chilling prophecy which echoed that of Eli and his wicked sons. "And the Lord has done just as He spoke through me; for the Lord has torn the kingdom from your hand and given it to your neighbor, to David. Just as you did not obey the Lord and did not execute His fierce wrath on Amalek, so the Lord has done this thing to you this day. Furthermore, the Lord will also hand Israel along with you over to the Philistines; so tomorrow you and your sons will be with me. Indeed, the Lord will hand the army of Israel over to the Philistines!" (1 Samuel 28:17–19).

This realization crashed on Saul like a thunderbolt from above. The prophecy flattened him like the millstone which cracked Abimelech's skull (Judges 9:52–53). He collapsed in anguish! "Then Saul immediately fell full length to the ground and was very afraid because of Samuel's words; there was no strength in him either, because he had eaten no food all day and all night" (1 Samuel 28:20). The dread from his impending doom left the king incapacitated for hours. His blasphemous appeal backfired. Despite his many feuds with David and the Philistines, Saul truly was his own worst enemy.

The Philistines prepared for their attack. Thousands of troops marched ahead, clearing the way for the king's royal court. David and his men marched close to Achish, guarding him as usual. Meanwhile, the Philistine commanders grew suspicious of David and his mighty men. Achish had grown fond of David's services, yet his commanders thought this battle was a conflict of interest: "Then the commanders of the Philistines said, 'What are these Hebrews doing here?' And Achish said to the commanders of the Philistines, 'Is this not David, the servant of Saul the king of Israel, who has been with me these days, or rather these years, and I have found nothing at all suspicious in him since the day he deserted to me to this day?'" (1 Samuel 29:3). This response infuriated the Philistine commanders. They refused to relent on this issue. They appealed to reason and Achish's self-interest. "But the commanders of the Philistines were angry with him, and the commanders of the Philistines said to him, 'Make the man go back, so that he will return to his place where you have assigned him, and do not let him go down to battle with us, or in the battle he may become an adversary to us. For how could this man find favor with his lord? Would it not be with the heads of these men?'" (1 Samuel 29:4).

The commanders concluded their retort by using David's larger-than-life reputation against him. Reminiscent of David's maniacal performance outside the gates of Gath, they once again referenced the songs of David's praise of slaying tens of thousands (1 Samuel 29:5). The Philistine commanders forced their leader's hand. Power is largely an illusion. A king is nothing without men willing to enforce his orders. A wise leader knows he must delegate tasks and duties. When done properly, the fruits of the kingdom are greater than the sum of its parts. When done poorly, chaos and corruption ensue in short order. It's called a war machine for good reason! Should individual components of the machine break down, the entire machine will follow suit. A chain only has the strength of its weakest link. Achish knew this and defaulted to the judgement of his men. His actions correlate to advice of Clausewitz, "In particular, you have to trust your subordinate commanders; therefore, the most trustworthy people should be selected for these posts, and this quality should count above all others" (Clausewitz 31).

Knowing he couldn't cross his men, Achish peacefully released David from service. Fortunately for David's sake, he was much more generous than Saul. The Philistine king summoned David to tell him the following. "As the Lord lives, you have indeed been honest, and your going out and your coming in with me in the army are pleasing in my sight; for I have not found evil in you since the day of your coming to me to this day. Nevertheless, you are not pleasing in the sight of the governors. Now then, return and go in peace, so that you will not do anything wrong in the sight of the governors of the Philistines" (1 Samuel 29:6–7). The rapid turn of events stunned David. The king sang nothing but praises yet turned him away in spite of this. Such is the realm of politics; some things never change.

Reminiscent of his shock at Saul's early dismissal, David asked what wrong he had committed. Acting as the king's personal bodyguard is no small feat! Only a select few can fill that role. Now this very king forced him to leave. The king politely, yet firmly, stood by his order. He once again complimented David's service and reiterated the conflict of interest. Without the support of his subordinate commanders, Achish was dead in the water. The dismissal wasn't personal, merely a political reality. Yet David and his band of warriors hadn't lost all. Achish repaid David by allowing him to keep the territory of Ziklag, whereas Saul gave David the point of his spear for all his good deeds. Achish gave David the final order to depart at dawn (1 Samuel 29:10). David and his men acted accordingly and began their trek to Ziklag. The Philistine army marched onward to battle.

FOLLOWING THE FOOTSTEPS OF JOSHUA

There was little rest for David and his men. The seeds of Saul's disobedience bore bitter fruit. The Lord had ordered Saul to wipe out the Amalekites, to continue Joshua's conquest (1 Samuel 15). He had refused to do so. Now, the Amalekites were back with a vengeance! Only a few days later, David returned to a scene of unimaginable horror. Far from a joyous reunion, David felt nothing but dread. "Then it happened, when David and his men came to Ziklag on the third day, that the Amalekites had carried out an attack on the Negev and on Ziklag, and had overthrown Ziklag and burned it with fire; and they took captive the women and all who were in it, from the small to the great, without killing anyone, and drove them off and went their way" (1 Samuel 30:1–2).

A far cry from the usual welcome of songs and dancing, only silence and ashes greeted them. The shock was too much to bear. The men broke down amongst the soot and smoldering ruins, weeping until their strength left them (1 Samuel 30:4). Things were so dire that several of the men considered a mutiny against David. Fortunately, cooler heads prevailed. "Also, David was in great distress because the people spoke of stoning him, for all the people were embittered, each one because of his sons and his daughters. But David felt strengthened in the Lord his God" (1 Samuel 30:6).

This mirrors the trials of Moses, who also faced violent hostility from those he led. Not long after their escape from Egypt, the Israelites grew restless and turned on him. They sarcastically asked, had Moses liberated them from slavery only to have them die of thirst? (Exodus

263

17:3). Desperation poured through his veins: "So Moses cried out to the Lord, saying, 'What am I to do with this people? A little more and they will stone me!'" (Exodus 17:4). Once again, the Lord intervened. He told Moses to take his staff and strike the rock at Horeb. Moses obeyed—and water flowed (Exodus 17:5–6).

Fear and division are some of Satan's favorite tools. Satan is as evil as he is cunning. The Father of Lies understands warfare as well. The fallen angel wants to drag us all to Hell with him! He wants us divided and distracted. The more infighting amongst believers, the better for the Devil. *Divide and conquer* remains a timeless strategy for a reason. Just as Satan tried to separate Moses and his followers, he placed a wedge between David and his men. In fact, *De Re Militari* highlights this time-honored tactic. There is no greater enemy than the enemy from within: "A prudent general will also try to sow dissention among his adversaries, for no nation, though ever so weak in itself can be completely ruined by its enemies unless its fall be facilitated by its own distraction" (Vegetius 66).

On some level, David understood the stakes. Thankfully, his faith had prepared him for the fight—just as it had for Moses. He donned the Shield of Faith (Ephesians 6:16). The enemy struck when the opportunity arose, but the real battle was only beginning. David turned to spiritual guidance. He prayed through the priest Abiathar, and the Lord answered clearly—just as He had with Joshua. HE told David: "Pursue, for you will certainly overtake them, and you will certainly rescue everyone" (1 Samuel 30:8).

Much later, someone described David and his mighty men as fierce as a wild bear robbed of her cubs (2 Samuel 17:8). Perhaps this very incident elicited such a ferocious description. Following the Lord's orders to pursue the Amalekites, David and his men made haste. Unfortunately, about one third of his forces were too ill or weak to continue their pursuit (1 Samuel 30:9–10). One rarely has ideal conditions going into combat. Yet, as we have seen many times in Biblical combat, "Valor is superior to numbers" (Vegetius 88). David had been valorous since his humble youth. With his family on the line, he was all the more so! Well seasoned to combat, David and his troops had a fighting spirit like no other.

Both Saul and Achish banished David from their kingdoms. He was an outcast everywhere he went. With his family gone, he truly had nothing else to lose. However, there is an upside when you have nothing left to lose; you stand to gain everything in victory. When you are at rock bottom, even a small victory seems like an epic feat. This concept is further elaborated by the legendary Samurai, Musashi. "If you keep your spirit correct from morning to night, accustomed to the idea of death and resolved on death, and consider yourself as a dead body, thus becoming one with the way of the warrior, you can pass through life with no possibility of failure" (Musashi 2).

David and his men set forth with icy resolve. This mission was nothing short of do or die. Fortunately, they met an unexpected ally on the way. They stumbled upon an emaciated slave, left for dead in the harsh wilderness. "Now they found an Egyptian in the field and brought him to David, and gave him bread and he ate, and they provided him water to drink" (1 Samuel 30:11). You catch more flies with honey than vinegar, as the old adage goes. David was substantially more generous than the slave's former master. A far cry from slave rations, David gave the dying man luxurious foods such as figs and raisin cakes. Considering the man hadn't eaten or drank in three days and nights, this must have tasted like a slice of heaven (1 Samuel 30:12). Abigail presented the same style of cakes to David (1 Samuel 25:18). Given this, we can infer they were prized indeed.

After the Egyptian recovered, David questioned him. "'To whom do you belong? And where are you from?' And he said, 'I am a young man of Egypt, a servant of an Amalekite; and my master abandoned me when I became sick three days ago. We carried out an attack on the Negev of the Cherethites, and on that which belongs to Judah, and on the Negev of Caleb, and we burned Ziklag with fire'" (1 Samuel 30:13–14). Naturally, David asked if he could lead them to the raiders. Fearing for his life, the Egyptian said he would help them in exchange for his freedom, both from harm and his former master (1 Samuel 30:15). The Egyptian was wise to do so. According to Sun Tzu, "Peace proposals unaccompanied by a sworn covenant indicate a plot" (Tzu 18). A man of his word, David agreed.

The Chinese general would also approve of David's actions. Kings and warlords often make use of torture or coercion for information. The problem is that under duress most people will do anything to stop the pain. This includes giving false claims or nonsense simply to get a temporary reprieve from their agony. *The Art of War* sheds light on a more effective and scalable approach for gathering intel. "Hence it is that which none in the whole army are more intimate relations to be maintained than with spies. None should be more liberally rewarded.... They cannot be properly managed without benevolence and straightforwardness" (Tzu 28). There is no need to twist an arm when you share a common enemy.

Armed with this new intel, David and his men were hot on the trail. Guided by the former slave and protected by Israel's most formidable warriors, they moved with urgency and resolve. Before long, they reached the outskirts of the enemy camp. "Now when he had brought him down, behold, they were dispersed over all the land, eating and drinking and celebrating because of all the great plunder that they had taken from the land of the Philistines and from the land of Judah" (1 Samuel 30:16).

This offers a powerful combat lesson: never celebrate too early. The Amalekites were drunk on their own success—literally and figuratively. In their pride and gluttony, they forgot the first rule of warfare: the fight isn't over until the enemy is broken. As Sun Tzu wisely observed, "To secure ourselves against defeat lies in our own hands, but the opportunity of defeating the enemy is provided by the enemy himself" (Tzu 7). The deeper one falls into sin, the duller one's senses become. These raiders were so blinded by greed that they failed to maintain even basic security. Their arrogance made them vulnerable, ripe for the slaughter.

Strategically, this was a textbook ambush. Clausewitz emphasized that "Surprise plays a much greater role in strategy than in tactics; it is the most valid starting point for victory." David understood this implicitly. He didn't hesitate. "And David slaughtered them from the twilight until the evening of the next day; and not a man of them escaped, except four hundred young men who rode on camels and fled" (1 Samuel 30:17). Though outnumbered, David and his men fought like lions. The threat to their families lit a fire in their hearts. This time, it wasn't just another battle—it was personal. Their ferocity was unmatched. Fueled by faith and righteous fury, David achieved what seemed impossible. "So David recovered all that the Amalekites had taken, and rescued his two wives. And nothing of theirs was missing, whether small or great, sons or daughters, plunder, or anything that they had taken for themselves; David brought it all back" (1 Samuel 30:18–19).

Here we see a living example of the peace-through-strength paradox. David was able to rescue his people precisely because he was dangerous. Harmless men don't stand between their families and barbarism. Only skilled, disciplined men do. The Levite from Judges 19 stands as a haunting contrast. When evil came knocking, he gave up his concubine without a fight. His cowardice cloaked itself in civility—but under pressure, the truth was revealed. Cowards are more dangerous than the enemy because they pretend to be allies. When wolves approach, they trip you and flee. And when others are slaughtered, they act surprised.

This is why Scripture explicitly condemns cowardice. "But for the cowardly, and unbelieving, and abominable, and murderers, and sexually immoral persons, and sorcerers, and idolaters, and all liars, their part will be in the lake that burns with fire and brimstone, which is the second death" (Revelation 21:8). Of all the sins listed, cowardice is first and foremost. Cowardice is not merely weakness; it is betrayal in disguise. And it never ends well.

Fortunately, David was cut from a different cloth. His courage and martial skill secured complete restoration. Nothing was lost. In fact, they gained more than they had had before. The Lord did not just rescue; He also rewarded. Israel didn't simply recover what it had lost. The Amalekites had recently raided the Philistines. David claimed all this as well (1 Samuel 30:20). The good shepherd returned to camp, spoils of war in tow.

With the dust settled from battle, David made his way back to camp. He still had one third of his troops to check on. Things were going well. Emotions ran high as families reunited. With the conflict settled, they could finally recuperate and rebuild. Yet, in short order, a few rabble-rousers overreached as greed captured their hearts. "Then all the wicked and worthless men among those who went with David said, 'Since they did not go with us, we will not give them any of the spoils that we have recovered, except to every man his wife and his children, so that they may lead them away and leave'" (1 Samuel 30:22). Previously, we discussed the corrosive nature of envy. Greed has similar effects. The scarcity mindset has an obvious hallmark. One trades a short-term win for a long-term loss! Should we follow this reasoning to the logical conclusion, things quickly turn disastrous. The more soldiers they exclude, the smaller the fighting force becomes. The smaller the army, the easier it is to defeat.

David seized the initiative and restored good order and discipline. Rather than using brute force or coercion, he shifted the focus to God's glory. "But David said, 'You must not do so, my brothers, with what the Lord has given us, for He has protected us and handed over to us the band of raiders that came against us. And who will listen to you in this matter? For as is the share of the one who goes down into the battle, so shall be the share of the one who stays by the baggage; they shall share alike'" (1 Samuel 30:23–24). Some of his men had the mercenary mindset. David transformed their myopic views by giving them a purpose greater than themselves.

David's ruling for the plunder was wise and just. He followed the guidance of Moses. After his crushing defeat of the Midianites, the Lord told Moses to "divide the spoils between the warriors who went to battle and all the congregation" (Numbers 31:27). Whether David knew of this precedent, or this was his own intuition, we will never know. However, what we *do* know is that this ruling ensured group unity and cohesion: "So it has been from that day forward, that he made it a statute and an ordinance for Israel to this day" (1 Samuel 30:25). With this matter settled, they were all freed up to focus on the Lord's work.

David's generosity didn't stop with the two hundred men who were left behind. "Now when David came to Ziklag, he sent some of the spoils to the elders of Judah, to his friends, saying, 'Behold, a gift for you from the spoils of the enemies of the Lord'" (1 Samuel 30:26). Too often when people reach success, they pull up the ladder behind them. The scarcity mindset is all too common. But not for David! In his flight from Saul, he had more allies than just those in Judah. He repaid bountifully to the people who had previously aided him (1 Samuel 30:31). David extended grace and generosity beyond merited action or deeds. His repayment was far more than a transaction. In this way, David stood as a precursor to Christ. The Apostle Paul illustrates this concept in his letter to the

Ephesians. Through Jesus, God blessed us with mercy that we did not and cannot earn through works or deeds. "For by grace you have been saved through faith; and this is not of yourselves, it is the gift of God; not a result of works, so that no one may boast" (Ephesians 2:8–9).

SAUL'S DOWNFALL

Saul and David had inverse trajectories. David's spirit grew brighter like a blazing flame. Meanwhile Saul's dwindled and began to sputter out. Saul now lacked the protection of both God and David, and the massive Philistine army encroached ever closer. "Now the Philistines were fighting against Israel, and the men of Israel fled from the Philistines but fell fatally wounded on Mount Gilboa" (1 Samuel 31:1). It's no surprise many of them fled. Their Commander-in-Chief gave a poor example to follow. The walls closed in, suffocating Saul. Endless waves of Philistine soldiers flooded the battle space. Jonathan fought valiantly, yet not even his prowess slowed down the horde, and he fell in battle (1 Samuel 31:2). Since the Spirit had long departed, the King of Israel was little more than a sitting duck. With Jonathan dead, the Philistines advanced even faster. "The battle went heavily against Saul, and the archers found him; and he was gravely wounded by the archers" (1 Samuel 31:3).

The Philistine assault was strategically sound. They had the advantage, yet they didn't waste time, manpower, or resources. They knew they didn't have to slaughter every man. With so many fleeing, they stuck to the high-value targets. *Principles of War* advocates the type of approach the Philistines utilized. With so many soldiers fleeing, they could now concentrate their forces more efficiently. "Even with superior forces, the direct attack should still be placed on only one point, which allows us to concentrate large forces against it, since completely encircling the enemy's army is possible only in the rarest circumstances.... This yields a higher chance of success on the main point. This success will outweigh all other disadvantages" (Clausewitz 5, 12).

Saul knew he was doomed! Likely thinking of how the Philistines had abused and tortured Samson, he sought another way out. "Then Saul said to his armor bearer, 'Draw your sword and pierce me through with it, otherwise these uncircumcised Philistines will come and pierce me through, and abuse me.' But his armor bearer was unwilling, because he was very fearful. So Saul took his sword and fell on it" (1 Samuel 31:4). This final act was the perfect culmination of Saul's long history of self-destructive behavior. Terrified and alone, his armor-bearer followed his master's example. With the king dead, the local Israelites had little option but to flee (1 Samuel 31:7) .This highlights arguably the biggest downfall of a top-down centralized command structure. Once the viper is de-fanged, it's little more than a limp noodle! Strike down the shepherd and the flock scatters. It's all the easier to pick the enemy off one by one.

The next day, the dust had settled. The spilled blood congealed into black tar as the bodies began to rot. The carrion birds reveled at such a lavish feast. Their celebration was cut short as the Philistine troops began looting the fallen corpses. Amongst the dead, they found the fallen bodies of Saul and his three sons. Confirming their longtime rival was defeated, they celebrated demonically. They paraded around Saul's body in a similar fashion to Samson, as they mocked the fallen king. Giving the praise to their gods they spread word far and wide. "They cut off his head and stripped off his weapons, and sent them throughout the land of the Philistines, to bring the good news to the house of their idols and to the people" (1 Samuel 31:9). The Philistines loved a good trophy, and they got their money's worth with Saul. They put the fallen king's armor in their temple and hung his body on the wall (1 Samuel 31:10).

In spite of Saul's many shortcomings, he had some admirable moments as well, especially early in his career. Saul's first real test occurred when the Ammonites besieged the Israelite territory of Jabesh Gilead. The first King of Israel led from the front as he staged an ambush and rescued his people. Saul's heroism was not forgotten by the people of Jabesh Gilead. Enraged at the mockery and disrespect from the Philistines, they braved the night for a rescue mission of their own.

They gave Saul kindness in the same way he had been kind to them in the past.

Under the cover of darkness, they recovered Saul's body (1 Samuel 31:11–12). After that, they buried the bones and they fasted seven days in mourning. In spite of the many terrible things Saul had done, David still honored Saul's divine appointment. Some time later, he gave Saul and Jonathan a proper burial with their family (2 Samuel 21:13–14).

Saul's downfall was as predictable as it was tragic. From the very beginning, he showed signs of insecurity and paranoia. Before Samuel even anointed him, Saul was already buckling under pressure, hiding among the supplies (1 Samuel 11:20–25). This echoed the early fear of Gideon, but unlike Gideon—who learned to cultivate faith and grow in courage—Saul never changed. He doubled down on fear, pride, and control. Some may argue he lacked confidence in himself. But at its core, Saul's failure stemmed from a lack of faith in the Lord. His actions made that clear early in his reign.

Faithful obedience was the central condition of kingship. Saul never truly grasped that. The prophet Samuel rebuked him sharply on three occasions. The first came when Saul disobeyed and performed religious rites he had no authority to conduct. Later, the Lord gave Saul a sacred mission: complete the conquest against the Amalekites. This command reached back to the Exodus, when the Amalekites had ambushed Israel in the wilderness. Saul's mission was clear. But once again, he failed—not from weakness, but from pride. He spared what God condemned. And when confronted, he blamed his troops and pleaded fear (1 Samuel 15:24). It was a pathetic display, completely at odds with the courage expected from Israel's king. This time, Samuel didn't just rebuke him—he renounced him.

From there, Saul spiraled further into rebellion. After Samuel's death, the king—desperate and unrepentant—resorted to necromancy, violating his own laws to summon Samuel's spirit. The prophet's final message sealed his fate. The next day, that prophecy was fulfilled exactly as spoken. Saul's long fall ended not with glory, but with ruin.

We've covered a lot of ground in the whirlwind story of Saul and David. Let's break down some of the important takeaway lessons.

SAUL AND DAVID CONCLUSION

Fundamentally, the book of 1 Samuel is a story of contrasts. Israel stood at a fork in the road. Embrace the ways of the world or have faith in God's plan. Despite the stern warning of Samuel, they opted for the former. Perhaps God allowed Saul's folly to teach us a lesson on what *not* to do. This book highlights a familiar Biblical theme: God often raises up the seemingly least likely among us. Earthly displays of wealth and prosperity are *not* indicative of leadership skills or capabilities. As the Lord told Samuel, HE looks at the heart when people look at outward appearance (1 Samuel 16:7). Saul was tall, rich and handsome. David on the other hand, was a no-name peasant boy. Saul came from a family of high standing. Meanwhile, David single-handedly fought off lions and bears to protect his flock.

The pair had inverse trajectories. David came from nothing and steadily rose to the top. Conversely, Saul did nothing to *earn* his kingship. After his crowning, he slid deeper and deeper into paranoia, insecurity, and disobedience. He lacked the discipline and skillset to rule effectively, whereas David's service as a shepherd forged him into the man he became. Saul represents human desire. David represents divine purpose and resilience in the face of adversity. David's faith led him to his strong and courageous actions, which the Lord favors.

Saul's story serves as a cautionary tale to all of us. As examined previously, all sin stems from the pridefulness of the heart. That's why it's often referred to as the mother of all sins. Despite his theocratic obligations, Saul truly thought he knew better than the Lord himself. He continuously prioritized his image and personal vendettas over the Lord's commands. He embraced the worst form of idolatry, *the idolatry of the self.* This was Satan's fatal flaw as well. Just as Satan envied the Lord, Saul envied David. The irony is that Saul truly could have had it all, if only he had honored God's commands. Saul's envy offers an important lesson: *Jealousy is corrosive and self-destructive.* It blinded Saul to his duties and alienated his allies. His descent into madness serves as a stark reminder of how sin erodes sound judgment and reason.

David's rise to power could not contrast with Saul further! His journey illustrates the power of faith, discipline, courage, and honing one's skills. God could have effortlessly ended Goliath with a bolt of lightning. Yet the simple matter is, He didn't need to! David had been practicing for years prior to this. Critics may shrug off David's shot as lucky. However, luck is when skill meets opportunity! Once we internalize this, we can train more confidently and trust in God's timing. David spent countless hours honing his technique and battling wild beasts! God simply offered an opportunity to use these skills for a grander purpose. David never prayed for an easy day in the fields. He prayed for the strength to endure and overcome challenges. In David's own words, "The Lord is my strength and my shield; My heart trusts in Him, and I am helped; Therefore my heart triumphs, And with my song I shall thank Him" (Psalm 28:7). Always remember, God helps those who help themselves. This attitude of humility contrasts sharply with Saul. Shortly after Saul blatantly disregarded the Lord's commands, he erected a statue of this apostasy, in his own honor.

The first book of Samuel is rich not only in spiritual teaching, but also in combat lessons. It offers a bevy of tactics and lessons that still resonate thousands of years later. David was truly a well-rounded warrior. David's many challenges highlight the need for adaptability and resourcefulness. The human mind is the most dangerous weapon known to humankind, for it *creates* all weapons. A stone is merely a deposit of minerals, until it's loaded into a sling. *Then* it becomes a lethal projectile! This is due to the mind's eye, from which all creativity and skill derive. David understood this as a boy. Saul saw Goliath and thought he needed heavy armor and weapons. Yet it's a fool's errand to attempt to beat a specialist at his own game. David was the only one to see an opportunity for victory. He declined the offer of Saul's armor and sword. This represented David rejecting the ways of the world.

Additionally, it's always unwise to try new equipment or techniques when the stakes are high. The practice room is for experimenting; the battlefield requires your A-game! Armed with faith and his trusted tools, he made short work of the giant.

David's feat was so shocking that its effects rippled throughout the Philistines! With their champion dead, the Philistines ran for the hills! Which brings us to the second point: David was a master of psychological warfare. By striking down one man, David filled thousands of hearts with pure terror! This had immediate effects on the battlefield as the Philistine soldiers fled. Moreover, the downstream effects aided David long after this. David leveraged this larger-than-life reputation when he found himself isolated and vulnerable. He utilized a strategy of deterrence by embodying a rabid dog. Confused and scared, the Philistines dismissed him with haste. David employed a similar strategy against Saul, not once, but twice!

Finally, the book of 1 Samuel underscores the importance of genuine connections and allies. The book of Ecclesiastes describes this concept elegantly. "And if one can overpower him who is alone, two can resist him. A cord of three strands is not quickly torn apart" (Ecclesiastes 4:12). Again we see the inverse trajectories at play. Saul started with everything and burned bridges and alienated allies. Conversely, David started from nothing and gained friends and allies organically. Saul's rash decision making and myopic view of his adversaries drove people away, whereas David's feats and character drew in friends like a magnet.

Too often powerful people opt for a Machiavellian-style friendship. They have a public and private face. Every interaction is quid pro quo. To them, it all boils down to transactions and political calculus. This was not the case with David. His friends risked their lives for him because he *inspired* them. One of the best examples is Jonathan. He courageously defended David, at the risk of his own life. Likewise, his sister Michal disobeyed Saul to aid in David's escape. Her quick thinking and ingenuity bought David a priceless head start. Additionally, David's interactions with King Achish provide fascinating insight. Even the fiercest of adversaries can shift to allies when the circumstances inevitably change. Achish was so impressed that he wanted David around for life. Yet his military commanders didn't approve of this appointment. The Philistine king ended things amicably and allowed David to keep Ziklag.

Despite all his allies, David was about to face a series of new challenges. His journey to the crown was nothing short of tumultuous. Yet, the challenges in his later career were even more insidious and treacherous.

David's story is far from over.

PART SIX:
DAVID AFTER SAUL—THE WEIGHT OF THE CROWN

DAVID AFTER SAUL

While Saul faced his impending doom, David followed the footsteps of Joshua and Moses. Where Saul chose disobedience, David revered the Lord's commands. He faithfully continued Joshua's conquest of Canaan as he put the Amalekites to the sword. Moses made the Lord's will crystal clear in his farewell address. He told Joshua and the Israelites entering the Promised Land, "Remember what Amalek did to you on the way when you came out of Egypt, how he confronted you on the way and attacked among you all the stragglers at your rear when you were tired and weary; and he did not fear God. So it shall come about, when the Lord your God has given you rest from all your surrounding enemies in the land which the Lord your God is giving you as an inheritance to possess, that you shall wipe out the mention of the name Amalek from under heaven; you must not forget" (Deuteronomy 25:17–19).

David and his men killed the lion's share of raiders, despite being vastly outnumbered. Only a few hundred escaped and fled Israelite territory. David crippled their efficacy and reach. Their days of raiding the Israelites were in the past. During the time of King Hezekiah, they were completely wiped out, approximately three hundred years later (1 Chronicles 4:42–43). David's stunning victory marks a significant turning point in the story. His valor and actions fulfilled Samuel's prophecy against Saul. This proved that David was the rightful King of Israel, not Saul.

A few days after his overwhelming victory, David received an unexpected visitor in his camp, his face marred with dust and his garments torn and frayed (2 Samuel 1:2). David quickly learned that this man had come from Saul's camp. Alarmed at his disheveled appearance, David asked what had happened to Saul's army. The newcomer gave a response that shook David to his core. He said that many of the Israelites had fallen in battle, including Saul and Jonathan (2 Samuel 1:4), David didn't want to believe it! He inquired as to how this mysterious man knew such information. Some simple questioning showed that his account was flimsy.

Surprisingly, the newcomer was an Amalekite! Clearly, he had misjudged David's relationship to Saul and Jonathan. The Amalekite claimed he had just happened to stumble upon a dying Saul. He painted himself as a good Samaritan who bestowed mercy upon Saul and put him out of his misery. "So I stood next to him and finished him off, because I knew that he could not live after he had fallen. And I took the crown which was on his head and the band which was on his arm, and I have brought them here to my lord" (2 Samuel 1:10).

Of course, no one is ever the villain in their own story. However, a more likely scenario is that this man was simply a looter seeking a generous reward. Not only was this Amalekite the Lord's sworn enemy, but also David still had respect for the position of king. "And David said to him, 'How is it you were not afraid to reach out with your hand to destroy the Lord's anointed?'" (2 Samuel 1:14). The looter expected rewards of silver and gold. He received sharpened iron instead! Not skipping a beat, David ordered his men to execute the Amalekite. "And David said to him, 'Your blood is on your head, because your own mouth has testified against you, saying, "I have finished off the Lord's anointed"'" (2 Samuel 1:16). Perhaps King Solomon thought of this incident when he wrote the following proverb. "The acquisition of treasures by a lying tongue Is a fleeting vapor, the pursuit of death" (Proverbs 21:6).

After years of persecution and exile, David was finally free from Saul's wicked pursuit. Regardless of his newfound freedom, he was distraught at the death of Jonathan, his closest friend. This grief extended beyond friendship. David still held the kingdom of Israel in high regard, undeterred by Saul's maltreatment of him. Saul's death left the kingdom in jeopardy! "Then David took hold of his clothes and tore them, and so also did all the men who were with him. And they mourned and wept and fasted until evening for Saul and his son

Jonathan, and for the people of the Lord and the house of Israel, because they had fallen by the sword" (2 Samuel 1:11–12). Jonathan had been David's greatest ally. With him and Samuel dead, David was now in uncharted territory.

After some time, David prayed to the Lord for guidance. The Lord told him to go to the capital city of Judah, Hebron (2 Samuel 1–2). David brought his wives with him, along with his troops and their families. They relocated from Ziklag, yet the city still remained in their control. To no one's surprise, the Judahites made David their king. "Then the men of Judah came, and there they anointed David king over the house of Judah" (2 Samuel 2:4). This was a public recognition of David's earlier anointing by Samuel (1 Samuel 16:13). Countless things had changed in those many years since his anointing by Samuel. Yet David's faith and reverence from the Lord had remained steadfast throughout the highs and lows.

However, not all in the realm accepted David as king. This was a pivotal point in David's story. It's only after seeing the shortfalls of the monarchy that we begin to understand God's intended system. He instituted judges and prophets for a reason. Arguably the best reason is that judges and prophets inspire unity. When the Lord raised up those heroes, all Israel could rejoice and rally around to support them. Judges were *instruments* of divine judgement, *not* divine themselves! They didn't stay in power for generations; they rose up when the people needed them the most. Contrast this to the system of monarchy. In those days, many cultures worshiped mortal kings like they were gods. Yet the Lord is the only one who is truly sovereign. Governments may be necessary for maintaining law and order but *worshipping* them is an affront to God himself!

The Lord knew the pitfalls of such a system of governance. Yet, He refused to trample the free will of the people. Samuel warned the people of this, to no avail (1 Samuel 8:19–22). Power abhors a vacuum. A kingdom without a ruler is akin to a body without a head. Sure enough, the Israelites were about to get their wish of being like the other nations on Earth. A bloody civil war was bubbling just under the facade of civility. Whenever the king falls, a *Game of Thrones* style power struggle is sure to emerge. Unfortunately, that's exactly what happened to the nation of Israel. Sometimes the most painful thing in life is getting *exactly* what you wished for.

After David's crowning in Judah, he learned that the people of Jabesh Gilead had buried Saul's remains. He sent messengers to bless them and express his gratitude for their heroism. He also extended a warm welcome to his kingdom. "And now may the Lord show kindness and truth to you; and I also will show this goodness to you, because you have done this thing. Now then, let your hands be strong and be valiant, since Saul your lord is dead, and also the house of Judah has anointed me king over them" (2 Samuel 2:6–7). In spite of this, they remained loyal to house Saul.

Around that same time, Abner, the commander of Saul's army, pulled the strings behind the throne. After the vicious battle with the Philistines, Saul only had one surviving son, Ish-Bosheth. In all likelihood, it took several years for the people to recognize Ish-Bosheth as king. "Ish-bosheth, Saul's son, was forty years old when he became king over Israel, and he was king for two years. The house of Judah, however, followed David" (2 Samuel 2:10). The fractured kingdom couldn't last for long.

Both house Saul and house David claimed the throne. Moreover, the Philistines occupied large territories while the Hebrews squabbled. The Philistines understood a timeless axiom of combat: *never interrupt your opponent while he's making a mistake!* The more Hebrews who fell in battle, the easier to swoop in later for loot and land. Perhaps the Messiah thought of this incident as He said the following: "Every kingdom divided against itself is laid waste; and a house divided against itself falls" (Luke 11:17).

Despite their contradictory claims to the throne of Israel, there was at least one attempt at diplomatic resolution. Perhaps they understood the looming Philistine threat. Or maybe they were just eager for an undisputed claim to royalty. Notwithstanding the above motivations, there is another likely motivating reason from David. Hark back to the first time David spared Saul's life. When the gravity of the situation dawned on Saul, he pleaded with David for mercy. Most importantly, he begged for the safety of his family (1 Samuel 24:21). David agreed and affirmed the oath. We know that David was a man of his word. The following incident is likely David's attempt to honor his oath to the fallen king.

We can't be completely certain of their motivations. What we know for certain is that things quickly escalated. Two opposing generals met in neutral territory. Abner represented house Saul. Representing David was his nephew, Joab. He was as effective as he was ruthless! At times, not even David himself could control him. However, for the time being, the opposing generals tried to de-escalate the situation. "And Joab the son of Zeruiah and the servants of David went out and met them by the pool of Gibeon; and they sat down, Abner's men on the one side of the pool and Joab's men on the other side of the pool" (2 Samuel 2:13).

Likely, they met in the pool as a sign of good faith and trust. Reflect back to Ehud. You can't smuggle weapons when your robe remains on dry ground. Abner suggested representative combat as a test to the throne. This method of conflict resolution echoed back to David and Goliath (1 Samuel 17:8–9).

Modern readers may scoff at such a method of diplomacy. It's worth noting that this method was arguably the most humane way to proceed forward. For this incident, only two dozen men willingly risked their lives. Contrast this with full-scale civil war in which every life hangs in the balance. Each army supplied a dozen men for this feat. The fighters squared off for this vicious fight to the death. The results were nothing short of disastrous! "And each one of them seized his opponent by the head and thrust his sword in his opponent's side; so they fell down together. Therefore that place was called Helkath-hazzurim, which is in Gibeon" (2 Samuel 2:16). Helkath Hazzurim roughly translates to *field of daggers*.

This fight offers many lessons, both spiritual and martial. First, the complete loss of these soldiers underscores the heavy price of civil war. In a civil war, *everyone* loses to some degree! Collateral damage is higher due to the blurry lines between friends and enemies. Even if you survive in one piece, you will almost certainly lose loved ones in the fight. Whoever emerges as the victor is often still in a weakened position. Troops are dead, supplies used up, and morale at rock bottom. Additionally, civil conflicts ensure that both sides are more vulnerable to outside threats. When Israel wanted to be like all other nations, it got exactly what it wished for (1 Samuel 8:20). Satan wants to sow division amongst believers. The more distracted we are fighting each other, the less we can defend against the Devil's evil deeds.

This bloody battle offers a bevy of combat lessons as well. The topic of knife defense is one of the most fiercely debated topics in martial arts. There are no silver bullet techniques, despite what some charlatans claim. Yet there are broader principles and concepts from which we can draw. In the words of Sun Tzu, "Water shapes its course according to the nature of the ground over which it flows; the soldier works out his victory in relation to the foe whom he is facing" (Tzu 12). Additionally, contemporary case studies demonstrate viable pathways to this thorny problem. Before we discuss remedies, let's discuss what *not* to do first. For unknown reasons, Abner proposed that young men fight this battle. Their youthful naivete and inexperience were on full display.

In every form of combat, it's wise to seek an asymmetrical advantage over your opponent. This applies equally to both one-on-one fighting and to full-scale war. The methodology may change, but the underlying reason *why* never does. Previously, we discussed the importance of armies seizing the high ground. The reason is simple. Due to relative positioning, you can inflict more damage on your opponent than he can inflict on you! This is the essence of combat.

We should always embrace this martial concept. A quick thought experiment: Think how easy it is to throw a rock downhill. Now picture yourself throwing the same rock *up* a hill. The rock remains the same, while the outcomes diverge dramatically. The soldiers in the makeshift arena did the exact opposite! They grabbed symmetrical ties on each other. Picture this scenario. Two fighters faced each other; they both placed their left hands on the opponent's head and neck. Commonly referred to as a 'collar tie' this grip is ubiquitous amongst grappling arts. While it's excellent for controlling someone, this was *not* the right tool for the job. This symmetrical tie up left their left flanks *entirely* exposed! Their right hands wielded a deadly dagger with an icepick grip. In a flash, they sank their blades into the unprotected flesh of their counterparts. According to *The Art of War,* "You can be sure of succeeding in your attacks if you only attack places which are undefended" (Tzu 11). To this effect, they prevailed, at the cost of their own lives. Unfortunately for them, they *both* left a vital point undefended.

Now that we have discussed what *not* to do, let's examine more effective approaches. Filipino martial arts (FMA) offer us the concept of defanging the snake. The concept is simple, when two bladesmen are fighting, one renders the other incapable of wielding the blade. You first attack the hand or arm holding the blade and then the body is wide open. It is also much easier to reach this target simply because it is closer.

Typically, practitioners perform this by slashing at their opponent's hand and fingers. Additionally, you can defang the snake by slashing the inside of your opponent's wrist or forearm. The reasoning is twofold. First, there are many blood vessels with little protection. One little slice is all that's needed to bleed out. Moreover, the flexor muscles in the forearm are what give the hand its gripping strength. Should these muscles or tendons sever, they render the hand useless; physically incapable of grasping anything. With blunt weapons like clubs or batons, you can crush the bones of the hands or wrist for similar effect. FMA practitioners would scoff at the idea of pulling your opponent, and his blade, closer to you. In the words of Sun Tzu, "If he is taking his ease, give him no rest. If his forces are united, separate them" (Tzu 4). Defanging the snake separates the forces like no other! Keep in mind that this is not conjecture or speculation. Filipino warriors have utilized this method of combat for centuries.

Later on, David's personal bodyguard, Benaiah, utilized this approach at least once. David's mightiest soldier squared off with a gargantuan Egyptian soldier. The Egyptian stood a towering seven and a half feet tall (1 Chronicles 11:23). He wielded a massive spear. Yet Benaiah didn't have a sling like his predecessor. He had to close the distance a different way. "And he killed an Egyptian, an impressive man. Now the Egyptian had a spear in his hand, but he went down to him with a club and snatched the spear from the Egyptian's hand, and killed him with his own spear" (2 Samuel 23:21).

One of the biggest advantages of spears is their range. A lone spearman can keep multiple swordsmen at bay simultaneously due to this. But how could Benaiah pull off such an epic feat? By defanging the snake of course! He almost certainly deflected the giant spearpoint with his club. With the blade out of the way, he now had a free shot at the giant's hand! Swinging as his life depended on it, he closed the distance to crush the massive hand under his club. With the hand incapacitated, he snatched the spear free and impaled his foe. Armed with only faith, a fighting spirit and modest tools, he too slayed a giant!

CASE STUDY

C uriously enough, a warrior can utilize this concept *without* weapons. Should an adversary have the upper hand with weaponry, it's advisable to disengage. However, that's not always possible. What if your loved ones are with you? You can't simply abandon them to the wolves! What if you are trapped and *can't* leave? Then your only option is to fight. Fortunately for us, a contemporary case study sheds light on this complex issue.

In November 2023, security camera footage captured the following incident occurring in South Florida. A knife wielding maniac viciously attacked a former professional MMA fighter and Christian, Javier Baez. As stated by a local news report, "According to a police report, 50-year-old Omar Marrero first tried to attack the MMA fighter while he was sitting inside his car in the parking lot of his Cutler Bay apartment" (Vallejo). What we know is that Baez emerged unscathed, but how? All the more surprising considering how Baez described his assailant, "He was out there screaming, going crazy. I think he was drugged up."

Baez escaped the car, yet the nightmare was far from over! While he made his escape, the assailant turned around to grab an even larger knife! The madman charged at Baez, and this is where Baez's years of training kicked in. His fighting instincts proved flawless! Like Benaiah, Baez first got out of the blade's way. Like a matador sidestepping the bull's horns, Baez utilized head movement and footwork to evade the razor-sharp blade. Baez embraced the way of the samurai. According to *The Book of Five Rings*, "When the enemy attacks, remain undisturbed but feign weakness. As the enemy reaches you, suddenly move away indicating that you intend to jump aside, then dash in attacking strongly as soon as you see the enemy relax" (Musashi 22).

According to Sun Tzu, "Security against defeat implies defensive tactics; ability to defeat the enemy means taking the offensive" (Tzu 7). Baez understood this principle as he implemented his grappling skills! With the blade offline, he grabbed his attacker with a 'claw' grip, a staple of wrestling and jiu jitsu. For context, they faced each other, and Baez trapped the attacker's knife-wielding arm and neck between his locked hands and arms. The beauty of this hold is its asymmetrical nature. Most importantly, it kept the attacker's knife-wielding hand dangling uselessly in the air. This embodies another maxim from *The Art of War*, "You can ensure the safety of your defense if you only hold positions that cannot be attacked" (Tzu 11). Baez understood this intuitively as he kept the knife far away.

The only viable option the attacker had was to attempt to grab the knife with his free arm. Baez's swift response shut this down long before he had the opportunity. The moment Baez secured the claw grip he hoisted his attacker off the ground for a vicious body slam to the concrete! This slam knocked the breath out of him. The legendary samurai, Musashi, would applaud Baez's quick thinking: "The important thing in strategy is to suppress the enemy's useful actions but allow his useless actions. However, doing this alone is defensive. First, you must ... suppress the enemy's techniques, foiling his plans, and thence command him directly. When you can do this you will be a master of strategy" (Musashi 23).

Yet Baez couldn't stop there! All it takes is one slip up to bleed out in a knife fight. Baez seamlessly strangled his attacker straight off the takedown. This is another benefit to the claw grip, it leads directly into submission attacks. Before fully securing the choke, Baez tucked the knife hand away where it couldn't harm him. With the snake defanged, Baez rendered his assailant unconscious with an arm triangle choke, also known as kata gatame. As stated previously, the chokehold is the undisputed king of combat techniques. This is due to a simple fact. The brain needs a constant stream of oxygen via the bloodstream. With the bloodstream occluded, consciousness fades in a matter of seconds. It doesn't matter how tough or strong you are. Not even stimulant drugs can help this. *Everyone* goes to sleep with a well applied stranglehold!

Or in the words of Mr. Baez himself, "I put him in a chokehold and he kinda just let it go. No one is worried about anything else but breathing when you are losing air."

Amazingly, all this occurred at lightning speed! From the time the attacker charged to getting strangled unconscious was only a few seconds! Sun Tzu would certainly praise Javier Baez! According to the Chinese general, "What the ancients called a clever fighter is one who not only wins, but excels in winning with ease" (Tzu 8). Javier Baez is a clever fighter indeed!

His response wasn't a fluke or dumb luck; he was simply prepared. This was nothing more than a day in the office to the seasoned fighter. According to him, "When you train so many times, it becomes just a reflex." Despite his retirement from professional fighting, Baez is still actively training. This is the way of the samurai! "You must train day and night in order to make quick decisions. In strategy it is necessary to treat training as a part of normal life" (Musashi 4). Additionally, "Timing in strategy cannot be mastered without a great deal of practice" (Musashi 7). This is not unique to the samurai, however. The Roman strategist who wrote *De Re Militari* concurs as well. "The old maxim is certain that the very essence of an art consists in constant practice" (Vegetius 43). Some paths might be smoother than others. But there are no shortcuts in fighting. Victory is the fruit of sweat and blood. Baez reaped his just reward.

As the old adage goes, when seconds count, police are only minutes away! The term first responder is a misnomer. It's far more accurate to describe the police as back up or secondary responders. This is not meant to denigrate law enforcement, but rather to emphasize the reality of self-defense. We must take ownership of our safety and realize that *we* are truly our own first responders! Baez continued to pin his assailant as he called the police. As stated in his interview, "I was able to hold him down with my knee and call the cops. He woke up and cops came and it was good. Easy." Few people would describe such a harrowing encounter as easy, but Javier Baez is far from average. He's a true hero!

The most surprising thing about this encounter is that neither of them had serious injuries. This underscores the unique nature of jiu jitsu. Jiu jitsu offers the complete spectrum of effective violence. From

288

low to high, the spectrum goes something like this. On the lowest level, a jiu jitsu practitioner can break an assailant's grips and run away. The next level is taking someone down to pin them. This may be uncomfortable for the person being pinned. But if done correctly, it keeps both parties safe. This is the approach Baez utilized. The next level is joint locks. Jiu jitsu practitioners can snap an attacker's arm or leg at their leisure. This damage ranges from hyper extensions to full blown orthopedic surgeries. If you doubt the efficacy of joint locks, there are countless video examples online. Be warned, they're liable to make your stomach churn. On the far end of the jiu jitsu spectrum lies death! A skilled practitioner can strangle you to death or break your neck. Dealer's choice. The above examples are all empty hand techniques. Yet jiu jitsu also mixes well with weaponry, as the samurai have demonstrated countless times on the battlefield. For example, after pinning a foe, it's much easier to stab them with a knife or dagger. If performed correctly they can't do the same to you!

In another interview a reporter asked Baez how he navigated such a dangerous situation. Baez said, "I stayed calm, kept my balance, kept my composure. Didn't bring him down and start beating on. Just took him down" (Gorchow) If Carl von Clausewitz were still alive, he'd surely acclaim Baez. According to the Prussian general, "We must not lose composure and resolve in these conditions; loss of these qualities is often among the first losses in a war. Without these qualities, even the most brilliant gifts of mind go to waste" (Clausewitz 2).

I personally reached out to Javier to get his take on what happened before the cameras began rolling. The full story is even more miraculous than what the news reported! Javier stated that he'd seen Marrero around his apartment complex, but he didn't know the man well. "He was outside screaming at some woman for quite some time. He came to my car upset trying to talk and let his anger out on me. I just ignored him and that's when he got mad" Realizing he had only made things worse, Javier calmly spoke to the man in an attempt to defuse the situation. This is when Javier fully grasped that Marrero was drunk and on cocaine.

If you've ever dealt with people in this state, you know it's about as productive as talking to a brick wall. When words fail, action becomes necessary. With no other options, Javier "acted real calm and quickly snatched his wrist and disarmed him." This enraged Marrero, who was fueled by liquid courage and stimulants. He turned around and grabbed a larger blade. *This* is what the security cameras captured. The parallel to David and Saul is truly uncanny! Javier showed mercy to his attacker not just once, but *twice*! Just like David did to Saul (1 Samuel 24/26). This incident truly is a modern day miracle.

The takeaway is that jiu jitsu gives you options! Javier Baez understood this. Under Florida law, he would have been justified in killing his attacker. Yet he chose not to. This stemmed from his Christian faith. Just as David spared the life of Saul, Javier Baez chose mercy for his attacker. We need more Christian men like Javier Baez! His heroic actions reinforce the distinction between peacefulness and harmlessness. In the words of Javier Baez, "training helped, but without Jesus I wouldn't be here all." Javier is in contact with Marrero's brother, who says he is in substance abuse treatment for his problems. Please pray for his recovery.

Jesus is the perfect example. He is both the Sacrificial Lamb and the Lion of Judah! Modern Christians know all about the merciful side of Jesus. Yet, few realize this same figure is prophesied personally to annihilate Satan himself! Hark back to Genesis and the Garden of Eden. The Lord makes this point abundantly clear in his damnation of the Devil. "Then the Lord God said to the serpent, 'Because you have done this, cursed are you more than all the livestock, and more than any animal of the field; on your belly you shall go, and dust you shall eat all the days of your life; and I will make enemies of you and the woman, and of your offspring and her Descendant; He shall bruise you on the head, and you shall bruise Him on the heel'" (Genesis 3:14–15).

We need skilled Christian warriors to defend against the barbarians of the world. Yet, if we forget about Christ, we will become nothing more than the monsters who seek to destroy us. We need the balance of the two. To summarize, "He, therefore, who desires peace, should prepare for war. He who aspires to victory, should spare no pains to form his soldiers. And he who hopes for success, should fight on principle, not chance" (Vegetius 45).

Now back to the epic saga of David.

THE THRONE IN THE BALANCE

This era of the unsettled throne was as bloody as it was tumultuous. Political maneuvering and debauchery ran rampant. Alliances shifted like the sand dunes of the desert. Diplomacy via representative combat failed. Such an ambiguous outcome boiled over into a full-scale battle between the houses of Saul and David. The first battle in the civil war was bloody and led to more conflict down the line. David's men seized victory, but not without cost. "That day the battle was very severe, and Abner and the men of Israel were defeated by the servants of David" (2 Samuel 2:17).

While the troops fought tooth and claw, one of the higher ups from David's side sought to kill general Abner personally. This was none other than Joab's brother, Asahel. He was said to be as swift and agile as a gazelle (2 Samuel 2:18). He may have had the speed of a gazelle, yet he had the blood thirst of a lion! Targeting his prey, he locked onto Abner with laser focus and resolve. Abner attempted to de-escalate his pursuer, but to no avail. As Asahel closed in, Abner issued a final warning. Perhaps this was out of genuine respect for Joab. Or maybe, Abner knew of Joab's ruthlessness and didn't want to make things personal. Whatever the reason, Abner appealed to the opposing general. "Then Abner repeated again to Asahel, 'Turn aside for your own good from following me. Why should I strike you to the ground? How then could I show my face to your brother Joab?'" (2 Samuel 2:22).

Asahel's determination was his downfall. Spears have a sharp broadhead at their fore end. Many also have a pointed spike toward their opposite end. This adds to balance and crafty fighters can use them as a secondary weapon. Which is exactly what Abner did. He used

Asahel's speed against him. Abner ran away; as Asahel closed in, he timed the attack perfectly so his attacker impaled himself! "However, he refused to turn aside; so Abner struck him in the belly with the butt end of the spear, so that the spear came out at his back. And he fell there and died on the spot. And it happened that all who came thereafter to the place where Asahel had fallen and died, stood still" (2 Samuel 2:23). In this way, Abner utilized the concept of kuzushi, or off balancing. He used his attacker's momentum against him.

This derailed not only Asahel's momentum, but also Joab's army as a whole. The unexpected loss of one of its leaders threw it off. It continued its pursuit, yet it fell a step behind. Abner's army regrouped as it seized a defensible area.

Joab may have been ruthless, but he wasn't reckless. He understood that his army couldn't make meaningful progress while fighting uphill. Abner signaled for a temporary ceasefire. "Then Abner called to Joab and said, 'Should the sword devour forever? Do you not realize that it will be bitter in the end? So how long will you refrain from telling the people to turn back from pursuing their kinsmen?'" (2 Samuel 2:24–25).

Realizing the futility of the situation, Joab agreed. The general of Judah bellowed his trumpet as he ordered his troops to a halt. The two opposing generals went their separate ways; for now, that is. They both needed to regroup and recalibrate. They both knew they needed to build larger armies. Joab was wise to cut his losses and quit while he had the lead. "Then Joab returned from pursuing Abner; but he gathered all the people together, and nineteen of David's servants were missing, besides Asahel. However, the servants of David had struck and killed many of Benjamin and Abner's men; 360 men were dead" (2 Samuel 2:30–31). Joab won that battle; Abner's casualties were eighteenfold his. If Joab had pursued Abner uphill, the losses would have been dramatically higher. Unlike Asahel, Joab stayed calm and calculated the odds in the heat of battle. The battle was over, but the bloody war raged on for years!

THE CONTINUED CIVIL WAR

C ountless battles raged during this tumultuous time in Israel's history. We don't know how many lives were lost, only that momentum clearly favored David: "Now there was a long war between the house of Saul and the house of David; and David became steadily stronger, while the house of Saul became steadily weaker" (2 Samuel 3:1). While David expanded his kingdom and fathered many sons, Saul's legacy continued to unravel.

Saul's remaining son, Ish-Bosheth, alienated his most crucial asset—his top general, Abner. Though Ish-Bosheth held the crown, Abner was the one holding the real power. "Now it happened that while there was war between the house of Saul and the house of David, Abner was strengthening himself in the house of Saul" (2 Samuel 3:6). Trust in your commanders is vital—especially in wartime. But Ish-Bosheth apparently didn't get the memo.

He accused Abner of sedition by claiming he had slept with one of Saul's concubines. Whether or not this was true is unclear. But the accusation was a political hand grenade. Perhaps Ish-Bosheth feared a power grab. Like his father, he might've suffered from the *bucket of crabs syndrome*—dragging others down just to stay on top. Rather than leading with strength and vision, he lashed out at the very man keeping him afloat.

Abner had every reason to be furious. He had done all the heavy lifting while Ish-Bosheth was merely a paper tiger. "Then Abner became very angry over Ish-bosheth's question and said, 'Am I a dog's head that belongs to Judah? Today I show kindness to the house of Saul your father, to his brothers and to his friends, and have not let you fall into the hands of David; yet today you call me to account for wrongdoing with that woman?'" (2 Samuel 3:8).

Abner's next words sealed his shift in allegiance. Swearing by God, he vowed to deliver the kingdom to David: "May God do so to me, and more so, if as the Lord has sworn to David, I do not accomplish this for him: to transfer the kingdom from the house of Saul, and to establish the throne of David over Israel and over Judah, from Dan even to Beersheba!" (2 Samuel 3:9–10). Ish-Bosheth realized too late that loyalty is a two-way street. He stayed silent—likely stunned by his own blunder.

Just like Saul before him, Ish-Bosheth repaid loyalty with suspicion. Abner, fed up, changed sides. As is often the case in war, *the enemy of my enemy is my friend.* "Then Abner sent messengers to David at his place, saying, 'Whose is the land? Make your covenant with me, and behold, my hand shall be with you to bring all Israel over to you'" (2 Samuel 3:12). David saw an opportunity and used it to demand the return of his first wife, Michal. "And he said, 'Good! I will make a covenant with you, only I require one thing of you, namely, that you shall not see my face unless you first bring Michal, Saul's daughter, when you come to see me'" (2 Samuel 3:13).

Additionally, David sent word to Ish-Bosheth, reminding him that Saul had authorized the marriage and that he had paid the dowry accordingly. Surprisingly, Ish-Bosheth agreed. "Ish-bosheth sent men and had her taken from her husband, from Paltiel the son of Laish. And her husband went with her, weeping as he went, following her as far as Bahurim. Then Abner said to him, 'Go, return.' So he returned" (2 Samuel 3:15–16). Paltiel wasn't peaceful; *he was harmless.* The moment he was confronted, he folded. Mere words were enough to stop him from defending his wife. Don't be like Paltiel. It's worth noting: David got the deal he wanted because he negotiated from a position of strength, not weakness.

With Michal's return came the symbolic unification of Israel. But the Philistine threat still loomed. Both sides of the long civil war were exhausted. As Sun Tzu observed, "There is no instance of a country having benefited from prolonged warfare" (Tzu 5). The longer Israel's factions bled each other dry, the easier it would be for the Philistines to conquer them all.

Abner, ever the pragmatist, wasted no time rallying support for a peace treaty. He addressed the elders with frank urgency: "Now Abner had a consultation with the elders of Israel, saying, 'In times past you were seeking for David to be king over you. Now then, do it! For the Lord has spoken regarding David, saying, "By the hand of My servant David I will save My people Israel from the hand of the Philistines, and from the hands of all their enemies""" (2 Samuel 3:17–18). He even secured the backing of Saul's own tribe, Benjamin. With the groundwork laid, Abner approached David in peace.

A feast welcomed him, and for the first time in years, hope seemed possible. "Abner said to David, 'Let me set out and go and gather all Israel to my lord the king, so that they may make a covenant with you, and that you may be king over all that your soul desires.' So David let Abner go, and he went in peace" (2 Samuel 3:21).

But not everyone welcomed this truce.

Joab and his men had been out raiding and returned flush with plunder—only to find David entertaining their longtime rival. Joab exploded. His blood still boiled from the death of his brother Asahel at Abner's hand. Furious, he stormed into David's presence: "Then Joab came to the king and said, 'What have you done? Behold, Abner came to you; why then have you let him go, so that he is already gone? You know Abner the son of Ner, that he came to gain your confidence, and to learn of your going out and coming in and to find out everything that you are doing'" (2 Samuel 3:24–25). Joab was a deadly field commander, but he lacked emotional control and discipline. His thirst for revenge clouded everything else.

He concocted a treacherous plan; a trap Machiavelli would be proud of. In blatant insubordination, Joab sent messengers to lure Abner back under false pretenses. Abner, having received the king's blessing, had no reason to suspect danger. But the snake in the grass struck with venom. "So when Abner returned to Hebron, Joab took him aside into the middle of the gate to speak with him privately, and there he struck him in the belly, so that he died on account of the blood of his brother Asahel" (2 Samuel 3:27).

The murder lacked both justice and political strategy. It left David in a political minefield. Abner had come in peace under the king's word. To violate that promise, especially by family, made David appear duplicitous and untrustworthy. Who would make a deal after hearing about this? Worse, they'd just butchered the one man who could've unified Israel under David's rule. Abner was no ordinary soldier. He was a proven leader, battle-hardened, respected, and valuable. He was an *asset* in a time when Israel was outnumbered by enemies on all sides.

David's family ties made the optics worse. Joab wasn't just a rogue general; he was David's nephew. This wasn't a simple matter of justice, it was a crisis of legitimacy. How could David rule a fractured nation when his own house wasn't in order?

The political tightrope had begun. David would walk it for the rest of his reign.

Fortunately, scripture offers insight into these difficult questions— both in the Old Testament and the New. Moses shows us the way. If a friend or family member leads you astray, he says: "you shall not consent to him or listen to him; and your eye shall not pity him, nor shall you spare or conceal him" (Deuteronomy 13:8). Though Moses referred specifically to idol worship, the core principle remains unchanged: God above all.

Jesus echoed this same timeless principle. "The one who loves father or mother more than Me is not worthy of Me; and the one who loves son or daughter more than Me is not worthy of Me. And the one who does not take his cross and follow after Me is not worthy of Me. The one who has found his life will lose it, and the one who has lost his life on My account will find it" (Matthew 10:37–39).

At first glance, this may sound harsh—but there is firm reasoning behind these words. God is the sovereign King and Creator of the universe. He surpasses all earthly authority. Jesus is God incarnate. "For in Him all the fullness of Deity dwells in bodily form, and in Him you have been made complete, and He is the head over every ruler and authority" (Colossians 2:9–10). We can only love others rightly when we first love the Lord.

Jesus made this point crystal clear in the Gospel of Matthew. A Pharisee, an expert in the law, tried to trap Him with a gotcha question—one of those legal snags some people worship like idols. He asked, "Teacher, which is the greatest commandment in the Law?" (Matthew 22:36). Christ gave a decisive answer: "'You shall love the

Lord your God with all your heart, and with all your soul, and with all your mind.' This is the great and foremost commandment. The second is like it, 'You shall love your neighbor as yourself'" (Matthew 22:37–39). Then He summed it all up: "Upon these two commandments hang the whole Law and the Prophets" (Matthew 22:40).

Praise Jesus for freeing us from the false idol of legalism! Yet secular crowds and cultural Christians often cling to the second commandment while discarding the first. Jesus made it clear: these commandments are not interchangeable—*they are sequential.*

Unlike the Pharisees, David knew what it meant to put God first. Perhaps he was too soft on Joab, but his priorities were righteous. He swiftly condemned his nephew's actions and declared his own innocence. Then he pronounced a curse on Joab and ordered his entire court into mourning: "Then David said to Joab and to all the people who were with him, 'Tear your clothes and put on sackcloth, and mourn before Abner.' And King David walked behind the bier" (2 Samuel 3:31).

David wept publicly and fasted in sorrow. He mourned Abner with sincerity, and the people saw his heart. He closed the funeral by honoring Abner's legacy and calling on divine justice: "And I am weak today, though anointed king; and these men, the sons of Zeruiah, are too difficult for me. May the Lord repay the evildoer in proportion to his evil" (2 Samuel 3:39). The sons of Zeruiah were Joab and his brothers. David may have lacked the force to discipline them then—but he never lacked the clarity to call their sins what they were.

THE SHIFTING SANDS OF LOYALTY

Things rapidly declined for the house of Saul. Much like his father, Ish-Bosheth never possessed a strong or courageous spirit. With his loyal watchdog Abner now dead, he was more exposed than ever. His fragile kingdom began to erode like a sandcastle in the tide. "Now when Ish-bosheth, Saul's son, heard that Abner had died in Hebron, his courage failed, and all Israel was horrified" (2 Samuel 4:1).

Opportunists never let a good crisis go to waste! Two wicked scoundrels saw their chance, and took it. Shockingly, they were Ish-Bosheth's own men! (2 Samuel 4:2). With Abner gone, greed overtook any sense of loyalty. While tragic, this was not surprising. Abner had been the real power. Without him, Ish-Bosheth was little more than a figurehead. As *The Art of War* warns, "When the common soldiers are too strong and their officers too weak, the result is insubordination" (Tzu 20). When a strong leader falls, a power vacuum always forms.

Like sharks scenting blood, the two killers quietly circled in for the kill: "Now when they had come into the house, as he [Ish-Bosheth] was lying on his bed in his bedroom, they struck him and killed him, and they beheaded him. And they took his head and traveled by way of the Arabah all night" (2 Samuel 4:7). Compare this to David, who spared Saul's life when given the same opportunity. That's the difference between greed and honor.

Proverbs has plenty to say about men like these. "One who works his land will have plenty of bread, But one who pursues worthless things lacks sense" (Proverbs 12:11). Driven by fantasies of gold and glory, they wasted no time delivering their grisly prize.

Or so they thought.

"Then they brought the head of Ish-bosheth to David at Hebron, and said to the king, 'Behold, the head of Ish-bosheth the son of Saul, your enemy, who sought your life; so the Lord has given my lord the king vengeance this day on Saul and his descendants'" (2 Samuel 4:8). Their cluelessness was staggering. Indeed, "Even fools are thought wise if they keep silent, and discerning if they hold their tongues" (Proverbs 17:28). These fools couldn't have picked a worse man to impress. Had they already forgotten the fate of the Amalekite who brought news of Saul's death?

David wasted no time condemning their treachery: "As the Lord lives, who has redeemed my life from all distress, when the one who informed me, saying, 'Behold, Saul is dead,' also viewed himself as the bearer of good news, I seized him and killed him in Ziklag, which was the reward I gave him for his news. How much more, when wicked men have killed a righteous man in his own house on his bed, shall I not now require his blood from your hands and eliminate you both from the earth?" (2 Samuel 4:9–11).

They came seeking rewards. Instead, they found swift justice. "Then David commanded the young men, and they killed them and cut off their hands and feet, and hung them up beside the pool in Hebron. But they took the head of Ish-bosheth and buried it in the grave of Abner in Hebron" (2 Samuel 4:12).

This punishment was both symbolic and strategic. On a practical level, David had to reestablish order in the aftermath of the civil war. Political murders could reignite the flames of rebellion. Perhaps if David had dealt more harshly with Joab, these wolves wouldn't have dared to strike. But the execution still sent a crystal-clear message. As *De Re Militari* notes, "But if the height of the mutiny requires violent remedies ... punish the ring-leaders only in order that ... all may be terrified by the example" (Vegetius 52).

The mutilation fit the crime. The hands that wielded the blade, and the feet that fled with their trophy, were cut away—emblems of their betrayal.

THE NEW KING

After many years of turmoil and bloodshed, Israel had grown weary. The sudden loss of Abner left the people vulnerable to outside threats. They coalesced around their shining hero, David. Everyone knew of his epic feats and immense courage. He had demonstrated battlefield prowess like no other before him. With Ish-Bosheth dead, no legitimate successors remained from Saul's house. Given his record, it's no surprise the tribes turned to David. "Then all the tribes of Israel came to David at Hebron and said, 'Behold, we are your bone and your flesh. Previously, when Saul was king over us, you were the one who led Israel out and in. And the Lord said to you, 'You will shepherd My people Israel, and you will be a leader over Israel'" (2 Samuel 5:1–2).

Years earlier, Samuel had anointed David in private. He was just a boy then, but the Spirit of the Lord had been with him ever since. Now thirty years old, David stood before the elders of Israel. "So all the elders of Israel came to the king at Hebron, and King David made a covenant with them before the Lord in Hebron; then they anointed David king over Israel" (2 Samuel 5:3).

David was now the undisputed ruler, and the people rejoiced. He had survived Saul's persecution and a brutal civil war. Yet he knew the Lord's work was not finished. Many leaders grow complacent after victory. Not David. He picked up where Joshua and the judges left off. His first mission as King of Israel was ambitious and strategic: he marched toward Jerusalem.

At that time, the city was called Jebus and was controlled by the Canaanite Jebusites. Though the Israelites had made attempts, none had succeeded in fully occupying it. Capturing this city served several purposes. First, it was a continuation of the conquest of Canaan. Though it was located on the border of Judah and Benjamin, the Jebusites had held it for generations.

During Joshua's campaigns, its king was killed, but the city held firm (Joshua 15:63). After Joshua's death, the Judahites made another push. "Then the sons of Judah fought against Jerusalem and captured it, and struck it with the edge of the sword, and set the city on fire" (Judges 1:8). Even so, they couldn't hold it. The Jebusites reclaimed control. "But the sons of Benjamin did not drive out the Jebusites who lived in Jerusalem; so the Jebusites have lived with the sons of Benjamin in Jerusalem to this day" (Judges 1:21).

Beyond fulfilling Joshua's mission, Jerusalem held strategic and symbolic value. The city was perched on Mount Zion, naturally defensible with deep valleys on three sides and strong fortifications. More importantly, it lay between the tribal territories of Benjamin and Judah—Saul's line and David's. By seizing and establishing Jerusalem as his capital, David could unify the kingdom under a new banner. No longer Judah versus Benjamin. It would be Israel together.

Still, how could David succeed where so many warriors had failed? The answer lay in his ingenuity, and in the Jebusites' arrogance. Their terrain had made them complacent.

The Jebusites taunted the Israelites from their elevated fortress. "Now the king and his men went to Jerusalem against the Jebusites, the inhabitants of the land; and they said to David, 'You shall not come in here, but even those who are blind and those who limp will turn you away,' thinking, 'David cannot enter here'" (2 Samuel 5:6). The Jebusites violated Clausewitz's maxim of defense: "The main principle of defense—never remain completely passive ... never rely entirely on the favorable conditions of the terrain" (Clausewitz 4, 10). They had rested on their laurels for far too long. Due to the failed attempts of Joshua and the judges they had grown apathetic. There's a saying in the military: *complacency kills*. Some rules are written in blood.

King David found the perfect man for this special operation. "Now David had said, 'Whoever is first to kill a Jebusite shall be chief and commander.' Joab the son of Zeruiah went up first, so he became chief" (1 Chronicles 11:6). Despite Joab's moral shortcomings, he was a brilliant warrior and strategist. We see another leadership lesson here: *become so good at your job that you're irreplaceable*. Joab had gotten off the hook easily for his assassination of Abner. Familial relations aside, he bought his freedom with his brutal warfighting efficacy. This is *not* advocating for abusing positions of power. It's betting on yourself in a competitive market, where only results matter. Joab knew he couldn't simply besiege the fortified city. He understood the following principle of Sun Tzu, "Let your plans be dark and impenetrable as night, and when you move, fall like a thunderbolt" (Tzu 14).

The Jebusites remained completely ignorant to the brewing storm on the horizon. Somehow, the Israelites obtained crucial intel. *Loose lips sink ships!* They found a weak point to the hilltop fortress. Outside the reinforced walls was the Gihon spring. This spring was the lifeblood of Jerusalem. David learned of a network of tunnels for drinking water. One of these secret tunnels allowed the residents to draw fresh water should the city fall under siege. He assigned Joab the daring mission. "And David said on that day, 'Whoever strikes the Jebusites is to reach those who limp and those who are blind, who are hated by David's soul, through the water tunnel.' For that reason they say, 'People who are blind and people who limp shall not come into the house'" (2 Samuel 5:8).

The translation of water shaft comes from the Hebrew word *tsinnor*. Curiously enough, this is the only time that specific term is mentioned in the entire Bible. In the late 1800s this secret tunnel was discovered by British archeologist Charles Warren (Warren et al. 191). Visitors can see this historical tunnel today with their own eyes! There are not many details surrounding this daring mission. From context, we can infer several things about this daring clandestine operation.

The following is a depiction of how this operation *likely* occurred; *it is an estimation based on contextual clues and tactical reasoning*.

After obtaining intelligence about these tunnels, David formed the infiltration plan. David and his men distracted the sentries at the main entrance. Under the cover of darkness, Joab and a small elite crew set forth behind enemy lines. The Gihon spring was almost certainly guarded, yet it remained outside the formidable city walls. Likely, they used their hand-to-hand skills to dispatch the guards quietly. Or perhaps they simply snuck by them. The Jebusites had long grown aloof towards security threats. After that, they gained entrance to the main pool. From there, they used the intel to point them in the right direction amongst the network of tunnels.

Imagine their dread. They didn't have flashlights or night vision goggles. They braved the terrain without knowing what lay just around the corner. Would they encounter armed guards patiently waiting to trap them? Would they slip and fall into some deep chasm? Maybe they would become hopelessly lost and die of the elements, their cries unanswered. Fortunately, their faith guided them. After some time, their claustrophobic expedition led them into the soft underbelly of the hardened city. Yet their harrowing encounter was far from over! One slip up meant certain death! Stealthier than a black cat under a new moon, they crept towards the city gates. They truly embodied the maxim, "Attack him where he is unprepared, appear where you are not expected" (Tzu 4). While the guards focused on David's camp outside, Joab opened the gate from *within*! With the gates wide open, David's men stormed through the city, ransacking it with ease.

THE CITY OF DAVID

D avid's army did what no other could. Truly a historic moment, this was one of the crowning jewels of his lengthy career. Wasting no time, David expanded this city into something more glorious. "David became greater and greater, for the Lord God of armies was with him" (2 Samuel 5:10). The people rallied around David as he reunited the broken Kingdom of Israel.

Moreover, even outsiders gave David recognition. A Phoenician king reached out to establish trade with Israel. "Then Hiram king of Tyre sent messengers to David with cedar trees, carpenters, and stonemasons; and they built a house for David" (2 Samuel 5:11). This ambitious undertaking and overwhelming success solidified David's undisputed place on the throne. "And David realized that the Lord had appointed him as king over Israel, and that He had exalted his kingdom for the sake of His people Israel" (2 Samuel 5:12).

Jerusalem may have earned the moniker 'City of David,' yet David wasted no time dedicating this victory to the Lord. David set forth on another ambitious mission: to relocate the Ark of the Covenant to the new capital of Jerusalem. He couldn't risk such a valuable treasure again falling into the hands of the Philistines. King David assembled a formidable protection detail. He leveraged his officers to rally the troops (1 Chronicles 13:1). The officers honored the king's wishes as they spread the word far and wide across Israel. The message was clear. It was no longer about Saul versus David, or Benjamin against Judah. All Israel stood united to honor the Lord.

A staggering thirty thousand came to support this holy undertaking (2 Samuel 6:1). It was a joyful celebration—music filled the air as the people sang, danced, and rejoiced. But in their excitement, they neglected key commands from the Law of Moses, with disastrous results. Long ago, the Israelites had been warned: only specific individuals were permitted to handle the Ark and only in particular ways (Numbers 4:15). Further instruction had been given on exactly how to transport it. "You shall put the poles into the rings on the sides of the ark, to carry the ark with them. The poles shall remain in the rings of the ark; they shall not be removed from it" (Exodus 25:14–15). The Ark was not to be touched.

Ignoring this, they placed the Ark on an ox-drawn cart. When the cart stumbled, a man named Uzzah reached out to steady it. "And the anger of the Lord burned against Uzzah, and God struck him down there for his irreverence; and he died there by the ark of God" (2 Samuel 6:7). To human eyes, this may seem harsh. But it illustrates a deeper truth: when we place our ways above the Lord's ways, *disaster is soon to follow*. Uzzah's intentions may have been noble, but obedience matters more than intent. Good intentions alone do not justify disobedience. In other words: "There is a way which seems right to a person, But its end is the way of death" (Proverbs 14:12).

As the old adage goes: *The path to Hell is paved with good intentions*. David was shaken, confused, and frustrated. He paused the journey and sent the Ark to a nearby town, where it brought great blessings (1 Chronicles 13:13–14). This period gave David time to reflect. The blessings proved the Ark wasn't the problem—it was how they had treated it. Just as the Philistines suffered for treating the Ark like a war trophy, Israel suffered for disregarding God's instructions. In response, David turned to Scripture and prayer. He reassembled Israel and this time and did things the right way.

Before the people, he acknowledged the previous failure: "Then David said, 'No one is to carry the ark of God except the Levites; for the Lord chose them to carry the ark of the Lord and to serve Him forever'" (1 Chronicles 15:2). Many leaders fear that admitting failure shows weakness. But the opposite is true. When leaders own mistakes and take corrective action, it builds trust. It proves they value the mission and the people more than their own ego.

David continued, taking full responsibility as king: "Because you did not carry it at the first, the Lord our God made an outburst against us, since we did not seek Him according to the ordinance. So the priests and the Levites consecrated themselves to bring up the ark of the Lord God of Israel" (1 Chronicles 15:13–14). Notice David said "we." He didn't pass the buck or shift the blame. He understood his duty to lead from the front. This time, the Ark was carried exactly as commanded. "The sons of the Levites carried the ark of God on their shoulders with the poles on them, just as Moses had commanded in accordance with the word of the Lord" (1 Chronicles 15:15).

The Ark's arrival in Jerusalem was a joyous occasion—proof of God's blessing upon David and Israel. Worship music filled the air, and the streets swelled with dancing and praise. Once again David led from the front. This wasn't a performance, the Spirit filled his heart. "And David was dancing before the Lord with all his strength, and David was wearing a linen ephod. So David and all the house of Israel were bringing up the ark of the Lord with joyful shouting and the sound of the trumpet" (2 Samuel 6:14–15). While all Israel celebrated this historic occasion, one person saw it as distasteful. None other than David's wife, Michal. "Then it happened, as the ark of the Lord was coming into the city of David, that Michal the daughter of Saul looked down through the window and saw King David leaping and dancing before the Lord; and she was contemptuous of him in her heart" (2 Samuel 6:16). It appears she had about as much reverence for the Lord as her father, Saul.

She couldn't believe the king would deign to celebrate with the commoners in such a fashion. David clearly understood something she missed: *we are all children of God*! If you're looking down your nose at someone you're not looking up towards God! David had risen up from a humble peasant's station in life. Conversely, Michal had been born into royalty. She did absolutely nothing to earn her lofty station in life. People rarely appreciate things they did not earn. Unaware of his wife's disapproval, David continued celebrating the Lord's grace and glory. On the other hand, Michal stewed in bitter resentment. "Now they brought in the ark of the Lord and set it in its place inside the tent which David had pitched for it; and David offered burnt offerings and peace offerings before the Lord" (2 Samuel 6:17). After the offerings, David blessed the people in the name of God. David had plenty of notable things he could brag about. But this wasn't about him or his campaigns. The focus was solely on worshiping the Lord. David closed the festivities with gifts for his people. "When David had finished offering the burnt offering and the peace offerings, he blessed the people in the name of the Lord of armies. Further, he distributed to all the people, to all the multitude of Israel, both to men and women, a cake of bread, one of dates, and one of raisins to each one. Then all the people left, each to his house" (2 Samuel 6:18–19).

The King of Israel received quite the shock upon his arrival home. Michal's sour demeanor stood quite the contrast to the sweetness of raisin cakes and joyful celebration. She wasted no time chastising him. Bitterly, she confronted him. "But when David returned to bless his own household, Michal the daughter of Saul came out to meet David and said, 'How the king of Israel dignified himself today! For he exposed himself today in the sight of his servants' female slaves, as one of the rabble shamelessly exposes himself!'" (2 Samuel 6:20).

David stood firm. He knew God's approval mattered far more than human pride or courtly decorum. "But David said to Michal, 'I was before the Lord, who preferred me to your father and to all his house, to appoint me as ruler over the people of the Lord, over Israel. So I will celebrate before the Lord! And I might demean myself even more than this and be lowly in my own sight, but with the female slaves of whom you have spoken, with them I am to be held in honor!'" (2 Samuel 6:21–22).

King David shows us an important lesson. *It's impossible to please everyone, so you might as well please God.* Modern culture remains obsessed with people-pleasing. This fruitless endeavor is as useful as chasing a mirage in an endless desert. Such a wild goose chase is sure to bring nothing but despair, fatigue, and heartache. It is worth reiterating, people-pleasing is *not* Biblical. Both the Old Testament and New Testament support this concept. King Solomon elaborated on this timeless idea. "The fear of man brings a snare, But one who trusts in the Lord will be protected" (Proverbs 29:25).

Pleasing people doesn't deliver us to salvation; only honoring the Lord does. Moreover, the Apostle Paul underscores this concept several times. Here is one notable example in a letter to early church leaders. "But just as we have been approved by God to be entrusted with the gospel, so we speak, not intending to please people, but to please God, who examines our hearts. For we never came with flattering speech, as you know, nor with a pretext for greed—God is our witness—nor did we seek honor from people, either from you or from others, though we could have asserted our authority as apostles of Christ" (1 Thessalonians 2:4–6).

God gave us discernment for a reason. In David and Michal's conflict, we can see which path bore the sweeter fruit. Jesus Christ made this point clear in His famous Sermon on the Mount: "You will know them by their fruits. Grapes are not gathered from thorn bushes, nor figs from thistles, are they? So every good tree bears good fruit, but the bad tree bears bad fruit. A good tree cannot bear bad fruit, nor can a bad tree bear good fruit. Every tree that does not bear good fruit is cut down and thrown into the fire. So then, you will know them by their fruits" (Matthew 7:16–20). David continued to have overwhelming success and victory, because the Lord was with him. Conversely, Michal faded into irrelevancy as she fell by the wayside. "And Michal the daughter of Saul had no child to the day of her death" (2 Samuel 6:23). Likely, this was God's final judgement on the house of Saul. Saul would never have a dynasty on the throne of Israel due to his repeated disobedience.

A DYNASTY LIKE NO OTHER

avid's dynasty, on the other hand, was beyond even his comprehension. Chapter 7 of 2 Samuel is one of the most important passages in the entire Bible. It introduces the Davidic Covenant—a divine promise that David's kingdom would reign eternally. But how could a mortal bloodline last forever? The answer: the Messiah, the One True King, would come from David's lineage to rule for all time.

At this point, David had settled into his palace in Jerusalem and enjoyed a temporary reprieve from war. Then a sense of guilt overtook him. "Now it came about, when the king lived in his house, and the Lord had given him rest on every side from all his enemies, that the king said to Nathan the prophet, 'See now, I live in a house of cedar, but the ark of God remains within the tent'" (2 Samuel 7:2–3). David wanted to build a temple for the Lord. Nathan initially agreed—it seemed like a noble idea. But that very night, God gave the prophet a powerful vision that would alter history forever.

Though David's intent was sincere, the Lord made it clear this was not his role. God began by reminding David of His dwelling since the time of the Exodus. "Go and say to My servant David, 'This is what the Lord says: Should you build Me a house for My dwelling? For I have not dwelt in a house since the day I brought up the sons of Israel from Egypt, even to this day; rather, I have been moving about in a tent, that is, in a dwelling place. Wherever I have gone with all the sons of Israel, did I speak a word with one of the tribes of Israel, whom I commanded to shepherd My people Israel, saying,' 'Why have you not built Me a house of cedar?'" (2 Samuel 7:5–7).

Then God reminded David just how far he had come, not by his own strength, but by divine providence. "Now then, this is what you shall say to My servant David: 'This is what the Lord of armies says: I Myself took you from the pasture, from following the sheep, to be leader over My people Israel. And I have been with you wherever you have gone, and have eliminated all your enemies from you; I will also make a great name for you, like the names of the great men who are on the earth'" (2 Samuel 7:8–9).

David's earthly feats were incredible. His rise from shepherd boy to warrior king is the ultimate underdog story. Yet, he would never live to see his greatest legacy fulfilled. God gave him just a glimpse of the glory to come.

And here, the Lord's wordplay speaks volumes. "The Lord also declares to you that the Lord will make a house for you. When your days are finished and you lie down with your fathers, I will raise up your descendant after you, who will come from you, and I will establish his kingdom" (2 Samuel 7:11–12).

David offered to build God a house—but instead, God promised to build David a house: not made of cedar and stone, but a royal lineage that would lead to the Savior of the world. This was David's true crowning achievement—not the sling, not the sword, not the throne, but the bloodline of Jesus Christ. Matthew 1:1 proclaims "The record of the genealogy of Jesus the Messiah, the son of David, the son of Abraham."

God elaborated further: "He shall build a house for My name, and I will establish the throne of his kingdom forever. I will be a father to him and he will be a son to Me; when he does wrong, I will discipline him with a rod of men and with strokes of sons of mankind" (2 Samuel 7:13–14).

These verses carry a powerful dual meaning. On one level, they speak of Solomon and the future kings of Israel, who would face discipline for disobedience. Yet the deeper, messianic layer looks forward to Christ's suffering. Though He was without sin, Jesus took on the punishment meant for us—flogged by human hands and struck with a Roman rod. This might sound bleak, but God's message is one of enduring hope. His love remains, even in the face of judgment. "But My favor shall not depart from him, as I took it away from Saul, whom I removed from you. Your house and your kingdom shall endure before

Me forever; your throne shall be established forever" (2 Samuel 7:15–16). A thousand years later, Jesus ascended to that eternal throne.

Nathan relayed this stunning revelation to David. Overwhelmed, the king approached the Lord with reverence and humility. "Then David the king came in and sat before the Lord, and he said, 'Who am I, Lord God, and who are the members of my household, that You have brought me this far? And yet this was insignificant in Your eyes, Lord God, for You have spoken also of the house of Your servant regarding the distant future. And this is the custom of mankind, Lord God'" (2 Samuel 7:18–19).

David's reaction mirrors Moses when God called him to the burning bush. This is when the Lord appointed him to liberate the Hebrews from slavery in Egypt: "But Moses said to God, 'Who am I, that I should go to Pharaoh, and that I should bring the sons of Israel out of Egypt?'" (Exodus 3:11). Both men recognized that divine missions are not earned, they're bestowed.

David poured out his gratitude before the Lord and embraced the covenant with joy. "For You, Lord of armies, God of Israel, have given a revelation to Your servant, saying, 'I will build you a house'; therefore Your servant has found courage to pray this prayer to You. Now then, Lord God, You are God, and Your words are truth; and You have promised this good thing to Your servant. And now, may it please You to bless the house of Your servant, so that it may continue forever before You. For You, Lord God, have spoken; and with Your blessing may the house of Your servant be blessed forever'" (2 Samuel 7:27–29).

And so, the Davidic Covenant was born! God's promise did not just shape Israel's monarchy—it also became the foundation of Christianity itself. David's kingdom became the soil from which the Messiah would rise. Scripture confirms David was a prophet, and Peter affirmed this in his sermon after Pentecost: "Brothers, I may confidently say to you regarding the patriarch David that he both died and was buried, and his tomb is with us to this day. So because he was a prophet and knew that God had sworn to him with an oath to seat one of his descendants on his throne, he looked ahead and spoke of the resurrection of the Christ, that He was neither abandoned to Hades, nor did His flesh suffer decay" (Acts 2:29–31).

David's connection to Jesus Christ is truly incredible. In fact, David prophesied the execution of the Messiah in stunning detail—approximately a thousand years before it happened! Let's take a closer look. For clarity, Psalm 22 is presented chronologically.

The Psalm opens with dread and isolation. David, from his years of persecution, was no stranger to such anguish: "My God, my God, why have You forsaken me? Far from my help are the words of my groaning" (Psalm 22:1). Now compare this with Christ's words on the cross. An ominous darkness had covered the sky for hours when Jesus cried out: "And about the ninth hour Jesus cried out with a loud voice, saying, 'Eli, Eli, lema sabaktanei?' that is, 'My God, My God, why have You forsaken Me?'" (Matthew 27:46). This was no coincidence. God was honoring His word. The exact words and the deep sense of abandonment match precisely.

The parallels continue: "All who see me deride me; They sneer, they shake their heads, saying, 'Turn him over to the Lord; let Him save him; Let Him rescue him, because He delights in him'" (Psalm 22:7–8). David described the contempt of the crowd, and that same venom was hurled at Christ: "In the same way the chief priests also, along with the scribes and elders, were mocking Him and saying, 'He saved others; He cannot save Himself! He is the King of Israel; let Him now come down from the cross, and we will believe in Him. He has trusted in God; let God rescue Him now, if He takes pleasure in Him; for He said, "I am the Son of God"'" (Matthew 27:41–43).

The mob's cruelty was matched only by its arrogance. It misunderstood Christ's mission. Of course He could stop them—He had the power. But He chose not to. Jesus stayed on the cross for one reason: He was paying our sin debt in full.

Still not convinced? The prophecy gets more specific. David wrote, "My strength is dried up like a piece of pottery, And my tongue clings to my jaws; And You lay me in the dust of death" (Psalm 22:15). Christ had been tortured for hours. His wounds bled steadily, His body weakened, and dehydration set in. At His breaking point, He cried out for mercy: "After this, Jesus, knowing that all things had already been accomplished, in order that the Scripture would be fulfilled, said, 'I am thirsty.' A jar full of sour wine was standing there; so they put a sponge full of the sour wine on a branch of hyssop and brought it up to His mouth" (John 19:28–29). David confirmed this detail elsewhere: "They also gave me a bitter herb in my food, And for my thirst they gave me vinegar to drink" (Psalm 69:21).

And then comes the most astonishing part: David described crucifixion—*centuries before it was invented.* "For dogs have surrounded me; A band of evildoers has encompassed me; They pierced my hands and my feet" (Psalm 22:16). This was long before the Romans ever existed and refined crucifixion techniques. How could David possibly know? Only through divine revelation. He must have seemed like a madman at the time; describing a methodology that had yet to be invented. Time would validate his claims.

David's vision of the cross didn't stop there. "I can count all my bones. They look, they stare at me" (Psalm 22:17). This too aligns with crucifixion. The victim's body hangs in such a way that bones become visible beneath stretched flesh. Death by crucifixion typically comes through asphyxiation. To speed up death, Roman soldiers would sometimes break the victims' legs. Once their legs collapsed, they could no longer lift their chest to breathe. It was agonizing. And if that wasn't cruel enough, it also prolonged the pain just long enough to crush both body and spirit.

The Gospel validates what David prophesied a millennium earlier. The Pharisees, in coordination with Roman governance, schemed to break the legs of the crucified. Their gruesome request had nothing to do with mercy—it was all about optics. They didn't want mutilated bodies defiling the Sabbath. "Now then, since it was the day of preparation, to prevent the bodies from remaining on the cross on the Sabbath (for that Sabbath was a high day), the Jews requested of Pilate that their legs be broken, and the bodies be taken away. So the soldiers came and broke the legs of the first man, and of the other who was crucified with Him; but after they came to Jesus, when they saw that He was already dead, they did not break His legs" (John 19:31–33).

There was more at play than religious posturing. According to John's Gospel, this also fulfilled Mosaic Law concerning the Passover lamb. "For these things took place so that the Scripture would be fulfilled: 'Not a bone of Him shall be broken.' And again another Scripture says, 'They will look at Him whom they pierced'" (John 19:36–37). This traces back to the command Moses gave regarding the sacrificial lamb: "They shall not leave any of it until morning, nor break a bone of it; they shall celebrate it in accordance with the whole statute of the Passover" (Numbers 9:12). Jesus is the Lamb of God—the ultimate and final sacrifice.

David also prophesied another specific detail: "They divide my garments among them, And they cast lots for my clothing" (Psalm 22:18). And that's exactly what happened: "And they crucified Him, and divided up His garments among themselves, casting lots for them to decide what each man would take" (Mark 15:24).

Despite the depravity and horror, both Psalm 22 and the Gospel end on a note of triumph. David closed his Psalm with praise for the coming Messiah: "A posterity will serve Him; It will be told of the Lord to the coming generation. They will come and will declare His righteousness To a people who will be born, that He has performed it" (Psalm 22:30–31). These final words echo directly through time, culminating in Christ's final breath on the cross: "Therefore when Jesus had received the sour wine, He said, 'It is finished!' And He bowed His head and gave up His spirit" (John 19:30).

With those words—"It is finished"—Christ declared the debt paid in full. He bore the punishment *we* deserved. He was pierced so we could heal. By accepting His sacrifice, we are granted eternal salvation.

Now, back to David's story.

THE GROWING KINGDOM

With the Lord's blessing, David was unstoppable. He crushed Israel's longstanding enemies, reclaimed territory, and asserted dominance throughout the region. "Now it happened afterward that David defeated the Philistines and subdued them; and David took control of the chief city from the hand of the Philistines" (2 Samuel 8:1). He then imposed a decisive defeat on the Moabites. The survivors paid tribute to Israel (2 Samuel 8:2).

Clausewitz would be pleased. One of his foundational principles of warfare was to "obtain the material means of combat and other resources of the hostile army" (Clausewitz 11). David understood this well and employed it to stunning effect during a battle near the Euphrates. "And David captured from him 1,700 horsemen and twenty thousand foot soldiers; and David hamstrung almost all the chariot horses, but left enough of them for a hundred chariots" (2 Samuel 8:4). This marked a significant upgrade to Israel's arsenal, as it was the first time they had chariots at their disposal.

Throughout the land, David's influence spread like wildfire. "When the Arameans of Damascus came to help Hadadezer, king of Zobah, David killed twenty-two thousand men among the Arameans. Then David put garrisons among the Arameans of Damascus, and the Arameans became servants to David, bringing tribute. And the Lord helped David wherever he went" (2 Samuel 8:5–6). David's growing empire brought in tremendous wealth, all of which he dedicated to the Lord. (2 Samuel 8:7)

This was just the beginning. Precious metals flowed in like a river of riches. Foreign rulers honored David and paid tribute accordingly.

Clausewitz emphasized another strategic aim of warfare: "Capture public opinion" (Clausewitz 11). David mastered this as well. His reputation became a weapon in itself.

When news of David's victories reached King Tou of Hamath, he didn't send an army—he sent his son bearing gifts of precious metals (2 Samuel 8:9–10). David's stunning reputation paralyzed many foes before a sword was ever drawn. For those stubborn or foolish enough to resist, the results were catastrophic (2 Samuel 8:13–14).

David's leadership, charisma, and battlefield experience—combined with Joab's ruthless execution—created an unstoppable force.

Yet it wasn't all blood and conquest. David was powerful, but not a tyrant. He remembered those who had helped him to rise to the throne. Chief among them was Jonathan, son of Saul. Jonathan had risked his life to save David more than once, asking only one thing in return: mercy for his descendants (1 Samuel 20:14–15).

Years later, David honored that oath. He learned there was still one descendent of Jonathan; his son, Mephibosheth (2 Samuel 9:1–3). David summoned him at once. Imagine Mephibosheth's fear: the grandson of a former king, now face to face with the reigning one. Moreover, he was crippled from a childhood accident; he stood no chance of escape, should things go wrong. But David quickly reassured him. "Then David said to him, 'Do not be afraid, for I will assuredly show kindness to you for the sake of your father Jonathan, and I will restore to you all the land of your grandfather Saul; and you yourself shall eat at my table regularly'" (2 Samuel 9:7).

Mephibosheth was stunned. This wasn't strategy or politics. It was personal. A solemn promise kept. David went even further. He appointed a servant and his household to manage the land and provide for Mephibosheth (2 Samuel 9:10). David honored his friend by caring for his son. He was a king who remembered loyalty, and repaid it in full.

This act of kindness begot another. Nahash, king of the Ammonites, had died. Scripture offers little detail about their relationship, but David clearly thought highly of him: "Then David said, 'I will show kindness to Hanun the son of Nahash, just as his father showed kindness to me.' So David sent some of his servants to console him about his father. But when David's servants came to the land of the Ammonites, the commanders of the Ammonites said to their lord Hanun, 'Do you think that David is simply honoring your father since he has sent you servants to console you? Has David not sent his servants to you in order to explore the city, to spy it out and overthrow it?'" (2 Samuel 10:2). Perhaps Nahash had aided David during his time under the Philistine king, Achish. We know David had previously sent plunder and gifts to those who helped him to flee Saul's persecution (1 Samuel 30:26–31). While the exact reason is unknown, David's sympathy was sincere, yet it backfired spectacularly.

The Ammonites made the age-old mistake of shooting the messenger. They thought David was sending spies and responded viciously (2 Samuel 10:3). What began as an act of goodwill was met with cruelty. "So Hanun took David's servants and shaved off half of their beards, and cut off their robes in the middle as far as their buttocks, and sent them away" (2 Samuel 10:4). Deeply humiliated, the men remained in Jericho until their beards regrew (2 Samuel 10:5).

David and his warriors would later be described as fierce as a bear robbed of her cubs (2 Samuel 17:8). *Never poke the bear.* The Ammonites escalated matters with no plan and no tactical reasoning— only arrogance. They antagonized a superior fighting force, which painted a massive bullseye on themselves.

Their rashness flies in the face of Sun Tzu's advice: "Move not unless you see an advantage; use not your troops unless there is something to be gained; fight not unless the position is critical" (Tzu 27). Realizing their blunder, the Ammonites scrambled for support and hired mercenaries: "Now when the sons of Ammon saw that they had become repulsive to David, the sons of Ammon sent messengers and hired the Arameans of Beth-rehob and the Arameans of Zobah, twenty thousand foot soldiers, and the king of Maacah with a thousand men, and the men of Tob with twelve thousand men" (2 Samuel 10:6).

David responded decisively, dispatching Joab to confront the threat. According to *De Re Militari*, "If he finds himself in many respects superior to his adversary, he must by no means defer bringing on an engagement" (Vegetius 64).

The fight was on! Joab mustered the entire army. The enemy presented two fronts: The Ammonites drew a battle formation at the entrance of their city gate. Meanwhile, the Aramean mercenaries were in open country (2 Samuel 10:8). Multiple fronts can overwhelm even a seasoned force. Prioritization is essential. Not all threats are equal—and Joab knew it.

This is where Joab's years of experience shone through. He calculated the odds and delegated troops accordingly. The Aramean mercenaries had a fearsome reputation, so he prioritized them. "And the sons of Ammon came out and lined up for battle at the entrance of the city, while the Arameans of Zobah and of Rehob and the men of Tob and Maacah were stationed by themselves in the field" (2 Samuel 10:9). For the Ammonite front, he assigned his brother Abishai. There's an old military saying: *No plan survives first contact with the enemy*. Joab understood how quickly battle can turn. So he made the plan dynamic. "And he said, 'If the Arameans are too strong for me, then you shall help me; but if the sons of Ammon are too strong for you, then I will come to help you'" (2 Samuel 10:11).

Clausewitz would certainly approve: "We must never lose sight, for a moment, of the following principle, the great importance of which I cannot stress enough: do not engage all your forces in battle at once and without a plan" (Clausewitz 8). Joab's plan offered flexibility and maneuverability—hallmarks of real-world tactics. Sun Tzu echoed this in his classic work: "Water shapes its course according to the nature of the ground over which it flows; the soldier works out his victory in relation to the foe whom he is facing" (Tzu 12).

Joab concluded with a rallying cry that showed both courage and faith: "Be strong, and let's show ourselves courageous for the sake of our people and the cities of our God; and may the Lord do what is good in His sight" (2 Samuel 10:12). He embodied the same warrior spirit that once drove Joshua.

Israel seized the initiative and struck first. The acclaimed Arameans fled like rabbits before a fox! (2 Samuel 10:13). All their bluster proved hollow. All the talk of their reputation was nothing but hot air; like the desert breeze. This once again highlights the crucial difference between a warrior and a mercenary. Warriors fight for cause and country. Mercenaries fight only for coin. As Machiavelli warned, "Mercenaries and auxiliaries are useless and dangerous; and if one holds his state based on these arms, he will stand neither firm nor safe" (Machiavelli 66). The Ammonites learned this the hard way. They burned through a fortune only to watch their hired swords scatter. "When the sons of

Ammon saw that the Arameans had fled, they also fled from Abishai and entered the city. Then Joab returned from fighting against the sons of Ammon and came to Jerusalem" (2 Samuel 10:14).

But the fight wasn't over. The Arameans regrouped for another attempt. Joab reported to David, who decided to lead personally. "Now when it was reported to David, he gathered all Israel together and crossed the Jordan, and came to Helam. And the Arameans lined up against David and fought him" (2 Samuel 10:17). Once again, with the Lord on his side, David routed them. "But the Arameans fled from Israel, and David killed seven hundred charioteers of the Arameans and forty thousand horsemen, and struck Shobach the commander of their army, and he died there" (2 Samuel 10:18).

David restored order through strength. He was peaceful, not harmless. As the timeless adage goes, "He, therefore, who desires peace, should prepare for war" (Vegetius 45). The results were undeniable: "When all the kings, servants of Hadadezer, saw that they had been defeated by Israel, they made peace with Israel and served them. So the Arameans were afraid to help the sons of Ammon anymore" (2 Samuel 10:19).

As Vegetius said, "No one dares to offend or insult a power of known superiority in action" (Vegetius 45). The Ammonites made that mistake once. Their crushing defeat remains an eternal reminder: Never side against the Lord's battles.

DAVID AND BATHSHEBA— A CAUTIONARY TALE OF IDLE HANDS

In terms of warfare accomplishments, David remains unparalleled. From single-handedly killing Goliath to restoring Jerusalem, he was truly one of a kind. He rose from humble shepherd boy to undisputed King of Israel. With such a meteoric rise, it's no wonder he might have felt invincible. Yet as we've seen time and again, David credited his victories to the Lord's blessing. The Spirit was with him. Despite his lofty kingship, he was still a peasant in this fallen kingdom—where Satanic forces still roam. Just like Adam and Eve, the silver-tongued serpent continues to whisper in our ears. David was no exception.

When David operated in his natural element—combat—he was unstoppable. But every man has his strengths and weaknesses. And David's greatest challenges were still ahead. As the old adage goes, *heavy is the head that wears the crown*. In his youth, David's trials were direct: kill the giant, win the battle, take the city. There were clear goals and obvious metrics for success. But the next phase of his life would test him in more subtle, insidious ways.

For decades, David had led from the front. But now, he sent Joab to war while he lingered in comfort: "Then it happened in the spring, at the time when kings go out to battle, that David sent Joab and his servants with him and all Israel, and they brought destruction on the sons of Ammon and besieged Rabbah. But David stayed in Jerusalem" (2 Samuel 11:1). This is the first warning sign. The ambition and grit that once defined him had dulled. As any seasoned fighter knows, staying on top is far harder than climbing to the top.

Why is that? First, the up-and-comer is hungry. He plays to win, while the reigning champ plays not to lose. There's a world of difference in that mindset. Second, once you're on top, your every move is studied. Your enemies probe for weaknesses. Last, success often breeds complacency. That's why we say, "don't rest on your laurels." Or as champion boxer Marvin Hagler once put it, "It's tough to get out of bed to do roadwork at 5 a.m. when you've been sleeping in silk pajamas" (Hagler).

And so we see a powerful truth: everyone needs a mission. Without one, even a warrior like David can falter. He had conquered every battlefield. He had wealth, status, and a legacy already etched in stone, yet it wasn't enough. In fact, it left him vulnerable. As the old saying goes, *idle hands do the Devil's bidding.*

It all began on a seemingly mundane evening: "Now at evening time David got up from his bed and walked around on the roof of the king's house, and from the roof he saw a woman bathing; and the woman was very beautiful in appearance" (2 Samuel 11:2). Not only was she married, but her husband, Uriah, was also one of David's top warriors. He was a member of the elite Mighty Men, also known as *The Thirty* (2 Samuel 23:39/1 Chronicles 11:41).

David should have turned away. Instead, temptation took root. When it comes to sin, people become master justifiers. In the heat of the moment, we become defense attorneys, pleading not before God's justice, but in favor of our selfish desires. We tell ourselves it's not really that bad. And this is where everything begins to unravel.

This chapter of David's story highlights the slippery nature of sin. He knew he shouldn't look lustfully at this mysterious woman—but he did anyway. He knew she was married, yet he pursued her regardless. David sent messengers to her. She came to his palace and they had an affair. After, she went back home and they tried to sweep the matter under the rug (2 Samuel 11:4).

Then came the consequences. Bathsheba told David she was pregnant.

David scrambled to cover up his misdeeds. Adultery was strictly prohibited under the Law (Exodus 20:14). If she were unmarried, he could have taken her as a wife. But with her pregnancy and her husband still alive, there was no easy way out. Still, David tried. Unfortunately, instead of coming clean, he doubled down on his shameful behavior. His panicked mind spun up a plan so devious that the Devil himself was surely grinning.

David summoned Uriah from the battlefield—under false pretenses. He disguised his wicked intent by asking routine military questions, then sent Uriah home to see his wife. David hoped Uriah would sleep with her, creating plausible deniability for the child's paternity.

But David miscalculated.

Uriah never went home. Instead, he slept at the entrance of the palace (2 Samuel 11:9). Confused, David asked why. Uriah's response was nothing short of honorable: "The ark and Israel and Judah are staying in temporary shelters, and my lord Joab and the servants of my lord are camping in the open field. Should I then go to my house to eat and drink and to sleep with my wife? By your life and the life of your soul, I will not do this thing" (2 Samuel 11:11).

David had no good answer. Uriah displayed nothing but loyalty, integrity, and professionalism—the very qualities every leader should aspire to.

Now David's plot grew more desperate. He invited Uriah to dinner and got him drunk, hoping to weaken his resolve. "Now David summoned Uriah, and he ate and drank in his presence, and he made Uriah drunk; and in the evening Uriah went out to lie on his bed with his lord's servants, and he still did not go down to his house" (2 Samuel 11:13).

Here we see the first glimpse of David's Machiavellian side. He hoped alcohol would compromise Uriah's principles, but it didn't work. With pressure mounting and no way to cover his tracks, David concocted a plan more akin to *Game of Thrones* than Godly leadership.

His poor decisions had backed him into a corner—and nothing is more dangerous than a man backed into a corner. Fear clouded his judgment. In one final attempt to erase his sin, David escalated dramatically. He sent Uriah back to the battlefield carrying his own death warrant: "So in the morning David wrote a letter to Joab and sent it by the hand of Uriah. He had written in the letter the following:

'Station Uriah on the front line of the fiercest battle and pull back from him, so that he may be struck and killed'" (2 Samuel 11:14–15).

Rather than face the truth, David chose to kill a loyal and valuable officer. Joab may not have known the true reason for the order, but he executed it nonetheless. "So it was as Joab kept watch on the city, that he stationed Uriah at the place where he knew there were valiant men. And the men of the city went out and fought against Joab, and some of the people among David's servants fell; and Uriah the Hittite also died" (2 Samuel 11:16–17). With this vile plan, David thought he had solved his problem. But life rarely works that way. Most of the time, we don't solve our problems; we just trade them out for new ones. The echoes of this act reverberated throughout David's life. Rather than simplify his troubles, he set into motion complications he couldn't foresee; troubles that would haunt him for the rest of his days.

Joab, having finished the dirty job, sent a message back to David. But he wasn't naive. He anticipated David might lash out, especially considering how tactically foolish the maneuver had been. Joab coached the messenger accordingly: "He ordered the messenger, saying, 'When you have finished telling all the events of the war to the king, then it shall be that if the king's wrath rises and he says to you, "Why did you move against the city to fight? Did you not know that they would shoot from the wall?"'" (2 Samuel 11:19–20).

Funnily enough, this exchange references a well-known story from the Book of Judges—Abimelech's downfall. Apparently, that military blunder had become common knowledge in Israel. Joab, clever as ever, anticipated the comparison and prepared a scripted reply. Should David reference this account, the messenger was to deflect by informing him that Uriah had fallen (2 Samuel 11:21),

Coded message in hand, the messenger went to David. He relayed Joab's report in full. David responded with a chilling calm. He masked everything under diplomatic coldness: "Then David said to the messenger, 'This is what you shall say to Joab: "Do not let this thing displease you, for the sword devours one as well as another; fight with determination against the city and overthrow it"; and thereby encourage him'" (2 Samuel 11:25).

David dismissed his envoy. Soon after, word reached Bathsheba. She mourned for her husband. Then she moved in with David, married him, and bore him a son. Their sins were quietly swept under the rug, for now. Yet they couldn't hide from the Lord's disapproval (2 Samuel 11:27).

The Lord sent a messenger. His name was Nathan.

Nathan was no ordinary prophet. He was trusted by David and known for his moral clarity, but this task tested everything. David had ordered the death of one man to hide his shame. What would he do to another who called it out publicly? "Don't shoot the messenger" is a cliché for good reason! Nathan had to walk a tightrope. He needed to speak the truth without provoking wrath.

So, he used strategy. Like Christ centuries later, he spoke in parables.

Nathan opened with a story: Two men lived in the same town (2 Samuel 12:2–4). One was rich beyond measure, with herds and flocks to spare. The other was destitute; he had only a single lamb. But that one lamb was cherished. The poor man raised it like a child. He fed it from his own table, held it in his arms, and loved it like family.

Then came a traveler.

The wealthy man, eager to entertain his guest, refused to sacrifice one of his own many animals. Instead, he stole the poor man's lamb. He butchered it and fed it to his guest. They ate well. The poor man was left with nothing.

To the rich man, it was just another meal.

To the poor man, *it was everything*.

This blatant greed and callousness outraged the former shepherd. "Then David's anger burned greatly against the man, and he said to Nathan, 'As the Lord lives, the man who has done this certainly deserves to die! So he must make restitution for the lamb four times over, since he did this thing and had no compassion'" (2 Samuel 12:5–6). Now that the issue was no longer abstract, David finally saw it clearly. But then Nathan dropped the hammer: David was the man in the story.

With the truth laid bare, David had nowhere to run. The Lord spoke through Nathan in no uncertain terms: "I also gave you your master's house and put your master's wives into your care, and I gave you the house of Israel and Judah; and if that had been too little, I would have added to you many more things like these! Why have you despised the word of the Lord, by doing evil in His sight? You have struck and killed Uriah the Hittite with the sword, you have taken his wife as your wife, and you have slaughtered him with the sword of the sons of Ammon. Now then, the sword shall never leave your house, because you have despised Me and have taken the wife of Uriah the Hittite to be your wife'" (2 Samuel 12:8–10).

David had been blessed arguably more than any figure in the Old Testament. The Lord had raised him from shepherd to king. We all fall short of God's glory—but in David's case, the fall was especially grievous. As God's anointed, he bore the sacred responsibility of leading by example.

Nathan pressed on: "This is what the Lord says: 'Behold, I am going to raise up evil against you from your own household; I will even take your wives before your eyes and give them to your companion, and he will sleep with your wives in broad daylight. Indeed, you did it secretly, but I will do this thing before all Israel, and in open daylight'" (2 Samuel 12:11–12).

Here we see one of the clearest contrasts between Saul and David. When the prophet Samuel confronted Saul, he made excuses. He blamed others. But God sees through every excuse and rationalization. David, by contrast, did something far more admirable—he owned it.

David could have deflected. He could have blamed Bathsheba for not hiding herself more carefully. He could have blamed Uriah for refusing the king's command to go home. He could have blamed his wives for not fulfilling him. There are always excuses—but excuses don't make a man innocent, only irresponsible. David didn't run from the truth. He didn't justify his behavior. He stood up and admitted it like a man. He accepted responsibility and the consequences that came with it.

Nathan's wisdom was on full display here. He didn't charge in shouting. He disarmed the king with a parable. Had Nathan led with accusation, he might have ended up like many prophets before him.

Scripture makes this clear: calling out sin in high places has always been dangerous work. Consider the reign of Ahab and Jezebel. They slaughtered God's prophets en masse. "For when Jezebel killed the prophets of the Lord, Obadiah took a hundred prophets and hid them by fifties in a cave, and provided them with bread and water" (1 Kings 18:4).

The letter to the Hebrews adds further context. The lives of the prophets were often marked by hardship and suffering: "And others experienced mocking and flogging, and further, chains and imprisonment. They were stoned, they were sawn in two, they were tempted, they were put to death with the sword; they went about in sheepskins, in goatskins, being destitute, afflicted, tormented (people of whom the world was not worthy), wandering in deserts, on mountains, and sheltering in caves and holes in the ground" (Hebrews 11:36–38).

Nathan's courage placed him in this lineage. But his strategy spared him. And David's contrition preserved the line of Christ.

This phenomenon isn't exclusive to Christianity or Judaism. Dissidents are almost always first on the chopping block—often quite literally. That's why teaching through parables is so brilliant. It's no accident Jesus used this method so frequently. People are quick to become defense attorneys at the first sign of discomfort. One brief encounter with sin or truth is enough to trigger elaborate justifications. Even the illiterate and uneducated possess this reflex. It's built into us from birth.

Just think of a child caught sneaking cookies from the jar. Watch how fast they launch into a passionate defense of their alleged crimes. That instinct never leaves us. The strength of a parable lies in its ability to bypass this defense mechanism entirely. Our egos act like ruthless gatekeepers, guarding against any truth that might wound our pride. Parables sneak past that gate like a Trojan horse. Once inside, the truth has room to take root and grow. That's exactly what happened to David. Once he connected the dots for himself, the truth landed with unforgettable force.

Parables also simplify. They strip away distractions and get to the heart of a matter. That clarity allows for better understanding, and more importantly, action. This doesn't mean watering down the message; it means communicating it more effectively. Jesus didn't just preach to the educated legalists. He taught anyone who would listen. That's precisely why the elites hated Him. He undercut their bloated authority and basically rendered their role obsolete. By teaching the people *how* to think, instead of merely what to believe, He broke their monopoly on spiritual interpretation.

Another key advantage of parables is their adaptability. A wise reader can apply them across a range of situations. The parable of the poor man's lamb, for instance, speaks to far more than just David's sin. It's a timeless warning against power abused and innocence exploited. That's why it still resonates three thousand years later.

Nathan's strategy worked. His words broke through. David saw the truth and didn't run from it. "Then David said to Nathan, 'I have sinned against the Lord.' And Nathan said to David, 'The Lord also has allowed your sin to pass; you shall not die. However, since by this deed you have shown utter disrespect for the Lord, the child himself who is born to you shall certainly die'" (2 Samuel 12:13–14).

The consequence came swiftly. "Then Nathan went to his house. Later the Lord struck the child that Uriah's widow bore to David, so that he was very sick" (2 Samuel 12:15). David was devastated. He fasted and pleaded with the Lord for days. "The elders of his household stood beside him in order to help him up from the ground, but he was unwilling and would not eat food with them" (2 Samuel 12:17).

On the seventh day, the child died.

David's servants were terrified to tell him. They feared his grief might explode into violence. But David saw the fear on their faces and understood what had happened. He asked them directly if the child was dead. They confirmed his worst fears.

Then something remarkable happened.

David didn't weep or scream. He didn't lash out. He simply got up, washed, changed clothes, and went to the house of the Lord to worship. Afterward, he returned home and asked for food (2 Samuel 12:20).

The royal court stood perplexed. "Then his servants said to him, 'What is this thing that you have done? You fasted and wept for the child while he was alive; but when the child died, you got up and ate food'" (2 Samuel 12:21).

This is where David's faith steadied him. He had pleaded with the Lord for mercy and grace. But once the Lord rendered His judgment, David understood there was no reversing it. When God speaks, it is final. David recognized that the tragedy was a direct result of his own actions. If he hadn't pursued Bathsheba, none of this would have happened. He accepted the Lord's discipline with humility and strength. Speaking to his attendants, David offered a quiet but profound expression of hope: "But now he has died; why should I fast? Can I bring him back again? I am going to him, but he will not return to me" (2 Samuel 12:23).

It's also worth remembering David's reaction to Nathan's parable: He must pay for that lamb four times over. This marked the first death of a son. That raises a morbid but pressing question—if David was to pay four times over, that meant three more of his sons would die prematurely. That dreadful possibility haunted him for the rest of his life.

Even in the midst of profound sorrow, hope began to stir again. Despite the loss of their first child, David and Bathsheba tried once more to build a family. "Then David comforted his wife Bathsheba, and went in to her and slept with her; and she gave birth to a son, and he named him Solomon. Now the Lord loved him, and sent word through Nathan the prophet, and he named him Jedidiah for the Lord's sake" (2 Samuel 12:24–25).

For context, Jedidiah means *loved by the Lord*. That name was more than a blessing; it was a sign of grace. David couldn't undo the damage he had done, but through repentance and obedience, he could still move forward. Neither he nor Bathsheba could have known then, but their son Solomon would one day become one of the most renowned figures in all of Scripture. Even today, Solomon is widely regarded as the wisest man in the Bible. Mortal man, that is.

Meanwhile, Joab continued the war against the Ammonites—the very campaign in which Uriah had fallen. A shrewd strategist, Joab seized control of critical infrastructure in the city of Rabbah. With the water supply captured, the Ammonites were running out of time. The dry climate would do what arrows could not. In the harsh desert conditions, dehydration kills kings and peasants all the same. Confident in his position, Joab sent a clever message to David. "Then Joab sent messengers to David and said, 'I have fought against Rabbah, I have even captured the city of waters. Now then, gather the rest of the people and camp opposite the city and capture it, or I will capture the city myself and it will be named after me'" (2 Samuel 12:27–28).

The challenge was playful, but clear: finish what was started, or let someone else get the glory. Reinvigorated with purpose, David responded. "So David gathered all the people and went to Rabbah, and he fought against it and captured it" (2 Samuel 12:29). With Joab having already broken the enemy's defenses, the final blow came easily.

What happened next was remarkable. David took the crown from the defeated king, not an ordinary trophy, but something almost mythical in description. "Then he took the crown of their king from his head; and its weight was a talent of gold, and it had a precious stone; and it was placed on David's head. And he brought out the plunder of the city in great amounts" (2 Samuel 12:30). For context, a talent weighs about seventy-five pounds (~34 kilos). Imagine bearing that on your head! At first glance, the glittering gold and precious gems seemed like a triumph. But moments later, the crushing weight would dig into your scalp, strain your neck, and remind you of an uncomfortable truth: the crown may dazzle, but it also burdens.

Perhaps God allowed this crown to come into David's possession as a symbolic warning. This was a *Sword of Damocles* moment. The old parable tells of Damocles, who envied a king's wealth and power. He got his wish for a day, enjoying a royal feast—until he noticed a sword suspended above his head by a single horsehair. His appetite vanished; soon after, so did Damocles. The lesson? Leadership may look desirable from a distance, but it comes with hidden dangers and burdens.

Everyone thinks they want the crown—until they feel its weight.

This heavy crown foreshadowed the trials David would soon face. He had triumphed over external enemies with ease. But his greatest struggles were yet to come—and they would not come from Philistines or Ammonites. They would come from within his own kingdom.

DAVID'S NEW CHALLENGES

Previously, we examined Abimelech, one of Gideon's many sons. Fathering children with numerous women is a surefire way to invite internal conflict. The reason is simple: too many competing interests. In matters like inheritance or royal succession, the stakes are zero-sum; there can only be one king. The more hands that reach for the crown, the more vicious the contest becomes. Abimelech is the textbook example. He butchered dozens of his half-brothers to seize Gideon's throne.

Now, in 2 Samuel 13, we see the same dark dynamic unfold in David's household. His military victories meant nothing if his kingdom collapsed from within. Tragically, his sons imitated his worst behaviors. One followed David's sexual immorality, but in a darker and more violent way. Another would mirror his bloodshed.

The trouble began with David's firstborn son, Amnon—heir to the throne. "Now it was after this that Absalom the son of David had a beautiful sister whose name was Tamar, and Amnon the son of David was in love with her" (2 Samuel 13:1). But this was no love. It was an obsession—sick and twisted. Tamar was Amnon's half-sister, and the full sister of Absalom. Instead of turning from this depravity, Amnon brooded over it. Eventually, his advisor Jonadab—described as a very shrewd man—asked what was wrong (2 Samuel 13:3). Amnon shared his wicked thoughts, and Jonadab supplied an equally wicked plan. Like Satan in the Garden of Eden, he whispered a vile idea into his ear. Amnon, the snake in the grass, acted on it.

Feigning illness, Amnon deceived his father and lured Tamar to his chamber (2 Samuel 13:5–10). Using the goodwill of his family, he set his trap. "Then Amnon said to Tamar, 'Bring the food into the bedroom, so that I may eat from your hand.' So Tamar took the pastries which she had made and brought them into the bedroom to her brother Amnon. When she brought them to him to eat, he took hold of her and said to her, 'Come, sleep with me, my sister'" (2 Samuel 13:10–11).

Tamar was horrified as she attempted to save herself. "But she said to him, 'No, my brother, do not violate me, for such a thing is not done in Israel; do not do this disgraceful sin!'" (2 Samuel 13:12). But Amnon didn't care. "However, he would not listen to her; since he was stronger than she, he violated her and slept with her. Then Amnon hated her with a very great hatred; indeed, the hatred with which he hated her was greater than the love with which he had loved her. And Amnon said to her, 'Get up, go away!'" (2 Samuel 13:14–15). Having used her and discarded her, he locked her out. Tamar, devastated and humiliated, tore her robe and fled, wailing (2 Samuel 13:19). This horrific abuse and betrayal surely haunted her for the rest of her days.

Her full brother Absalom took her in, offering refuge—though the damage was done. She was never the same again. "Now when King David heard about all these matters, he became very angry. But Absalom did not speak with Amnon either good or bad; for Absalom hated Amnon because he had violated his sister Tamar" (2 Samuel 13:21–22). Absalom said nothing, but his silence was deadly. He didn't seek scandal—he wanted justice. And when justice didn't come, he settled for vengeance.

What happened next is left unclear. David was angry, but no punishment was recorded. This echoes his prior leniency with Joab. David had a pattern of being soft on family. And after his own moral collapse—adultery, murder, and cover-up—he may have felt unqualified to discipline others.

David's failure to lead by example became a festering wound in his household. The fallout from his sin rippled outward. Though he was forgiven by the Lord, David's credibility was shattered. And the people around him, especially his own sons, were not so quick to forget. David's infidelity sparked the domino effect. The kingdom snowballed downhill, gaining speed and destructive power. The Lord had forgiven him, but people are much harder to convince.

ABSALOM'S VENGEANCE

S ome two years passed without justice. This only deepened Absalom's hatred for his half-brother. At first, he may have given David the benefit of the doubt. But as time dragged on with no punishment, Absalom lost respect for his father's leadership. In his eyes, David had forfeited the moral high ground. So, Absalom decided to act.

He turned to deception, just as warfare often demands. Absalom planned a sheep-shearing event and invited all his brothers. This gathering appeared innocent and even generous. Amnon had no reason to suspect a thing, which is exactly what Absalom wanted. Then came the order: "See now, when Amnon's heart is cheerful with wine, and I say to you, 'Strike Amnon,' then put him to death. Do not fear; have I not commanded you myself? Be courageous and be valiant" (2 Samuel 13:28).

This strategy echoes Sun Tzu: "A clever general avoids an army when its spirit is keen, but attacks it when it is sluggish and inclined to return" (Tzu 14). Amnon, drunk and defenseless, never stood a chance. As Sun Tzu also wrote: "You can be sure of succeeding in your attacks if you only attack places which are undefended" (Tzu 11). Absalom understood this intuitively. The assassination went off without a hitch. Absalom's men fled the scene, and Absalom vanished with them. This marked the beginning of his personal rebellion.

There's a hard lesson here: when people lose faith in the system, they take justice into their own hands. Wise leaders must understand this danger. This is not to justify tyranny or censorship—but to stress the importance of trust. Leaders must walk the walk. They must act with integrity. Justice delayed is justice denied, and that denial erodes authority.

David's silence on Amnon spoke louder than any decree. Moses warned of this very pitfall: "You shall not distort justice, you shall not show partiality; and you shall not accept a bribe, because a bribe blinds the eyes of the wise and distorts the words of the righteous. Justice, and only justice, you shall pursue, so that you may live and possess the land which the Lord your God is giving you" (Deuteronomy 16:19–20). David's judgment was clouded by family ties, and possibly by guilt. He failed to discipline Amnon, and that failure turned fatal. Perhaps Solomon reflected on this tragedy when he later wrote: "He who withholds his rod hates his son, But he who loves him disciplines him diligently" (Proverbs 13:24).

David tried to spare Amnon, but ended up destroying him.

News of the assassination rocked David to his core. Another wave of anguish crashed upon him (2 Samuel 13:31). Early reports claimed that all of David's sons had been slaughtered. But one official clarified: "Let my lord not assume that they have put to death all the young men, the king's sons, for only Amnon is dead; because this has been set up by the intent of Absalom since the day that he violated his sister Tamar" (2 Samuel 13:32).

Then David knew. There was no mystery here, just a reckoning. Imagine the emotions that tore through his heart: Grief for his firstborn. Shame over his inaction. Fear for his other sons. And above all, guilt. David had set a poor example, and his sons took his poor example even further. Amnon mirrored his lust. Absalom mirrored his violence. As the theocratic leader of Israel, David bore ultimate responsibility. The buck stopped with him.

The only question now: Could he stop the bleeding before the kingdom tore itself apart?

At that moment, the watchman reported that David's sons, save for Absalom and Amnon, had returned home (2 Samuel 13:34–35). Then, a surprising emotion overtook the king: joy. Not all was lost. There was still room for hope. The other sons wept for their shattered family. Absalom, much like his father, fled into exile. His self-imposed banishment lasted three years. And all the while, David's heart ached for his safe return (2 Samuel 13:39).

But Absalom's assassination of Amnon had another major effect; it cleared the throne. Absalom was now the heir to Israel. The question lingered: Would he inherit the throne ... or take it by force? That thought surely kept King David up at night. How could such a fractured kingdom survive?

As usual, Joab had a plan. The seasoned general held great influence in David's court. For reasons we're not told, Joab opted for an indirect approach to reconcile the king with his exiled son. Perhaps he had tried a direct appeal and failed. We don't know. What we do know is that Joab concocted a theatrical ruse to steer David's heart. Ever the tactician, he recruited a woman from Tekoa—an actress with a sympathetic story. Playing the role of director, Joab gave her a script with clear intent (2 Samuel 14:2–3).

Joab's strategic mind orchestrated the scene from start to finish. His knack for manipulating court dynamics rivaled his brilliance on the battlefield. He played both arenas with equal mastery. With the stage set, the woman put on a heartfelt performance: "Now when the woman of Tekoa spoke to the king, she fell on her face to the ground and prostrated herself, and said, 'Help, O king!'" (2 Samuel 14:4). David asked her what was troubling her. She claimed to be a widow with two sons. One day, the brothers fought, and no one was there to intervene. One son was killed in the scuffle. The tragedy tore her family apart. Now her clan demanded justice, blood for blood.

This left her in an impossible position: if they killed her remaining son, her family line would be wiped out forever. She pleaded with the king: "Now behold, the entire family has risen against your servant, and they have said, 'Hand over the one who struck his brother, so that we may put him to death for the life of his brother whom he killed, and eliminate the heir as well.' So they will extinguish my coal which is left, so as to leave my husband neither name nor remnant on the face of the earth" (2 Samuel 14:7).

David, moved by her story, promised to protect her family. Pleased with his response, she asked the king to invoke the Lord's name. David gladly complied. Little did he know his answer was prophetic with a cruel sense of irony. "Then she said, 'May the king please remember the Lord your God, so that the avenger of blood will not continue to destroy, otherwise they will destroy my son.' And he said, 'As the Lord lives, not one hair of your son shall fall to the ground'" (2 Samuel 14:11).

David walked right into Joab's ruse.

This promise, meant for a fictional son, was really about his real one. It was a foreshadowing of Absalom. Yet sometimes, the most painful judgment in life is getting *exactly* what you asked for.

THE DANGER OF HALF MEASURES

Now the actress asked to speak plainly with David, and he granted her request. This is where her carefully rehearsed performance began to unravel. In doing so, we see David's more merciful side on display—even as she challenged him directly. "The woman said, 'Why then have you planned such a thing against the people of God? For in speaking this word the king is like one who is guilty, in that the king does not bring back his banished one. For we will surely die and are like water spilled on the ground, which cannot be gathered up. Yet God does not take away life, but makes plans so that the banished one will not be cast out from Him'" (2 Samuel 14:13–14).

Few commoners would dare speak to a monarch so bluntly. But she held her ground. Her reasoning was sound, yet her approach was extremely risky. She claimed the people feared for the future of the kingdom, and this time, they weren't wrong. David's years of discernment tipped him off to something peculiar. The entire interaction felt staged.

He confronted her directly: "So the king said, 'Is the hand of Joab with you in all this?' And the woman replied, 'As your soul lives, my lord the king, no one can turn to the right or to the left from anything that my lord the king has spoken. Indeed, it was your servant Joab who commanded me, and it was he who put all these words in the mouth of your servant'" (2 Samuel 14:19).

The jig was up. She confessed the entire scheme and complemented David's sharp perception.

Begrudgingly, David went along with Joab's plan. He ordered Joab to bring Absalom back to Jerusalem. Joab, delighted, fell to the ground in thanks. This seemed like a step toward healing, but there was a twist. David allowed Absalom back into the city on one condition: "However, the king said, 'He shall return to his own house, but he shall not see my face.' So Absalom returned to his own house and did not see the king's face" (2 Samuel 14:24).

We can't be certain of David's reasoning behind this partial pardon. It was likely a mix of unresolved pain: anger over Amnon's murder, guilt for never punishing him in the first place, embarrassment over his own misdeeds with Bathsheba, and dread over what Israel's future might hold. Whatever the case, David chose a half measure, and *half measures double the problems*. This is a lesson he would learn the hard way.

Meanwhile, Absalom thrived in the royal city. As heir to the throne and a man of striking appearance, he quickly gained favor (2 Samuel 14:25). During this time, Absalom fathered several children and lived in luxury. Yet resentment simmered beneath the surface. He had been allowed back, but not fully restored (2 Samuel 14:28). The ambiguity of his position gnawed at him. Was he the crowned prince of Israel ... or a disgraced exile in disguise?

Leaders must be resolute, even if that causes temporary discomfort. Wishy-washy back and forth is a recipe for disaster! Mixed messages act like a festering wound. Painful though it may be, infected flesh must be purified. It stings at first, but true healing follows. David's handling of Absalom, however, was more like picking at a stubborn scab. The constant irritation only worsened the wound. Absalom wasn't suffering from a physical injury, but his heart was wounded. And that wound festered.

Crestfallen, Absalom reached out to Joab for help. But Joab gave him the cold shoulder (2 Samuel 14:29). When words fail, action becomes necessary—and Absalom understood this. Fed up with years of neglect, he took dramatic action. He ordered his servants to burn down Joab's barley fields! He made himself impossible to ignore. The squeaky wheel gets the grease.

With Joab now forced to listen, Absalom laid his demands bare. The half measures had to end. He would accept either full restoration or execution, nothing in between (2 Samuel 14:32). After all, King Saul had once prepared to execute his son Jonathan. Would David follow that same path? Instead, he likely recalled the woman's "burning coal" metaphor and chose mercy. "So when Joab came to the king and told him, he summoned Absalom. Then Absalom came to the king and prostrated himself with his face to the ground before the king; and the king kissed Absalom" (2 Samuel 14:33).

David forgave him—but was the damage already done? Was this reconciliatory gesture too little, too late? *Never push a loyal person too far; they make the most dangerous enemies.*

Initiative is like fire. When contained, it can cook a meal or forge metal into useful tools. But if left untamed, it destroys everything in its path. Initiative is one of the greatest qualities good leaders possess. Yet prudence and reason must temper this virtue. Without these virtues in harmony, things quickly deteriorate into chaos. Absalom had initiative in spades. When David was asleep at the wheel, Absalom had taken matters into his own hands. Years later, sensing weakness again, he did the same. He had lost respect for his father's rule, and now, he began his quiet rebellion.

In fact, Absalom may have been the first Israelite leader to field horse-drawn chariots. "Now it came about after this that Absalom provided for himself a chariot and horses, and fifty men to run ahead of him" (2 Samuel 15:1). This royal display dazzled the public. Absalom looked like a king out of legend. "Now in all Israel there was no one as handsome as Absalom, so highly praised; from the sole of his foot to the top of his head there was no impairment in him" (2 Samuel 14:25).

He was young, strong, and bore a mane of hair like a lion's. That hair was so thick he only cut it once a year when it became too heavy. The cuttings weighed about five pounds! (2 Samuel 14:26). He radiated vitality and power.

But this wasn't just for show. Absalom knew how to work a crowd. He rose early and met people outside the city gate, intercepting those seeking an audience with the king. Like a seasoned politician, he presented *himself* as the solution to all of society's problems, not the Lord. "Then Absalom would say to him, 'See, your claims are good and right, but you have no one to listen to you on the part of the king.' Moreover, Absalom would say, 'Oh that someone would appoint me judge in the land, then every man who has a lawsuit or claim could come to me, and I would give him justice!'" (2 Samuel 15:3–4).

We can't be entirely sure of Absalom's motivations. Did he truly believe he was a purveyor of justice? Given the lengths he went to in avenging his sister, perhaps. Or was this a calculated political coup, crafted to usurp his father's throne? We cannot say for certain. Whatever the case, we know one thing: the people loved it. Imagine a peasant's surprise when this handsome prince stepped down from his royal chariot to lend an ear.

Absalom didn't just listen; he greeted them warmly as friends and equals (2 Samuel 15:5). Anyone can fake sincerity for a short time. But Absalom carried on like this for years! He steadily courted favor and built popular support. "Absalom dealt this way with all Israel who came to the king for judgment; so Absalom stole the hearts of the people of Israel" (2 Samuel 15:6). The prince became the face of the kingdom. He gave the people someone they could believe in, someone they trusted.

This politicking continued for four years. At that point, Absalom asked David's permission to travel to Hebron, the capital city of Judah. He claimed he had made a vow to the Lord. Thinking nothing of it, David approved the trip. Once in Hebron, Absalom prepared to strike. He clearly understood the principle from *The Art of War*: "Let your plans be dark and impenetrable as night, and when you move, fall like a thunderbolt" (Tzu 14). Years of careful, patient planning culminated in this decisive moment. Earlier, he had demanded either full restitution or the death penalty. Now we see that same all-or-nothing mentality come to life. Safely out of David's reach, he made his move: "But Absalom sent spies throughout the tribes of Israel, saying, 'As soon as you hear the sound of the trumpet, then you shall say,' 'Absalom is king in Hebron!'" (2 Samuel 15:10). With him were two hundred men, unaware of the true purpose of their journey.

Absalom also recruited key allies to strengthen his hand and weaken David's. Absalom secretly recruited Ahithophel, David's advisor, to conspire against the king (2 Samuel 15:12). This defection was devastating. The loss of Ahithophel crushed morale and dealt a serious blow to David's chances of survival. "Now the advice of Ahithophel, which he gave in those days, was taken as though one inquired of the word of God; so was all the advice of Ahithophel regarded by both David and Absalom" (2 Samuel 16:23). As it turns out, Ahithophel was Bathsheba's grandfather (2 Samuel 23:34). Like Absalom, he bided his time. He waited patiently for his chance at vengeance. Apparently, he subscribed to the old axiom: keep your friends close, and your enemies closer.

The Kingdom of Israel erupted into chaos. An envoy caught David flat-footed with grim news: "Then a messenger came to David, saying, 'The hearts of the people of Israel are with Absalom'" (2 Samuel 15:13). David understood there could be no negotiation. The longer he waited, the more innocent lives would be caught in the crossfire. This was do or die. First, he had to escape the city. This was more than just a personal flight—it was a necessary move to spare the people of Jerusalem from a brutal siege. "So David said to all his servants who were with him in Jerusalem, 'Arise and let's flee, for otherwise none of us will escape from Absalom. Go quickly, or he will hurry and overtake us, and bring disaster on us and strike the city with the edge of the sword'" (2 Samuel 15:14).

With his loyal followers in tow, David departed. He stood by the city gates, making sure everyone evacuated safely. Some of the Levitical priests, led by Zadok, attempted to take the Ark of the Covenant with them. Surprisingly, David told them to leave it. Whatever the outcome, he respected the will of the Lord (2 Samuel 15:25–26). The priests obeyed the king's command.

Devastated by betrayal, David pressed onward. It was at this low point that David received more painful news: "Now someone informed David, saying, 'Ahithophel is among the conspirators with Absalom.' And David said, 'Lord, please make the advice of Ahithophel foolish'" (2 Samuel 15:31). Would the Lord answer his prayer? Or was this the Lord showing David just how foolish he had become?

Losing a trusted advisor was a gut punch to the embattled king. Fortunately, not all was lost. At the summit of the mount, David reunited with Hushai the Arkite, a loyal ally. Hushai offered to join him in exile. But David had a better plan. He deployed Hushai as a covert agent. Instead, he sent him back to Jerusalem with a mission: to infiltrate Absalom's court and counter Ahithophel's influence. David countered: "But if you return to the city and say to Absalom, 'I will be your servant, O king; even as I was your father's servant in time past, so now I will also be your servant,' then you can foil the advice of Ahithophel for me" (2 Samuel 15:34). Hushai was to act as a smokescreen, every word a calculated misdirection. True to his loyalty, Hushai returned to Jerusalem just as Absalom entered the city (2 Samuel 15:37).

With his strategic deception underway, David continued forward, and he soon encountered an unexpected ally. Ziba, the steward of Mephibosheth, met him with badly needed supplies. Years earlier, David had shown Mephibosheth kindness, restoring land and honor to Saul's crippled grandson. Now Ziba returned the favor. "Now when David had gone on a little beyond the summit, behold, Ziba the servant of Mephibosheth met him with a team of saddled donkeys, and on them were two hundred loaves of bread, a hundred cakes of raisins, a hundred summer fruits, and a jug of wine" (2 Samuel 16:1). These provisions sustained David's weary group for the journey ahead.

Then came a jarring encounter with a bitter relative of Saul named Shimei. His greeting was as hostile as it was theatrical. "He also threw stones at David and all the servants of King David; and all the people and all the warriors were on his right and on his left. This is what Shimei said when he cursed: 'Go away, go away, you man of bloodshed and worthless man! The Lord has brought back upon you all the bloodshed of the house of Saul, in whose place you have become king; and the Lord has handed the kingdom over to your son Absalom. And behold, you are caught in your own evil, for you are a man of bloodshed!'" (2 Samuel 16:6–8).

David knew there was some truth in the man's words. His men, however, were ready to silence the critic. Abishai, ever the warrior, offered to decapitate the troublemaker. David refused. He had larger problems to contend with. Once again, we see a striking parallel to his earlier dealings with Saul. And once again, David returned cursing with mercy: "Perhaps the Lord will look on my misery and return good to me instead of his cursing this day" (2 Samuel 16:12). The king told his guards to stand down. Shimei continued to hurl insults, stones, and dirt until they were out of range.

Eventually, they reached safe harbor at Bahurim. There, they refreshed themselves with the bread and raisin cakes provided by Ziba.

David had escaped Jerusalem with his life intact—but the battle for the kingdom was far from over.

A PROPHECY FULFILLED

Meanwhile, Absalom wasted no time claiming the royal city of Jerusalem. Ahithophel stood close beside the usurping prince. They believed they held the upper hand, exactly the illusion you want when running a PSYOP. Ever loyal to King David, Hushai performed his covert duties flawlessly. "Now it came about, when Hushai the Archite, David's friend, came to Absalom, that Hushai said to Absalom, 'Long live the king! Long live the king!'" (2 Samuel 16:16). Absalom, suspicious at first, questioned why Hushai hadn't left with David. But Hushai wisely appealed to Absalom's ego. When dealing with leaders, especially insecure ones, it helps to affirm their self-image. Hushai understood this dynamic well.

He answered, "No! For whomever the Lord, this people, and all the men of Israel have chosen, his I shall be, and with him I shall remain. Besides, whom should I serve? Should I not serve in the presence of his son? Just as I have served in your father's presence, so I shall be in your presence" (2 Samuel 16:18–19). The flattery worked. Egotistical people are easily swayed by confirming their self-image. Satisfied with the explanation, Absalom accepted Hushai into his inner circle. The covert operation continued without a hitch.

Eager to assert dominance, Absalom turned to Ahithophel for his next move. The advice he received was shocking: "Have relations with your father's concubines, whom he has left behind to take care of the house; then all Israel will hear that you have made yourself repulsive to your father. The hands of all who are with you will also be strengthened" (2 Samuel 16:21). This strategy wasn't about pleasure. It was about power, humiliation, and control. It resembled mating animal

343

behavior more than royal decorum. Like a lion overthrowing an older rival, the victorious male takes the pride as a symbol of dominance. Winner takes all.

Absalom needed no convincing. "So they pitched a tent for Absalom on the roof, and Absalom had relations with his father's concubines in the sight of all Israel" (2 Samuel 16:22). This vulgar act marked the point of no return. It wasn't just a power grab; it was a fulfillment of prophecy. The words of the prophet Nathan now rang with sobering clarity. "This is what the Lord says: 'Behold, I am going to raise up evil against you from your own household; I will even take your wives before your eyes and give them to your companion, and he will sleep with your wives in broad daylight. Indeed, you did it secretly, but I will do this thing before all Israel, and in open daylight'" (2 Samuel 12:11–12).

And let us not forget, Ahithophel was Bathsheba's grandfather. His brutal advice may have served a dual purpose: helping Absalom claim the throne while simultaneously exacting personal vengeance against David. A tactical two-for-one. Take the kingdom and crush David's spirit.

It was psychological warfare at its finest. Humiliate your opponent, erode his legitimacy, and weaken morale. This was more than a coup; it was a public ritual of humiliation. The question now loomed over Israel: Was David's reign truly finished? Could he possibly regain favor after such a scandalous display?

After Absalom's vulgar proclamation of kingship, they got back to business. Ahithophel, ever the cunning strategist, proposed a swift and decisive strike on the exiled king. "Furthermore, Ahithophel said to Absalom, 'Please let me choose twelve thousand men and let me set out and pursue David tonight. And I will attack him while he is weary and exhausted and startle him, so that all the people who are with him will flee. Then I will strike and kill the king when he is alone'" (2 Samuel 17:1–2).

Tactically speaking, Ahithophel's advice was solid. Both Sun Tzu and Miyamoto Musashi would likely approve. He understood that not all targets hold equal value. Cut off the head, and the body dies. War is costly. Israel had already endured one civil war after Saul's death; no one hungered for another. Each new conflict drained manpower and resources, weakening the kingdom's edge. Sun Tzu would have nodded in agreement: "In war, then, let your great object be victory, not lengthy campaigns.... There is no instance of a country having benefited from prolonged warfare" (Tzu 5).

Ahithophel's proposal also aligned with the principles of other strategists. He knew David's forces were fierce and battle-hardened. Their best chance was to strike when those men were weary and disoriented. He aimed to instill fear, break morale, and cause chaos. Musashi echoes this approach: "In large-scale strategy it is important to cause loss of balance. Attack without warning where the enemy is not expecting it, and while his spirit is undecided follow up your advantage and, having the lead, defeat him" (Musashi 26). Vegetius concurs: "If the enemy makes excursions or expeditions, the general should attack him after the fatigue of a long march, fall upon him unexpectedly" (Vegetius 66).

Every point Ahithophel made was tactically sound. But he forgot one critical factor: the Lord was on David's side.

Though Absalom and his advisors initially favored the plan, they decided to hear a second opinion. Hushai's moment had come. The PSYOP began to take effect. And ironically, Hushai sowed doubt not with lies, but with truth: "Then Hushai said, 'You yourself know your father and his men, that they are warriors and they are fierce, like a bear deprived of her cubs in the field. And your father is an expert in warfare, and he will not spend the night with the people'" (2 Samuel 17:8).

This was the brilliance of Hushai's deception. He leaned into David's fearsome reputation. From the day he had slayed Goliath to his daring conquest of Jerusalem, everyone in Israel knew David was not a man to underestimate. Then Hushai brought up David's guerrilla warfare experience. "Behold, he has now hidden himself in one of the ravines, or in another place; and it will be that when he falls on them at the first attack, whoever hears it will say, 'There has been a slaughter among the people who follow Absalom!' And even the one who is valiant, whose heart is like the heart of a lion, will completely despair; for all Israel knows that your father is a mighty man, and those who are with him are valiant men" (2 Samuel 17:9–10).

The seeds of doubt had been sown.

But Hushai didn't stop at discrediting Ahithophel. Good leaders don't just point out problems, they offer solutions. And Hushai knew how to appeal to Absalom's vanity. He proposed a dramatic display of royal power: "But I advise that all Israel be fully gathered to you, from Dan even to Beersheba, like the sand that is by the sea in abundance; and that you personally go into battle. Then we will come to him in one of the places where he can be found, and we will fall on him just as the dew falls on the ground; and of him and of all the men who are with him, not even one will be left" (2 Samuel 17:11–12).

It was a grand vision. Absalom leading the charge, basking in glory. Hushai's plan was tailor-made to his ego.

It worked. "Then Absalom and all the men of Israel said, 'The advice of Hushai the Archite is better than the advice of Ahithophel.' For the Lord had ordained to foil the good advice of Ahithophel, in order for the Lord to bring disaster on Absalom" (2 Samuel 17:14).

David's psychological warfare was working. The Lord had answered his prayer. The trap was set.

Now that Absalom had bought into Hushai's plan, the operation moved to its next phase. Hushai sent a secret message to David, warning him of Absalom's intentions and urging an immediate relocation of his forces. To transmit this critical intelligence, Hushai coordinated with loyalist priests who still supported the exiled king. In a time before electronic communications, messages were carried by hand—an inherently risky process. One misstep could collapse the entire network.

That nearly happened. "But a boy did see them, and he told Absalom; so the two of them left quickly and came to the house of a man in Bahurim, who had a well in his courtyard, and they went down into it. And the woman took a cover and spread it over the well's mouth and scattered barley meal on it, so that nothing was known" (2 Samuel 17:18–19).

Absalom's men arrived soon after, but the woman misled them, sending the pursuers in the wrong direction. Once the danger had passed, the messengers emerged and delivered their urgent warning to David. He responded without delay, moving his troops across the Jordan river, out of Absalom's immediate reach.

The woman's quick thinking was reminiscent of others in Israel's history. Recall how Michal helped David to escape Saul's men (1 Samuel 19), and how Rahab hid the Israelite spies from Jericho's authorities

(Joshua 2). These moments of tactical deception preserved lives and influenced the course of history.

Then came a stunning development. Ahithophel, the brilliant tactician and architect of the rebellion, withdrew entirely. "Now when Ahithophel saw that his advice had not been followed, he saddled his donkey and set out and went to his home, to his city, and set his house in order, and hanged himself; so he died and was buried in his father's grave" (2 Samuel 17:23).

It was a dramatic and grim exit. Disregarded and disillusioned, Ahithophel likely recognized the writing on the wall. Without the advantage of his rapid-strike strategy, the rebellion stood little chance. If David regained power, and all signs pointed that way, Ahithophel would be branded a traitor and executed.

Much like Saul, he feared the humiliation and consequences of defeat more than death itself. His suicide sent a chilling message to Absalom's camp. The rebellion's mastermind had lost confidence in its success. This dark omen cast a long shadow over the rebel army as it prepared for war. The tide was turning, and they could feel it.

THE FINAL BATTLE

David succeeded in his hasty crossing of the Jordan river. With an obstacle in between him and his adversary, he now recalibrated and planned a battle in his favor. David divvied up the troops amongst his commanders as he told them he would march alongside them in battle (2 Samuel 18:1–2). They were now at a fork at the road. A warrior at heart, David tried to enter the battle himself, yet his commanders unilaterally blocked him. Their reasoning was sound. They already knew Absalom's plans via secret intel. Never give your enemies easy ammo to use against you. They justified their tactical reasoning to the king: "But the people said, 'You should not go out; for if in fact we flee, they will not care about us; and if half of us die, they will not care about us. But you are worth ten thousand of us; so now it is better that you will be ready to help us from the city'" (2 Samuel 18:3).

The irony was bitter! Think back, all this trouble started when David *chose* not to go to battle in Rabbah (2 Samuel 11:1). His idle hands did the Devil's bidding. He lustfully sought out a married woman. When things spiraled out of control, he implemented a murderous conspiracy to cover up the mess. The long term fallout led to his son Absalom's rebellion. Now, when the multiyear feud reached a fever pitch, he *couldn't* fight!

David understood the plan from his commanders. He added one simple, yet fanciful request before they marched to the front lines: "But the king commanded Joab, Abishai, and Ittai, saying, 'Deal gently with the young man Absalom for my sake.' And all the people heard when the king commanded all the commanders regarding Absalom" (2

Samuel 18:5). Combat isn't gentle and these commanders were *far* from gentlemen! Yet as a father, David at least had to try.

David's army set forth to take the offensive. The battle took place in the forest of Ephraim. For David's battle-hardened troops, this was a one-sided affair. "The people of Israel were defeated there by the servants of David, and the slaughter there that day was great, twenty thousand men" (2 Samuel 18:7). As fierce as Joab was, the terrain proved even more treacherous! "For the battle there was spread over the whole countryside, and the forest devoured more people that day than the sword devoured" (2 Samuel 18:8). David's army continued slaughtering the traitors with ease.

Eventually, they caught their real target, Absalom. David's men happened upon a once-in-a-lifetime opportunity to attack their foe. "Now Absalom encountered the servants of David. Absalom was riding on his mule, and the mule went under the branches of a massive oak. Then his head caught firmly in the oak, and he was left hanging between the sky and earth, while the mule that was under him kept going" (2 Samuel 18:9). Perhaps King Solomon thought of this incident when he wrote the following. "Before destruction the heart of a person is haughty, But humility goes before honor" (Proverbs 18:12).

Absalom had no one else to blame. His arrogance led him to overstep and bite the hand that feeds. The rebellious prince grew up in a life of luxury, whereas David and his men grew up fighting for their lives. Moreover, Jesus discussed this type of attitude in the Gospel of Luke. "For everyone who exalts himself will be humbled, and the one who humbles himself will be exalted" (Luke 14:11). There are few things more humiliating than dangling helplessly while no one comes to your aid.

Oh, the irony! His thick hair, once a symbol of power and vigor, was now the very thing which trapped Absalom. Dangling in the air, he was totally at the mercy of Joab and his men. One of the men informed Commander Joab that Absalom was trapped in a tree. Confused, Joab asked the man why he didn't seize the easy victory. Essentially, the soldier answered that he feared David's wrath more than he desired wealth and rewards. A reasonable response given David's reputation and gravitas.

But Joab wasn't having it.

Despite his many flaws, Joab was a man of action who always led from the front. He lived his life by the motto: better to ask forgiveness than permission. Today was no exception. He sought to end the rebellion personally. "Then Joab said, 'I will not waste time here with you.' So he took three spears in his hand and thrust them through the heart of Absalom while he was still alive in the midst of the oak. And ten young men who carried Joab's armor gathered around and struck Absalom and killed him" (2 Samuel 18:14–15). Joab understood that you can't leave certain loose ends untied; traitors, especially so. *De Re Militari* concurs. "If the height of the mutiny requires violent remedies ... punish the ring-leaders only in order that, though few suffer, all may be terrified by the example" (Vegetius 52).

The bloody rebellion ended as abruptly as it started. After countless lives lost, it was finally over. The fighting stopped. The rebels scattered and fled after their leader died. Yet, Joab had one more matter to attend to. They dropped Absalom's body into a pit in the forest. Then, they piled up massive heaps of rocks over the body. At first, it appeared they were being respectful with the burial. Further examination shows the exact opposite. Joab was *mocking* the traitorous prince! Much like Saul, Absalom had constructed a monument not to the Lord, but *himself*. "Now Absalom in his lifetime had taken and set up for himself a memorial stone, which is in the King's Valley, for he said, 'I have no son to continue my name.' So he named the memorial stone after his own name, and it is called Absalom's Monument to this day" (2 Samuel 18:18). Of course, David would have preferred a proper burial for his fallen son. This displays the level of animosity between the two sides of the civil war.

Shortly after this victory, a messenger reached David. He immediately asked if Absalom was safe. The envoy confirmed every father's worst fear; his son was dead. Absalom was the third son David had lost. The death of Absalom hit him the hardest. He broke down and his royal composure crumbled. "Then the king trembled and went up to the chamber over the gate and wept. And this is what he said as he walked: 'My son Absalom, my son, my son Absalom! If only I had died instead of you, Absalom, my son, my son!'" (2 Samuel 18:33).

This tragic spectacle sowed the seeds of uncertainty in David's soldiers. What should have been a victory celebration felt more like a funeral. Rather than being proud of defending the kingdom, David's reaction made them feel shamed and confused. "So the victory that day was turned into mourning for all the people, because the people heard it said that day, 'The king is in mourning over his son.' And the people entered the city surreptitiously that day, just as people who are humiliated surreptitiously flee in battle" (2 Samuel 19:2–3).

While tragic, David's raw emotional state didn't exactly inspire faith and confidence in his soldiers. Mixed messaging never leads anywhere well. Joab read the room and knew disaster lay just around the corner. He confronted the king with a dire warning. He told David that he was humiliating their men, after they had risked their lives on his behalf. Joab continued the scolding: "Then Joab came into the house to the king and said, 'Today you have shamed all your servants, who have saved your life today and the lives of your sons and daughters, the lives of your wives, and the lives of your concubines, by loving those who hate you, and by hating those who love you. For you have revealed today that commanders and servants are nothing to you; for I know today that if Absalom were alive and all of us were dead today, then it would be right as far as you are concerned'" (2 Samuel 19:5–6).

Fortunately, he still had a chance. Joab proposed a solution: "Now therefore arise, go out and speak kindly to your servants, for I swear by the Lord, if you do not go out, no man will stay the night with you, and this will be worse for you than all the misfortune that has happened to you from your youth until now!" (2 Samuel 19:7). This was a wise decision, for, "Fidelity is seldom found in troops disheartened by misfortunes" (Vegetius 63). Leaders must lead by example. Heavy is the head which wears the crown. David met with his troops to congratulate them on their victory.

With Absalom gone, David turned his attention to securing the throne before another contender emerged. He acted swiftly, sending word through loyal priests to lobby the elders of Judah. To restore unity, David extended an olive branch—even to former enemies.

In a bold move, he offered military command to Amasa, his nephew and former general of Absalom's forces. It was a striking gesture of reconciliation. David said to a messenger: "And say to Amasa, 'Are you not my bone and my flesh? May God do so to me, and more so, if you will not be commander of the army for me continually, in place of Joab'" (2 Samuel 19:13).

Why replace Joab? Perhaps David still resented him for killing Absalom despite explicit orders. Or maybe Joab's long record of insubordination had finally become intolerable; he had previously murdered Abner in cold blood (2 Samuel 3:27). Whatever the reason, David saw Amasa as a symbol of peace, a bridge back to his tribe.

The strategy worked. The people of Judah rallied behind David and personally escorted him home. "The king then returned and came as far as the Jordan. And the men of Judah came to Gilgal in order to go to meet the king, to escort the king across the Jordan" (2 Samuel 19:15).

A NEW THREAT

L ike most journeys, there were a few bumps along the way.
David's army ran into Shimei, the man who had hurled rocks and insults during their flight. To Shimei's horror, he realized David was the victor in the bloody war. He wasted no time begging for the new king's forgiveness. He fell to the ground as he pleaded for mercy (2 Samuel 19:20). True to form, Joab's brother Abishai, wanted to execute the man on the spot. David understood this would *not* inspire unity in the tenuous Kingdom of Israel. King David gave his word that he would not harm Shimei. The caravan continued onward towards the royal city of Jerusalem.

David smoothed out this bump in the road, but their journey back to Jerusalem was far from over. One more opportunist saw his chance to seize the throne. "Now a worthless man happened to be there whose name was Sheba, the son of Bichri, a Benjaminite; and he blew the trumpet and said, 'We have no share in David, Nor do we have an inheritance in the son of Jesse; Every man to his tents, Israel!'" (2 Samuel 20:1). The men of Judah remained loyal to David while many others sided with Sheba.

Recall that Saul hailed from Benjamin. Apparently, Sheba felt more loyalty towards the previous king than the current ruler. Political alliances continued to shift like the desert sand. David made it safely back to Jerusalem, yet there was no rest with Sheba plotting yet another rebellion against him. He gave his new commander, Amasa, his first mission. Kill the traitor before he kills all of us. "Now the king said to Amasa, 'Summon the men of Judah for me within three days, and be present here yourself.' So Amasa went to summon the men of Judah, but he was delayed longer than the set time which he had designated for him" (2 Samuel 20:4–5).

David knew he needed to take out Sheba before he evaded his grasp. Time would only put more distance between them. Additionally, more time allowed Sheba greater opportunities to recruit rebels and amass resources. King David knew better than to underestimate a threat. "And David said to Abishai, 'Now Sheba the son of Bichri will do us more harm than Absalom; take your lord's servants and pursue him, so that he does not find for himself fortified cities and escape from our sight'" (2 Samuel 20:6).

David let Joab and his brother Abishai loose. Now off of their leashes, David's attack dogs hunted down their prey resolutely. While in hot pursuit, they encountered their old rival, Amasa. He never made it to Jerusalem. David had appointed Amasa as the head of his troops. But Joab wasn't the type of person who appreciated his toes stepped on. He never trusted Amasa. Yet again, Joab disregarded David's commands as he set the trap for Amasa: "And Joab said to Amasa, 'Is it going well for you, my brother?' And Joab took hold of Amasa by the beard with his right hand to kiss him. But Amasa was not on guard against the sword that was in Joab's hand, so he struck him in the belly with it and spilled out his intestines on the ground, and did not strike him again, and he died. Then Joab and his brother Abishai pursued Sheba the son of Bichri" (2 Samuel 20:9–10).

Here, we see a striking parallel to Ehud (Judges 3). Perhaps Joab learned about deception from stories of the early judge. Much like Ehud, Joab *also* used his left hand to eviscerate his rival. Was Joab motivated by envy of this man replacing him? Or was this more pragmatic? Was this to protect the kingdom from further rebellion? Whatever the reasoning, the soldiers understood who was calling the shots. "Now one of Joab's young men stood by him and said, 'Whoever favors Joab and whoever is for David, follow Joab!'" (2 Samuel 20:11). This was just another day in the office for Joab, but many of his followers were shocked! This brazen assassination left many of them stunned! "But Amasa was wallowing in his own blood in the middle of the road. And when the man saw that all the people stood still, he removed Amasa from the road to the field and threw a garment over him when he saw that everyone who came by him stood still" (2 Samuel

20:12). Out of sight, out of mind. With his competitor eliminated, Joab led his troops in pursuit of Sheba the traitor.

Sheba made the most of his head start. With haste, he made it all the way to the Northernmost territory. He *thought* he was out of reach of the King's grasp. In his impulsiveness, he severely misjudged David's capabilities. Sheba took refuge in the city of Abel Beth Maakah. The city had a reputation for its wisdom and faithfulness (2 Samuel 20:18). Joab wasn't there to play games. He immediately began displaying a show of force. He directed his sizable army to construct ramps and equipment to besiege the city.

Previously, we discussed the dangers of besieging a city directly. The defender has a significant advantage in this scenario. However, Joab was ready to quell this rebellion before it could gain proper footing. Musashi elaborates on this important concept, "In large-scale strategy, when we see that the enemy has few men, or if he has many men but his spirit is weak and disordered, we knock the hat over his eyes, crushing him utterly. If we crush lightly, he may recover" (Musashi 28). With determination in his heart, Joab began attacking the edges of the fortified walls. Fortunately for the residents of Abel, they lived up to their wise reputation. Before any bloodshed, they pleaded with Joab for a peace deal. "And they came and besieged him in Abel Beth-maacah, and they built up an assault ramp against the city, and it stood against the outer rampart; and all the people who were with Joab were wreaking destruction in order to topple the wall. Then a wise woman called out from the city, 'Listen, listen! Please tell Joab, Come here that I may speak with you'" (2 Samuel 20:15–16).

Joab may have been vicious and cold blooded, yet he wasn't unreasonable. He was ruthless, but there was always a logic to his aggression. He wasn't a raving lunatic. Rather, he was single minded in his goals and wasted no time cutting down anyone or anything in his way. He approached the wise woman with curiosity. Imagine her terror as she looked out and witnessed a massive army trying to break down the walls. Sometimes it only takes a singular act of courage to change the course of history.

She demonstrated great bravery as she negotiated for her city. She pleaded: "I am one of those who are ready for peace and faithful in Israel. You are trying to destroy a city, even a mother in Israel. Why would you swallow up the inheritance of the Lord?" (2 Samuel 20:19). Joab clarified matters. He wasn't after the city, only one man who personally rebelled against David's throne. He told the woman that he would gladly withdraw his forces on one condition: turn over Sheba. The wise woman promptly agreed. The city of Abel respected the Lord's will and David's rule. The residents rounded up Sheba and made an example of him. "Then the woman wisely came to all the people. And they cut off the head of Sheba the son of Bichri and threw it to Joab. So he blew the trumpet, and they were dispersed from the city, each to his tent. Joab also returned to the king at Jerusalem" (2 Samuel 20:22). As quickly as it started, Sheba's rebellion ended.

Perhaps king Solomon reflected on this incident and he wrote the following proverb. "One who says to the wicked, 'You are righteous,' Peoples will curse him, nations will scold him; But for those who rebuke the wicked there will be delight, And a good blessing will come upon them" (Proverbs 24:24–25). Additionally, this mission has a parallel to the Book of Judges. Hark back to the story of the Levite and his concubine (Judges 19–20).

After learning of the disgusting abuse and murder of the Levite's concubine, Israel's leadership demanded justice. Instead of turning over the few guilty offenders, the tribe of Benjamin doubled down. It intentionally covered up those heinous crimes and made them its hill to die on. As a result, a gruesome civil war erupted which killed over 65,000 men in just a few days (Judges 20). Their stubbornness greatly backfired on the Benjaminites. It also led to countless of their Israelite brothers dying alongside them. Fortunately, Abel lived up to its reputation of wisdom and holiness. Despite Joab's fearsome reputation, there was no collateral damage. Only guilty blood spilled that day.

Sun Tzu would certainly approve of the efficiency of this operation. According to *The Art of War*, "the skillful leader subdues the enemy's troops without any fighting; he captures their cities without laying siege to them; he overthrows their kingdom without lengthy operations in the field" (Tzu 6). It's worth noting that these negotiations went so smoothly because Joab negotiated from a position of strength. Besieging a city is very risky and costly. Yet Joab demonstrated he was willing to risk it all to accomplish the mission. It is wise to seek negotiations only after you have found a point of leverage. Even if you are bluffing, you are operating from a position of *perceived* strength. Think back to young David acting like a rabid dog foaming at the mouth (1 Samuel 21:10–15). The truth was that David was utterly alone and defenseless. Yet his performance was enough to convince his foes that he wasn't worth the risk.

Joab understood one thing clearly: you don't have to wipe out an entire city. You simply have to convince its leadership that you will! This is a *much* more efficient use of time, resources, and manpower. In effect, Joab embraced the jiu jitsu mindset. He identified a weakness and leveraged it to his full advantage. This example once again highlights the distinction between harmlessness and peacefulness. Joab could only negotiate peace because he was ready to fight and conquer. In other words, "He, therefore, who desires peace, should prepare for war" (Vegetius 45).

Sheba's death brought relative peace and stability to the kingdom, for now that is. As soon as the internal threat died off, the external threats came back in full force. Unfortunately, peace is fleeting like the wind. Just as the kingdom began to settle, a lingering threat reared its ugly head. David's longtime rivals, the Philistines, once again caused trouble for the Israelites. David had learned his lesson about leading from the sidelines. He accompanied his men on the battlefield yet again.

This time, however, things were different. Several decades had passed since David's first battle with Goliath. Then, he was a spry young man with a bright future. Now, his age had finally caught up with him. He still had his fighting spirit, yet his aged body couldn't keep up with his youthful heart. "Now when the Philistines were at war with Israel again, David went down, and his servants with him; and when they fought against the Philistines, David became weary" (2 Samuel 21:15). Another Philistine giant saw David's weakened state and took his shot. Armed with a massive spear and sword, the giant pounced on the opportunity.

Fortunately, David's warriors had earned the title of Mighty Men for a reason! Joab's brother, Abishai, heroically swooped in to his king's aid. "But Abishai the son of Zeruiah helped him, and struck the Philistine and killed him. Then David's men swore to him, saying, 'You shall not go out again with us to battle, so that you do not extinguish the lamp of Israel'" (2 Samuel 21:17). This analogy demonstrates how much

David was respected in his time. In spite of his flaws, he was the guiding light of Israel. This is the power of great leadership; it illuminates pathways that many are too afraid to investigate. People aren't *really* afraid of the darkness. They are afraid of the *unknown risks* hiding in the shadows.

Good leaders act courageously in this unknown territory rife with danger. Once the leaders clear the path, it's much easier for others to follow. That's what it means to lead from the front. These battles against the Philistines carried on for some time. David's Mighty Men lived up to their reputation and continued slaying their massive foes. In fact, one of them killed the brother of Goliath! (2 Samuel 21:19). Curiously enough, one of these giants had a remarkable feature. "And there was war at Gath again, where there was a man of great stature who had six fingers on each hand and six toes on each foot, twenty-four in number; and he also had been born to the giant" (2 Samuel 21:20). This time, King David's brother followed in his footsteps to slay the gargantuan warrior (2 Samuel 21:21).

DAVID STUMBLED

This brush with death forced David to confront a painful issue; he couldn't lead Israel for much longer. This King of Israel had much soul searching to do before he appointed his successor. David had more immediate concerns. God yet again punished Israel. Some have concluded it was because so many Israelites rose up to rebel against the Lord's anointed, David. Others conclude it was because David lacked faith. When he resumed his role as King of Israel, he demanded a census. As we have examined previously, when you are fighting the Lord's battles, you don't need much. An excellent example is Gideon. The Lord instructed him to prune down his troops from thousands to only a few hundred. With their unconventional tactics, they overcame incredible odds.

Despite Joab's reckless nature, even he thought David's plan was a bad idea (2 Samuel 24:3). Maybe it was fear motivating David. Perhaps it was pride. Both of them are among Satan's favorite tools for manipulation and control. David fell victim to one of the Great Serpent's many dirty tricks. "Then Satan stood up against Israel and incited David to count Israel" (1 Chronicles 21:1).

In spite of Joab's warning, David carried on with his plan. The census numbers revealed something incredible! "And Joab gave the number of the census of the people to the king: in Israel there were eight hundred thousand valiant men who drew the sword, and the men of Judah were five hundred thousand men (2 Samuel 24:9). David had a whopping 1.3 *million* troops fighting for the Kingdom of Israel! The Book of Chronicles denotes a slightly different number, 1.1 million (1 Chronicles 21:5). In all likelihood, the number was even higher. This discrepancy is likely because Joab failed to include Levi and Benjamin in the census (1 Chronicles 21:6).

We will never know the exact number of fighters. What we *do* know is that Israel had greatly increased in size, just as the Lord promised. For context, when Joshua led Israel into the Promised Land, there were approximately 600,000 fighting-age males (Numbers 26:51). In spite of the near-constant warfare and countless warriors falling in battle, Israel's army doubled in size! Once David realized how blessed they were, his heart sank like a stone. He had let fear get the better of him, and now he felt like a fool. One cannot honor fear and faith at the same time. They are mutually exclusive properties. This is likely why Satan uses fear so often. Upon learning the truth, David regretted his decision. "Now David's heart troubled him after he had counted the people. So David said to the Lord, 'I have sinned greatly in what I have done. But now, Lord, please overlook the guilt of Your servant, for I have acted very foolishly'" (2 Samuel 24:10).

The Lord sent David word through the prophet, Gad. The theocratic King of Israel knew he needed to atone for this faithless behavior. The Lord gave David three terrible choices to choose from as He spoke through the prophet. "So Gad came to David and said to him, 'This is what the Lord says: "Take for yourself three years of famine, or three months to be swept away before your foes while the sword of your enemies overtakes you, or else three days of the sword of the Lord: a plague in the land, and the angel of the Lord destroying throughout the territory of Israel." Now, therefore, consider what answer I shall bring back to Him who sent me'" (1 Chronicles 21:11–12).

While none of these choices are appealing, David's reasoning was sound. He knew the Lord's love and grace was greater than that of his enemies. David gave his answer, "I am in great distress; please let me fall into the hand of the Lord, for His mercies are very great. But do not let me fall into human hands" (1 Chronicles 21:13). When you have nothing but bad options to choose from, it's usually best to get through it hastily. Sometimes the only way out is through the pain.

The plague was vicious and brutal, reminiscent of the horrors God sent the Pharaoh of Egypt. The root cause was the same: humanity's hubris. Sometimes we need a humbling experience. The Sovereign King of the Universe has an infinite number of ways to do that. "So the Lord sent a plague upon Israel from the morning until the appointed time, and seventy thousand men of the people from Dan to Beersheba died" (2 Samuel 24:15).

The angel of death swept across Israel, far and wide. The wave of death and destruction rampaged towards Jerusalem with masses of bodies in its wake. Just before it reached David, something incredible happened. "When the angel extended his hand toward Jerusalem to destroy it, the Lord relented of the disaster and said to the angel who destroyed the people, 'It is enough! Now drop your hand!' And the angel of the Lord was by the threshing floor of Ornan the Jebusite" (2 Samuel 24:16). David saw the angel heading towards him. He was completely unaware that the Lord had ordered the angel to cease fire. The King of Israel looked up to an astonishing sight. "Then David raised his eyes and saw the angel of the Lord standing between earth and heaven, with his drawn sword in his hand stretched out over Jerusalem. Then David and the elders, covered with sackcloth, fell on their faces" (1 Chronicles 21:16).

A note to all pacifist Christians: *even angels have swords*, and they know how to use them! David begged and pleaded to the Lord. He took ownership of the problems and lobbied for the sake of his people. He could have blamed a million things, but he didn't. True leaders know the buck stops with them. "Then David spoke to the Lord when he saw the angel who was striking down the people, and said, 'Behold, it is I who have sinned, and it is I who have done wrong; but these sheep, what have they done? Please let Your hand be against me and against my father's house!'" (2 Samuel 24:17). This analogy of shepherd and sheep is yet another precursor to Christ. Both were willing to die to protect their people.

Scripture makes it abundantly clear that God knows us better than we know ourselves. The prophet Jeremiah elaborated on this important concept. "I, the Lord, search the heart, I test the mind, To give to each person according to his ways, According to the results of his deeds" (Jeremiah 17:10). We can come up with all kinds of justifications on the fly. It's less painful to rationalize bad behavior than to confront our misdeeds. Our inner defense lawyers love a good challenge! Just like Adam and Eve, we can attempt to lie to the Lord. In reality, we are only lying to ourselves. Most people will accept a comforting delusion over a painful truth. But not David! The Lord read his heart and knew his repentance was genuine. In His infinite grace, He relented. The significance of this wasn't lost on David; he knew this was a holy site. The Spirit spoke through the prophet Gad. "So Gad came to David that day and said to him, 'Go up, erect an altar to the Lord on the threshing floor of Ornan the Jebusite'" (2 Samuel 24:18). Ever faithful, David searched for the owner of the establishment.

The owner's name was Ornan. Imagine his surprise! To him, it was just another day in the office. He was performing his daily work with his sons as usual. With no warning, they witnessed a spectacle so great it was nearly incomprehensible. They hid in terror (1 Chronicles 21:20). Angels are so magnificent it's difficult for the human mind to comprehend their presence. This is a consistent pattern throughout the Old Testament and the New Testament. Encounters with angels often coincide with moments of divine mercy or judgment, turning points in the Biblical story.

A particularly noteworthy example is when an angel told Mary she would birth the Messiah: "And coming in, he said to her, 'Greetings, favored one! The Lord is with you.' But she was very perplexed at this statement, and was pondering what kind of greeting this was. And the angel said to her, 'Do not be afraid, Mary, for you have found favor with God'" (Luke 1:28–30). Ornan was also greatly troubled at this divine encounter. His reaction mirrored those of many who stood in the presence of divine messengers.

More than just spectacle, these moments often foreshadow deeper truths. The threshing floor, once a site of judgment and mercy, would later become the foundation for Solomon's temple. Here, David interceded to stop the plague, just as Christ would one day intercede to remove the penalty of sin. Both stood between the people and God's wrath, offering sacrifice in their stead.

David found the awestruck man, and he dropped and bowed at the king's presence. He knew without a doubt that the Lord was with King David. Ornan was committed to aiding the theocratic king in any way possible.

David got straight to business. "Then David said to Ornan, 'Give me the site of this threshing floor, so that I may build on it an altar to the Lord; you shall give it to me for the full price, so that the plague may be brought to a halt from the people'" (1 Chronicles 21:22). Ornan was still scared to death! It's hard to care about money when you're staring death in the face. He told David to take everything and use it as he saw fit—from the land itself to the livestock and wheat that came with it. Ornan was happy to trade it all to spare his life and protect his family. Fortunately, that wasn't necessary. David had the earthly authority to accept. But it wasn't about his rights as king, it was about honoring the Lord.

He made his reasoning crystal clear. "No, but I will certainly buy it for the full price; for I will not take what is yours for the Lord, nor offer a burnt offering which costs me nothing" (1 Chronicles 21:24). David paid six hundred gold shekels for the site. That's approximately 15 pounds of pure gold, an extravagant price, but a fitting one for atonement. He understood a profound truth: true sacrifice must be personal. If it doesn't cost us anything, it doesn't mean anything.

This example is yet another precursor to the ultimate sacrifice made by Christ. Here, paying the full price holds tremendous significance. Jesus used very similar language moments before He died on the cross. One of the final things He said was a simple, "It is finished" (John 19:30). The Greek word is τετέλεσται (tetelestai). In the ancient world, it was a commercial term used to indicate that a debt was completely paid.

But this was no financial transaction. It was Christ taking on the full burden of humanity's sin—past, present, and future. Where David paid in gold, Jesus paid in blood. The only thing we must do is accept Him as Lord and Savior—a far lighter burden than the cross He bore. Critics often ask, "Why would God send His Son to die for us?" The answer lies in the principle David already understood: *a sacrifice that costs nothing means nothing*. The Gospel of John drives this point home. "For God so loved the world, that He gave His only Son, so that everyone who believes in Him will not perish, but have eternal life. For God did not send the Son into the world to judge the world, but so that the world might be saved through Him" (John 3:16–17).

Even the way David's warriors once described him as the "lamp of Israel" (2 Samuel 21:17) foreshadows the coming Messiah. John described Jesus in similar terms. "And this is the judgment, that the Light has come into the world, and people loved the darkness rather than the Light; for their deeds were evil. For everyone who does evil hates the Light, and does not come to the Light, so that his deeds will not be exposed. But the one who practices the truth comes to the Light, so that his deeds will be revealed as having been performed in God" (John 3:19–21).

With Ornan now on board, King David wasted no time honoring God's grace and mercy. After "paying in full," work immediately began on the altar. He knew there was something sacred about this ground. "Then David built an altar there to the Lord, and offered burnt offerings and peace offerings. And he called to the Lord, and He answered him with fire from heaven on the altar of burnt offering" (1 Chronicles 21:26). The fire from heaven was a divine signature, God's seal of acceptance and forgiveness. Peace returned to the land. "The Lord commanded the angel, and he returned his sword to its sheath" (1 Chronicles 21:27).

PASSING THE TORCH

Through divine revelation, the Lord told David that this site would host the future temple. Though David longed to build it himself, the Lord had chosen Solomon for this sacred task. At the time, Solomon was still young and inexperienced, so David explained the divine mission entrusted to him: "But the word of the Lord came to me, saying, 'You have shed much blood and have waged great wars; you shall not build a house to My name, because you have shed so much blood on the earth before Me. Behold, a son will be born to you, who shall be a man of rest, and I will give him rest from all his enemies on every side; for his name will be Solomon, and I will give peace and quiet to Israel in his days. He shall build a house for My name, and he shall be My son and I will be his Father; and I will establish the throne of his kingdom over Israel forever'" (1 Chronicles 22:8–10).

This marked a turning point in Israel's history. The conquest of the Promised Land had required grizzled warriors like Joshua, the judges, and David; men hardened by war and sacrifice. Taking land is always harder than holding it. But now the time for swords was giving way to a time for stability. Solomon's reign would be defined by consolidation, peace, and prosperity. This peace afforded Solomon the freedom to pen much Scripture we treasure today. He wrote most of Proverbs, Ecclesiastes, and Song of Songs—books still studied for their insights into life, morality, and the nature of God. His kingdom, while less bloodied, would shine in other ways.

Though David would not build the temple, he poured himself into the preparation. The scope of his logistical effort was astonishing. "Now behold, with great pains I have prepared for the house of the Lord a hundred thousand talents of gold and a million talents of silver, and bronze and iron beyond measure, for they are in great quantity; I have also prepared timber and stone, and you may add to that" (1 Chronicles 22:14). This represents thousands of tons of precious metal; an offering of staggering value.

But the goal wasn't just scale, it was sacred excellence. The temple would be the finest creation of its kind. Unified under David's leadership, all of Israel was mobilized to assist Solomon. In this, we see one of the monarchy's enduring strengths: complete unity and focus. David led not for his own glory, but to bless future generations. There is an old adage: *A society grows great when old men plant trees whose shade they know they shall never sit in.* David embodied this timeless truism.

Unfortunately, David would not live to see the temple's completion. By now, he had ruled for decades. He knew time was short, but he used his remaining strength to lay the foundation—spiritually, politically, and materially. The warrior king was preparing the way for a peaceful kingdom—just as Christ, the greater Son of David, would one day prepare the way for a kingdom not built by human hands.

At this point, David was very old, and his health was failing. Although he had previously declared Solomon as his successor, he had not yet formally passed the crown. Sensing an opportunity, Adonijah, David's eldest surviving son, made his move. Whether motivated by entitlement or political ambition, he echoed Absalom as he seized the moment. "Now Adonijah the son of Haggith exalted himself, saying, 'I will be king.' So he prepared for himself chariots and horsemen, with fifty men to run before him" (1 Kings 1:5).

Once again, David's failure to discipline his household bore bitter fruit. There is a time for grace and a time for correction. Overlooking wrongdoing may feel compassionate in the short term, but it often fosters rebellion or resentment in the long run. Indeed, "Foolishness is bound up in the heart of a child; The rod of discipline will remove it far from him" (Proverbs 22:15).

Adonijah swiftly gathered allies behind the king's back. His scheming rivaled that of Absalom and would have made Machiavelli proud. Among the defectors was Joab, David's longtime commander. Like Adonijah, Joab had rarely been held accountable. He had a pattern of defiance, yet David had consistently let it slide. That leniency cost him. By refusing to confront disloyalty, David invited it.

Abiathar the priest also joined the conspiracy. Adonijah hosted a feast to celebrate his premature reign, deliberately excluding Solomon and David's loyal supporters (1 Kings 1:9–10). Perhaps Solomon had this episode in mind when he later wrote, "He will die for lack of instruction, And in the greatness of his foolishness he will go astray" (Proverbs 5:23).

Word of the rebellion reached David's trusted circle. The prophet Nathan, alarmed by the threat to God's promise and the nation's stability, acted quickly. He approached Bathsheba, Solomon's mother, urging her to intervene. Her maternal instincts ignited! If Adonijah took the throne, he would surely take out his rival, Solomon. This episode also exposes one of the darker consequences of polygamy: divided loyalties and deadly rivalries among half-brothers. As with Gideon's sons, when inheritance and power are at stake, blood is often spilled.

Bathsheba reminded David of his solemn oath. "So she said to him, 'My lord, you yourself swore to your servant by the Lord your God, saying, "Your son Solomon certainly shall be king after me, and he shall sit on my throne"'" (1 Kings 1:17–18). Nathan then entered and reinforced her plea. With wisdom and tact, he gave David an opportunity to act without losing face. Nathan asked, "Then Nathan said, 'My lord the king, have you yourself said, "Adonijah shall be king after me, and he shall sit on my throne?"'" (1 Kings 1:24).

Nathan concluded by drawing a line in the sand. "But me, even me your servant, Zadok the priest, Benaiah the son of Jehoiada, and your servant Solomon, he has not invited. Has this thing been done by my lord the king, and you have not let your servants know who shall sit on the throne of my lord the king after him?" (1 Kings 1:26–27).

They were at a fork in the road. David knew he needed to handle this matter with resolution. There were two brothers, but there could only be one King. David summoned Bathsheba and his other trusted allies. Adonijah operated in the shadows. David understood that sunlight is the best disinfectant. He needed all of Israel to know who his successor was. David ordered a trusted priest and prophet to anoint Solomon publicly, officially making him King of Israel (1 Kings 1:33–35).

This crisis left no room for personal pride or arrogance. Some political figures refuse ever to transfer power. Like shipwrecked sailors, they stubbornly cling on to what they can until the bitter end. Sun Tzu would certainly applaud King David's selflessness. According to *The Art of War*, "The general who advances without coveting fame and retreats without fearing disgrace, whose only thought is to protect his country and do good service for his sovereign, is the jewel of the kingdom" (Tzu 21). It was finally time for his son to rise to the occasion. No one wanted an ending like Sheba, with his decapitated head thrown out like garbage.

David's loyal followers rallied around him. They were eager to execute the King's final order. "And Zadok the priest then took the horn of oil from the tent and anointed Solomon. Then they blew the trumpet, and all the people said, 'Long live King Solomon!' And all the people went up after him, and the people were playing on flutes and rejoicing with great joy, so that the earth shook at their noise" (1 Kings 1:39–40). Israel rejoiced at their new King! Most of Israel, that is.

Word quickly spread to Adonijah's band of traitors. Realizing the futility of their situation, they fled like roaches exposed to light. "Then all the guests of Adonijah trembled and got up, and each went on his way" (1 Kings 1:49). There's an important lesson here. Fair-weather friends will stick with you just long enough to get you in trouble! The moment they see the storm clouds on this horizon they suddenly vanish, leaving you in the dust. If your so-called friends are willing to betray other people for you, they will betray you shortly after.

Adonijah realized he was utterly alone and vulnerable. His followers fled in desperation for their own skins. Moreover, he had directly challenged one of the most accomplished warriors in the Bible. David was too old to fight. Fortunately, his hand-picked warriors surrounded him. David's Mighty Men were well seasoned from countless battles. They had slain giants; this wannabe rebel posed no real challenge.

Realizing he had burned all his bridges, Adonijah appealed to the Lord's mercy. "Adonijah also was afraid of Solomon, and he got up, and went, and took hold of the horns of the altar" (1 Kings 1:50). This practice goes back to the time of Moses (Exodus 21:12–14). Officials told Solomon, "Behold, Adonijah is afraid of King Solomon, for behold, he has taken hold of the horns of the altar, saying, 'May King Solomon swear to me today that he will not put his servant to death with the sword'" (1 Kings 1:51). Adonijah hadn't resorted to violence yet; that likely saved him. Solomon didn't promise protection, rather, he gave his half-brother an opportunity to repent. "And Solomon said, 'If he is a worthy man, not one of his hairs will fall to the ground; but if wickedness is found in him, he will die'" (1 Kings 1:52).

There was a new sheriff in town! David let Adonijah's behavior slide, while Solomon delivered even-handed justice. "So King Solomon sent men, and they brought him down from the altar. And he came and prostrated himself before King Solomon, and Solomon said to him, 'Go to your house'" (1 Kings 1:53). Solomon could only choose peace because he negotiated from a position of strength.

The throne of Israel was now undisputed, or so it appeared. Knowing his time was limited, David gave King Solomon important advice: "I am going the way of all the earth. So be strong, and prove yourself a man" (1 Kings 2:2). David once again reiterated the importance of following God's Law. This simple rule was the key to the Kingdom of Israel. He again told Solomon of the Lord's promise. "Do your duty to the Lord your God, to walk in His ways, to keep His statutes, His commandments, His ordinances, and His testimonies, according to what is written in the Law of Moses, so that you may succeed in all that you do and wherever you turn, so that the Lord may fulfill His promise which He spoke regarding me, saying, 'If your sons are careful about their way, to walk before Me in truth with all their heart and all their soul, you shall not be deprived of a man to occupy the throne of Israel'" (1 Kings 2:3–4).

Aside from the spiritual lessons, there was tactical advice as well. He warned Solomon of the threat from within. He mentioned how Joab had now betrayed him three times. David knew the Kingdom could not flourish with such an insidious threat. Joab wasn't the kind of loose end you leave untied. David made this abundantly clear as he told Solomon the following: "So act as your wisdom dictates, and do not let his gray hair go down to Sheol in peace" (1 Kings 2:6).

Shortly after his advice to Solomon, David uttered his final proclamation. He began by praising the glory of God. "The Spirit of the Lord spoke through me, and His word was on my tongue. The God of Israel said it; the Rock of Israel spoke to me: 'He who rules over mankind righteously, who rules in the fear of God, is like the light of the morning when the sun rises, a morning without clouds, when the fresh grass springs out of the earth from sunshine after rain.' Is my house not indeed so with God? For He has made an everlasting covenant with me, properly ordered in all things, and secured; for will He not indeed make all my salvation and all my delight grow?" (2 Samuel 23:2–5).

David was a God-fearing man, unlike his predecessor, Saul, who only glorified himself. Moreover, David closed his speech calling for justice against evildoers. He yet again highlighted the importance of strong and capable men. "But the worthless, every one of them, are like scattered thorns, because they cannot be taken in hand; instead, the man who touches them must be armed with iron and the shaft of a spear, and they will be completely burned with fire in their place" (2 Samuel 23:6–7). Harmless men can't fight back against the wicked. Only strong and peaceful men are capable of such feats. And thus, David went the way of the world. His family buried him in Jerusalem, also called the *City of David*. "Now the days that David reigned over Israel were forty years: in Hebron he reigned for seven years, and in Jerusalem he reigned for thirty-three years. Then Solomon sat on the throne of his father David, and his kingdom was firmly established" (1 Kings 2:11–12).

ONE LAST SHOT

With his father gone, Adonijah made one last underhanded attempt to seize power. Accustomed to leniency, he failed to learn from Solomon's mercy. Some men only learn the hard way. He made a critical error. In the words of Sun Tzu, "He who exercises no forethought but makes light of his opponents is sure to be captured by them" (Tzu 19). That perfectly described Adonijah. Never underestimate your adversary.

This time, he used Solomon's mother, Bathsheba, to press his case. Adonijah requested to marry Abishag, one of King David's concubines. At the time, claiming a former king's concubine was viewed as a symbolic act of taking the crown. It is telling that he went through Bathsheba instead of asking Solomon directly—likely a way to test the young king's resolve.

Solomon, widely considered one of the wisest minds in the Bible, immediately saw through this maneuver. He knew other defectors were still lurking in the shadows. Solomon said to his mother, "And why are you requesting Abishag the Shunammite for Adonijah? Request for him the kingdom as well—since he is my older brother—for him, for Abiathar the priest, and for Joab the son of Zeruiah!" (1 Kings 2:22).

This was a point of no return. Adonijah repaid Solomon's mercy with treachery. The king had extended an opportunity for redemption; Adonijah responded with a veiled coup attempt. "Then King Solomon sent the order by Benaiah the son of Jehoiada; and he struck him so that he died" (1 Kings 2:25). Given that Benaiah had previously slain both lions and giants, this was a simple task.

The Art of War affirms the need for decisive action: "If, however, you are indulgent, but unable to make your authority felt; kind-hearted, but unable to enforce your commands ... then your soldiers must be likened to spoilt children; they are useless for any practical purpose" (Tzu 21). David had indeed spoiled several of his children. Now, Solomon was left to clean up the mess. It was time to set a new tone for the Kingdom of Israel.

But the treachery ran deeper than Adonijah alone. Solomon next dealt with Abiathar, the turncoat priest and descendant of Eli. Though Solomon had justification to execute him, he chose mercy, remembering Abiathar had once carried the Ark of the Lord. "So Solomon dismissed Abiathar from being priest to the Lord, to fulfill the word of the Lord, which He had spoken regarding the house of Eli in Shiloh" (1 Kings 2:27). The Lord's prophecies are never wrong, even if we do not understand them at the moment.

One final matter remained: a reckoning with Joab, arguably the most dangerous man in Israel. The disgraced commander realized Solomon was purging the threats to the throne. In desperation, he ran to claim asylum. "Now the news came to Joab because Joab had followed Adonijah, though he had not followed Absalom. So Joab fled to the tent of the Lord and took hold of the horns of the altar" (1 Kings 2:28). But asylum was reserved for those guilty of accidental killing (Exodus 21:14). Joab's murders were premeditated and political. His record of defiance and bloodshed had finally caught up to him. Joab refused to leave the tent, forcing Benaiah's hand.

Solomon issued the order: "Do just as he has spoken, and execute him and bury him, so that you may remove from me and from my father's house the blood which Joab shed without justification" (1 Kings 2:31). Solomon spoke the truth. Joab had murdered two Israelite commanders in cold blood, despite David's orders. Solomon added, "So their blood shall return on the head of Joab and on the head of his descendants forever; but for David and his descendants, and his house and his throne, may there be peace from the Lord forever" (1 Kings 2:33).

Ever loyal, Benaiah struck Joab down. The king rewarded him accordingly. "And the king appointed Benaiah the son of Jehoiada over the army in his place, and the king appointed Zadok the priest in place of Abiathar" (1 Kings 2:35). With the mutineers gone and order restored, King Solomon ushered in a new era of strength, justice, and prosperity for Israel.

DAVID AFTER SAUL
CONCLUSION

D avid's life was nothing short of tumultuous! His journey was a veritable whirlwind of ups and downs. Everything from the highest heights to tragedy and heartache. Much of his story is bittersweet. Despite Saul's years of persecution towards him, David still lamented Saul's death. Despite all the hardships and treachery, David still honored the position of the Lord's anointed. When an opportunist claimed to have killed Saul, David had him executed for his wanton disgrace. David's troubles during Saul's reign were quite overt, from the towering threat of Goliath to Saul hurtling spears at him. While quite lethal, at least these threats were easy to identify, and thus, combat. However, the threats David faced after Saul were far more insidious. Just like the proverbial snake in the grass, David was often unaware of the lurking danger that plagued his reign.

Power abhors a vacuum. After Saul's death, a brutal civil war erupted between the tribes of Judah (David) and Benjamin (Saul). To their credit, the rival commanders, Joab and Abner, attempted resolution via representative combat. Unfortunately, this ended disastrously. All the young men slaughtered each other in a brutal stalemate. This catastrophe highlights the need for asymmetrical advantages in combat. The civil war raged on for some time and countless Israelite soldiers fell in battle. Eventually, David brokered a peace deal with the tribe of Benjamin. David had the advantage by far. He could have wiped out their entire tribe, yet he chose mercy for the benefit of Israel as a whole. There is a reason Jesus said, "Every kingdom divided against itself is laid waste; and no city or house divided against itself will stand" (Matthew 12:25). David inspired unity within the kingdom.

David wasted no time as the undisputed King of Israel. Though it was thought impossible by naysayers, his first mission was to conquer the fortress city of Jerusalem. Fortunately, David had a track record of overcoming seemingly impossible feats! The conquest of Jerusalem offers several practical insights. First, the Jebusites made the timeless error of underestimating their opponents. Their taunts and mockery only fueled the Israelites' motivation. Additionally, we saw the importance of out-of-the-box strategic thinking. The Israelites fought asymmetrically as they deftly maneuvered the secret tunnel to enter the fortified city. By embracing the jiu jitsu mindset, they strategically leveraged a weak point and exploited it against their enemies.

After taking the city, David led by example in the worship ceremonies. When King David ordered the Ark brought into the city, he danced and celebrated with his people. This was in spite of his wife's protestations. David held the Ark in great reverence. When David prayed over how to treat the Ark, God told him something even greater. The Lord gave David an everlasting dynasty! Today we refer to this as the *Davidic Covenant*. The Lord promised that David's throne would endure through the ages (2 Samuel 7:16). This promise was fulfilled by Jesus Christ.

After this covenant, David had overwhelming success with his enemies in the land. Ironically, this is where his life became immensely more complicated. David was a natural fighter. He spent his youth protecting his flock from lions and bears! Yet his role as King of Israel offered far more insidious challenges. The saying, *heavy is the head that wears the crown,* applies perfectly. It's often much harder to stay at the top than to rise to the top. With David's long list of defectors and traitors, it's easy to see why.

A central theme of David's epic life is the fallibility of humankind. Cherished by his people and feared by his enemies, David's reputation was larger than life. Indeed, this is a very rare combination. Machiavelli elaborates on this balance in *The Prince*. "Upon this a question arises: whether it be better to be loved than feared or feared than loved? It may be answered that one should wish to be both, but, because it is difficult to unite them in one person, it is much safer to be feared than loved, when, of the two, either must be dispensed with" (Machiavelli 86). The love and admiration of his people combined with the terror of his enemies made David into a truly one-of-a-kind figure. He was one of the greatest warrior kings to ever live.

Yet even such a man has his flaws. David became too comfortable. He let Joab do all his fighting for him as he lounged around his royal palace. David's boredom led to his affair with Bathsheba. This unfortunate incident highlights the slippery nature of sin. It started with David accidentally seeing her naked as she bathed. He should have stopped and turned away. Yet he kept pushing the envelope. He tried to negotiate with temptation, which never ends well. It's a bit like wrestling with a pig. The swine drags you down to its level and beats you with experience! After that, you wind up dirty and humiliated. Meanwhile, the filthy pig loves every second of it! The only way to beat it at its game is to not play in the first place.

David's affair serves as a cautionary tale. A key takeaway is the danger of hero worship. We *all* fall short of the glory of God. David pulled off numerous seemingly impossible feats, but even the best amongst will inevitably stumble. Placing faith in heroes or celebrities is a fool's errand. This shameful encounter highlights the need for Jesus Christ, the only man without sin. HE is the only one worthy of worship.

Moreover, David's affair underscores the danger of unintended consequences. The prophet Nathan showed immense moral courage as he confronted David. Countless prophets have been executed for calling out royal transgressions. Nathan's parable of the greedy shepherd forced David to grapple with his misdeeds. Parables are effective in large part because they can bypass individuals' egos. The defense attorney in our minds never gets the chance to argue back. With the ego to the side, things often become much clearer. There is a reason Jesus used parables so frequently.

David repented to the Lord and HE cleared away his sin. Yet this was not without cost. David paid for his sin four times over, just as he responded to Nathan's parable. It's worth reiterating, genuine repentance requires action, *not* empty words. The Lord forgave him, but people are much harder to convince! After the affair, David lost respect from much of his family and his peers. This greatly contributed to the rebellions against him. Fortunately, there was still hope for the disgraced king.

We can learn a lesson from Absalom's rebellion. The rift in the father–son relationship began with David's weak response to Amnon's heinous crimes. This likely stemmed from David's lack of self-respect after his affair. It's hard to talk the talk after you fail to walk the walk. Perhaps King Solomon reflected on this incident when he wrote, "To show partiality to the wicked is not good, Nor to suppress the righteous in judgment" (Proverbs 18:5).

The fallout from Amnon's crimes highlights the need for even-handed justice. Leaders must enforce good order and discipline. If not, people take matters into their own hands. Absalom's vigilante justice is the perfect example. David's lax treatment of Amnon led to Absalom losing all respect for him. He didn't stop with just his revenge killing. He tried to usurp the Throne of Israel itself. Even though Absalom chased David from Jerusalem, David still had the upper hand. This is where his years of experience shone through. David recruited Hushai to deploy a PSYOP against the rebellious regime. This mission directly led to an overwhelming victory and the end of Absalom's rebellion.

David's earthly accomplishments are truly in a league of their own! Yet these pale in comparison to David's greatest feat, paving the way for the Messiah, Jesus Christ (Matthew 1). There are countless parallels between David and Jesus. One of the most beautiful is when David paid the full price in his sacrifice to the Lord (1 Chronicles 21:24). Despite all the trials and tribulations, David still honored the Lord above all else. Critics may focus on David's flaws or shortcomings. Yet these shortcomings prove why we need Jesus in the first place. We can't save ourselves; only the Lord can save us. We are living in a fallen kingdom (Genesis 3). People will make mistakes, but God doesn't. Before you throw stones towards David, remember this: without David, we wouldn't have Jesus! In God's infinite wisdom, He can redirect human sin and folly towards righteous ends (1 Corinthians 1:27–31).

PART SEVEN:
NEHEMIAH—THE REBUILDER

NEHEMIAH INTRODUCTION

Nehemiah was a man of virtue, honor, and courage. He sorted through the rubble of Jerusalem to build a better future for his people. Before we look at Nehemiah's contributions, we must first examine *why* he was rebuilding Jerusalem in the first place. King Solomon did many great things for the Kingdom, yet he had his flaws like anyone else. Solomon began his reign humbly. He knew he had big shoes to fill. Sons are almost always measured by the success of their fathers. How could the new king live up to such epic feats as David?

One night, the Lord showed himself to the new king and told him to ask for whatever he wanted. Solomon poured out his heart in gratitude. He knew the Lord was ultimately the reason for his father's successful leadership of Israel. With a humble heart, Solomon made a request to the Lord: "And now, Lord my God, You have made Your servant king in place of my father David, yet I am like a little boy; I do not know how to go out or come in. And Your servant is in the midst of Your people whom You have chosen, a great people who are too many to be numbered or counted. So give Your servant an understanding heart to judge Your people, to discern between good and evil. For who is capable of judging this great people of Yours?" (1 Kings 3:7–9).

The Lord was pleased at Solomon's selfless request: "Because you have asked this thing, and have not asked for yourself a long life, nor have asked riches for yourself, nor have you asked for the lives of your enemies, but have asked for yourself discernment to understand justice, behold, I have done according to your words. Behold, I have given you a wise and discerning heart, so that there has been no one like you before you, nor shall one like you arise after you'" (1 Kings 3:11–12).

Wisdom and discernment are gifts beyond measure. Yet the Lord had something else in store for the up and coming King. The Lord rewarded Solomon's selflessness yet again. HE said: "I have also given you what you have not asked, both riches and honor, so that there will not be any among the kings like you all your days" (1 Kings 3:13).

There's an important lesson here. There are no shortcuts to either prosperity or honor. One must earn them. Ironically, the quicker one tries to obtain them the quicker they crash and burn. The reason is simple: *both greed and pride are self-destructive emotions*. People with those traits often trade short term wins for long term losses. Greediness leads to ostracization. Pridefulness leads to humiliation and embarrassment. Yet, when one focuses on wisdom, one can navigate these tricky waters steadfastly. As the old adage goes: *Give a man a fish, and you feed him for a day. Teach a man to fish, and you feed him for a lifetime.* Solomon didn't ask the Lord for a handout or payday. He prayed for tools to provide for the Kingdom of Israel. Yet there was a catch! The Lord gave Solomon one simple caveat: obedience to the Lord (1 Kings 3:14).

King Solomon did magnificent things for the Kingdom of Israel. Yet everyone has their blind spots. Much like David, Solomon's blind spot was lust. Solomon possessed incredible wisdom and foresight, but lust has a way of blinding our faculties of reason. Solomon was no exception to this. Early into his reign, he made a political alliance with the Pharaoh of Egypt. Solomon married the Pharaoh's daughter as a result (1 Kings 3:1). This marked the beginning of Solomon's rift with God's Law. Moses had warned about this generations earlier. "Furthermore, you shall not intermarry with them: you shall not give your daughters to their sons, nor shall you take their daughters for your sons. 'For they will turn your sons away from following Me to serve other gods; then the anger of the Lord will be kindled against you, and He will destroy you quickly'" (Deuteronomy 7:3–4).

Meanwhile, Solomon continued building the Lord's temple in Jerusalem. After many years the temple and royal palace were finished. Solomon dedicated the holy site to the Lord with many sacrifices and festivities. Shortly after, the Lord again appeared to Solomon. The Lord reminded him to live faithfully and to obey His commands, just like David. So long as Solomon followed the Lord's laws, he would have the

Lord's blessing. But God gave him a stark warning that echoes the Book of Judges: "But if you or your sons at all turn from following Me, and do not keep My commandments and My statutes which I have set before you, and go and serve other gods and worship them, then I will cut off Israel from the land which I have given them; and this house which I have sanctified for My name I will cast out of My sight, and Israel will be a proverb and a byword among all peoples" (1 Kings 9:6–7).

Solomon continued his work for the Kingdom. He aged and his willpower diminished as the influence of his foreign wives increased. "Now King Solomon loved many foreign women along with the daughter of Pharaoh: Moabite, Ammonite, Edomite, Sidonian, and Hittite women" (1 Kings 11:1). Moses and the Lord had given the Israelites ample warning against marrying outside the faith. Solomon's example was particularly extreme. " He had seven hundred wives, who were princesses, and three hundred concubines; and his wives turned his heart away. For when Solomon was old, his wives turned his heart away after other gods; and his heart was not wholly devoted to the Lord his God, as the heart of David his father had been" (1 Kings 11:3–4).

This heresy greatly angered the Lord. There is a reason the First Commandment says, "You shall have no other gods before Me" (Exodus 20:3). Much like in the cycle of judges, Solomon participated in blasphemy and idolatry. Once again, this all stemmed from the pridefulness of the heart. Solomon *thought* he could pull it off. However, it's impossible to follow both the Lord and false idols. Ironically, Solomon proposed the idea of splitting the baby in two early into his career (1 Kings 3:16–18). This was a clever test to determine honesty and integrity. The lesson of that story was that half-measures lead to poor outcomes for all. In this same way, Solomon attempted to split the baby of these two competing faiths. "Then Solomon built a high place for Chemosh, the abhorrent idol of Moab, on the mountain that is east of Jerusalem, and for Molech, the abhorrent idol of the sons of Ammon. And he did the same for all his foreign wives, who burned incense and sacrificed to their gods" (1 Kings 11:7–8). Yet again, Israel embraced the ways of the world over the ways of God.

The Lord was furious at Solomon. In their arrogance, the Israelites *thought* a King would solve their problems, despite the warnings of Samuel (1 Samuel 8). However, they still had the same problems with idolatry as their predecessors. In the words of Solomon himself, "What has been, it is what will be, and what has been done, it is what will be done. So there is nothing new under the sun" (Ecclesiastes 1:9). All people are fallible; we are still tarnished by the shortcomings of Adam and Eve. That's why Jesus Christ is the only true King, He was the only man without sin. Due to our imperfect nature, we cannot reach salvation on our own. Attempts to do so result in disaster.

God condemned Solomon's actions. This also serves as foreshadowing to Nehemiah. "So the Lord said to Solomon, 'Since you have done this, and you have not kept My covenant and My statutes, which I have commanded you, I will certainly tear the kingdom away from you and will give it to your servant. However, I will not do it in your days, only for the sake of your father David; but I will tear it away from the hand of your son. Nevertheless I will not tear away the whole kingdom, but I will give one tribe to your son for the sake of My servant David and for the sake of Jerusalem, which I have chosen'" (1 Kings 11:11–13). None of the above is meant to denigrate the legitimate contributions from Solomon. Rather, it serves as a cautionary tale. If the wisest man in the Bible can fall victim to his basest desires, *so can you!*

Many generations passed from the time of the judges to the time of Solomon. Yet the familiar pattern arose. Recall the cycle of Judges: apostasy, oppression, crisis, and deliverance. It all started with Israel turning its back to the Lord: "So the Lord's anger burned against Israel, and He gave them over to plunderers who plundered them; and He sold them into the hands of their enemies all around, and they were unable to stand before their enemies. While they were in distress, they cried to the Lord, but He would not answer them, because they had forsaken the Lord and had served the Baals and the Ashtaroth" (Judges 2:13–14).

King Solomon committed apostasy by serving false gods. This meant that the next part of the cycle was imminent: *oppression.* Solomon reigned for forty years. After his death, his son Rehoboam ascended to the Throne of Israel. Where Solomon ruled with wisdom and foresight, his son led with an iron fist. King Rehoboam's authoritarianism and hubris led to strife within the kingdom. He ignored the advice of his elders and ruled as a tyrant. When the Israelites begged for mercy he responded callously: "And the king answered the people harshly, for he ignored the advice of the elders which they had given him. And he spoke to them according to the advice of the young men, saying, 'My father made your yoke heavy, but I will add to your yoke; my father disciplined you with whips, but I will discipline you with scorpions'" (1 Kings 12:13–14).

This maltreatment led to the split of the Northern and Southern kingdoms. Of course, this made it easier for invaders to attack effectively. The Northern kingdom was the first to fall. The Southern kingdom (Judah) lasted a bit longer, though it still had the same problems of idolatry and blasphemy. Eventually, it was exiled to the Kingdom of Babylon (Jeremiah 29:1–14).

A good while later, Cyrus King of Persia conquered Babylon. The good Lord moved his heart. The Persian King issued the following decree. "This is what Cyrus king of Persia says: 'The Lord, the God of heaven, has given me all the kingdoms of the earth, and He has appointed me to rebuild for Him a house in Jerusalem, which is in Judah. Whoever there is among you of all His people, may his God be with him! Go up to Jerusalem which is in Judah and rebuild the house of the Lord, the God of Israel; He is the God who is in Jerusalem. And every survivor, at whatever place he may live, the people of that place are to support him with silver and gold, with equipment and cattle, together with a voluntary offering for the house of God which is in Jerusalem'" (Ezra 1:2–4). Roughly seventy years later, King Artaxerxes sent Nehemiah to rebuild Jerusalem.

NEHEMIAH

After many years of bloodshed and conquest, Jerusalem was little more than rubble and ash. The once magnificent city was now a mere shadow of its former self. In fact, Jeremiah had prophesied the downfall of the City of David many years prior. The Lord spoke through Jeremiah: "Behold, I will send and take all the families of the north, declares the Lord, and I will send to Nebuchadnezzar king of Babylon, My servant, and will bring them against this land and against its inhabitants and against all these surrounding nations; and I will completely destroy them and make them an object of horror and hissing, and an everlasting place of ruins. This entire land will be a place of ruins and an object of horror, and these nations will serve the king of Babylon for seventy years" (Jeremiah 25:9–11). Just like in Judges, when the people turned their back to God they fell into the hands of an oppressive regime.

Yet it wasn't all doom and gloom. Our merciful God always provides a path towards redemption, but only for those who want it. Jeremiah also prophesied the rebuilding and restoration of Jerusalem: "And I will restore the fortunes of Judah and the fortunes of Israel, and will rebuild them as they were at first. And I will cleanse them from all their wrongdoing by which they have sinned against Me, and I will forgive all their wrongdoings by which they have sinned against Me and revolted against Me. It will be to Me a name of joy, praise, and glory before all the nations of the earth, which will hear of all the good that I do for them, and they will be frightened and tremble because of all the good and all the peace that I make for it" (Jeremiah 33:7–9).

Once again, *genuine repentance requires action.* Surely as the Lord lives, that's exactly what happened. This didn't happen overnight. It happened in waves. Many years later, the Persian King Cyrus the Great defeated the Babylonians and reconquered Jerusalem. The Lord swayed his heart. Cyrus issued a royal decree to rebuild the temple and for the Hebrews to return to Jerusalem (Ezra 1:1–4). Cyrus appointed Zerubbabel to lead the charge (Ezra 2:2–1). The rebuilding of the temple alone took over two decades of back-breaking labor! Yet there was still much work to do.

Later, King Artaxerxes commissioned Ezra to lead the second wave of returning exiles. The temple was ready, yet the people needed religious instruction and leadership. No matter how glorious a building's architecture may be, it is a mere husk without people inside. A church is nothing without the Holy Spirit in its parishioners' hearts. The Hebrews lacked spiritual education, leadership, and guidance. This is where Ezra enters the story. The king bestowed full royal authority on Ezra for this mission. "So this Ezra went up from Babylon, and he was a scribe skilled in the Law of Moses, which the Lord God of Israel had given; and the king granted him all he requested because the hand of the Lord his God was upon him" (Ezra 7:6).

King Artaxerxes, a God-fearing man, wrote the following in his letter to Ezra: "Whatever is commanded by the God of heaven, let it be done with zeal for the house of the God of heaven, so that there will not be wrath against the kingdom of the king and his sons" (Ezra 7:23). The Persian king concluded his letter with the following message: "And you, Ezra, according to the wisdom of your God which is in your hand, appoint magistrates and judges so that they may judge all the people who are in the province beyond the Euphrates River, that is, all those who know the laws of your God; and you may teach anyone who is ignorant of them. And whoever does not comply with the Law of your God and the law of the king, judgment is to be executed upon him strictly, whether for death or for banishment or for confiscation of property or for imprisonment" (Ezra 7:25–26).

Ezra worked diligently for many years rebuilding Jerusalem spiritually. He educated his fellow Hebrews and taught them how to cultivate their faith. In other words, "Train up a child in the way he should go, even when he grows old he will not depart from it" (Proverbs

22:6). Of course, we are all children of God. Ezra succeeded in this endeavor, yet the city still had major physical deficits. Though their souls were nourished, the flock lacked a shepherd to secure their physical safety. This is where Nehemiah comes in.

One day, he learned of the desolation of Jerusalem: "And they said to me, 'The remnant there in the province who survived the captivity are in great distress and disgrace, and the wall of Jerusalem is broken down and its gates have been burned with fire'" (Nehemiah 1:3).

The reality of the situation shook him to his core. Nehemiah broke down and wept as he fasted and prayed mournfully. He didn't make excuses or attempt to negotiate with God. Instead, he humbled himself and took ownership of the situation. The following is part of his prayer for forgiveness: "We have acted very corruptly against You and have not kept the commandments, nor the statutes, nor the ordinances which You commanded Your servant Moses" (Nehemiah 1:7).

The first step in fixing a problem is acknowledging the problem in the first place. Nehemiah clearly understood this. His humility and ownership display his leadership qualities. In other words, "One who is on the path of life follows instruction, but one who ignores a rebuke goes astray" (Proverbs 10:17).

Despite his sorrow, Nehemiah continued his royal duties honorably as cupbearer for the king. This was far more than the position of a simple servant. It was a position of utmost trust and confidence. The reason is simple. Royal cupbearers first tasted the king's drink as a test for poison. Whether he realized it or not, Nehemiah had already internalized the warrior mindset. According to Musashi, "the way of the warrior is resolute acceptance of death" (Musashi 1). He willingly put his life on the line for the good of the kingdom.

While sipping wine every day might sound like a walk in the park, for Nehemiah it must have been daunting. Every day might be his last. This surely made his everyday actions more meaningful. Staring death in the face daily made him the perfect candidate for his next assignment. Clausewitz concurs: "One must get used to the idea of dying with honor, to continually nurture that idea to get used to it" (Clausewitz 2). Perhaps this was divine providence at play? Young David is a great example of this. He fought ferocious beasts countless times before defeating Goliath. He didn't kill the giant by mere chance or coincidence; he had practiced for years prior. In this same way, Nehemiah grew accustomed to his own mortality as he put his life on the line for the kingdom.

Nehemiah performed his duties stoically, yet the king knew something wasn't right. The king asked his trusted servant why he looked so sad, as he concluded Nehemiah was heartbroken at something (Nehemiah 2:2). Nehemiah told the king how he had recently learned of the utter desolation of Jerusalem. The king asked him what he wanted. Nehemiah took a step back and prayed for guidance. He told King Artaxerxes the following. "If it pleases the king, and if your servant has found favor before you, I request that you send me to Judah, to the city of my fathers' tombs, that I may rebuild it (Nehemiah 2:5). He didn't ask for handouts. Rather, he took the initiative and volunteered to go into harm's way to rebuild the city himself! When appealing to superiors, never come to them with problems. They have enough problems already and will likely just kick the can down the road. Rather, *offer solutions to problems* and proceed from there. The king was pleased with his servant's initiative and motivation.

Victory loves preparation. Nehemiah asked for letters to the local governors to ensure safe passage on this mission. He also asked for building supplies so he could begin work immediately. Additionally, the king provided an armed escort of cavalry officers for the journey (Nehemiah 2:7–9). Of course, any ambitious endeavor is sure to have some naysayers mucking about. "And when Sanballat the Horonite and Tobiah the Ammonite official heard about it, it was very displeasing to them that someone had come to seek the welfare of the sons of Israel" (Nehemiah 2:10).

The trip to Jerusalem went without a hitch. After settling in, Nehemiah set out under the cover of darkness with a few trusted people to get a closer look. None of his compatriots knew about the mission they would soon undertake. Nehemiah's sorrow was well founded. The once strong walls were more akin to a graveyard than robust fortifications. They haunted Nehemiah, like a ghost of a long gone era. The ruins were a mere shadow of the glorious strength and security of the past. Inspecting the walls at night was quite wise. Nehemiah looked at the walls from his adversary's perspective. Recall that Joab had infiltrated these very walls, likely under the cover of darkness (2 Samuel 5:6–8).

After surveying the damage, Nehemiah returned to camp and rallied the troops. He said: "You see the bad situation we are in, that Jerusalem is desolate and its gates have been burned by fire. Come, let's rebuild the wall of Jerusalem so that we will no longer be a disgrace" (Nehemiah 2:17). He elaborated on the importance of this holy mission. He spoke of the king's support and of course gave all the glory to God. Nehemiah led from the front. His initiative and courage swayed the hearts of his men as they followed his lead.

His team saw his vision, but of course a few malcontents butted their way into the conversation. "But when Sanballat the Horonite and Tobiah the Ammonite official, and Geshem the Arab heard about it, they mocked us and despised us, and said, 'What is this thing that you are doing? Are you rebelling against the king?'" (Nehemiah 2:19). If you're catching flak, you're likely close to the target! Remember, people-pleasing isn't a Christian virtue. Pleasing God is!

These short sighted insults rolled off Nehemiah like water off a duck's back. He didn't let their words get under his skin. He kept his composure and let his faith guide him. He responded with the following: "The God of heaven will make us successful; therefore we His servants will arise and build, but you have no part, right, or memorial in Jerusalem" (Nehemiah 2:20). He understood something important about dealing with fools. *Never get into a wrestling match with a pig!* It will drag you down to its level and beat you with experience. You inevitably wind up covered in mud and the pig loves every second of humiliating you. Rather, Nehemiah took a lesson from Proverbs. "One who corrects a scoffer gets dishonor for himself, And one who rebukes a wicked person gets insults for himself. Do not rebuke a scoffer, or he will hate you; Rebuke a wise person and he will love you" (Proverbs 9:7–8).

THE REBUILDING

Thus, the real work began! Rebuilding an entire city's fortifications surely was a daunting task. There is a timeless adage for completing seemingly impossible tasks. *You eat an elephant one bite at a time.* Prioritize what is most important and execute tasking accordingly. Rinse and repeat. That's exactly what Nehemiah's team did! Their results were nothing short of astonishing! They completed the entire wall of Jerusalem, including ten gates, in just fifty-two days! (Nehemiah 6:15). Their results speak volumes about the power of leveraging habit and discipline. While they finished in great time, the project was far from a cakewalk.

The work was grueling. The hours were long and the threats never ending. The ever-looming threat constantly chipped away at the nerves. Yet Nehemiah's team didn't complain or shy away from this harsh reality and the back-breaking labor. It answered the call of duty stoically and proficiently. In fact, it faced quite a bit of opposition during this project.

It began with harsh words and insults. One naysayer derided them: "What are these feeble Jews doing? Are they going to restore the temple for themselves? Can they offer sacrifices? Can they finish it in a day? Can they revive the stones from the heaps of rubble, even the burned ones?" (Nehemiah 4:2). Tobiah the Ammonite, another cynical fool, chimed in from the peanut gallery: "Even what they are building—if a fox were to jump on it, it would break their stone wall down!" (Nehemiah 4:3). This was nothing short of typical bullying. First, bullies try to embarrass or intimidate. If that fails, violence usually follows. However, Nehemiah was a man of immense courage. He had long grown accustomed to staring death in the face. He worked steadfastly and uttered the following prayer. "Hear, O our God, how we are an object of contempt! Return their taunting on their own heads, and turn them into plunder in a land of captivity" (Nehemiah 4:4).

Nehemiah and his team understood their assignment. They understood that life isn't fair. They knew they would face insults and jeers. They knew their adversaries wanted them dead. Yet instead of succumbing to intimidation, they put their noses to the grindstone and carried on. Their work echoes that of Theodore Roosevelt's famous *Man in the Area* speech, which reads as follows.

"*It is not the critic who counts*; not the man who points out how the strong man stumbles, or where the doer of deeds could have done them better. *The credit belongs to the man who is actually in the arena*, whose face is marred by dust and sweat and blood; who strives valiantly; who errs, who comes short again and again, because there is no effort without error and shortcoming; but who does actually strive to do the deeds; who knows great enthusiasms, the great devotions; who spends himself in a worthy cause; who at the best knows in the end the triumph of high achievement, and who at the worst, *if he fails, at least fails while daring greatly, so that his place shall never be with those cold and timid souls who neither know victory nor defeat*" (Roosevelt 1910, Italics added for emphasis).

Nehemiah understood a valuable lesson that his critics failed to grasp. The only real failure is *failing to try in the first place!* He understood the real difference between winners and losers. Put simply, winners try and fall short more than losers ever bother to try in the first place. Fear paralyzes many people. We must remember, fear is a tool of Satan. It has no place in the Christian arsenal. Hebrews elaborates on this important concept and illustrates how Christ led by example. In fact, this is one of the main reasons why the Lord took the form of a man as Jesus Christ. "Therefore, since the children share in flesh and blood, He Himself likewise also partook of the same, so that through death He might destroy the one who has the power of death, that is, the devil, and free those who through fear of death were subject to slavery all their lives" (Hebrews 2:14–15).

With steady hands and steadfast hearts, the work continued. According to Nehemiah himself, "So we built the wall, and the entire wall was joined together to half its height, for the people had a mind to work" (Nehemiah 4:6). This is where things escalated! It finally dawned on the naysayers that Nehemiah was serious. Their attempts at intimidation had failed. This placed them at a fork in the road. They had to make a choice. Either swallow their pride and back down or double down. Like the fools they were, they chose the latter! Some lessons are only learned the hard way.

ENEMY RESISTANCE

Nehemiah's enemies stewed in anger and bitterness. Their pride wouldn't let this affront to their egos go unchecked. "So all of them conspired together to come to fight against Jerusalem and to cause confusion in it" (Nehemiah 4:8). Nehemiah's response gives us a roadmap forward. First, he sought guidance from the Lord. Always a good way to start any venture. Yet he didn't stop there; he posted guards around the clock to meet the threat head on! The work continued as planned, though the rough conditions began to take a toll. Some people began to lose faith while their enemies grew bolder as they plotted in the shadows: "And our enemies said, 'They will not know or see until we come among them, kill them, and put a stop to the work'" (Nehemiah 4:11). They plotted and schemed their evil deeds, waiting for an opportune time to pounce from the shadows. They embodied the prowling lion the Apostle Peter mentioned in one of his letters (1 Peter 5:8).

Whispers of this plot made their way to some of the local residents. Their hearts sank as fear crept in. Hands trembling with fear, they said to Nehemiah over and over again: "They will come up against us from every place where you may turn" (Nehemiah 4:12). They understood an important lesson about bullies: *you can't run from them!* There are no shortages of bad actors out in the world. They constantly lurk in the shadows, preying on those they perceive as weak. The only solution is to confront them head on. As Christians, we are called to a lifestyle of de-escalation. Yet the truly wicked don't care about such things. Should de-escalation fail, we *must* be the good shepherds of the flock. Did young David abandon his sheep at the first sign of a lion? No, he squared off and faced the threat head on. He was all alone out in the fields, armed with little more than sticks and rocks. Yet he knew he was the only thing standing in between the predator and its prey! Additionally, Christ exemplified the Good Shepherd's selfless sacrifice to rescue the flock.

Nehemiah understood this timeless principle as he followed in the footsteps of brave warriors like Joshua and David. He kept his wits about him as he surveyed the situation. After determining the most vulnerable points, he dispatched soldiers accordingly. They willingly placed themselves in harm's way to protect the most vulnerable amongst them, an embodiment of the warrior ethos. In Nehemiah's own words, "then I stationed men in the lowest parts of the space behind the wall, the exposed places, and I stationed the people in families with their swords, spears, and bows" (Nehemiah 4:13).

Jesus Christ himself approved self-defense. HE said to His disciples shortly before His arrest and execution, "But now, whoever has a money belt is to take it along, likewise also a bag, and whoever has no sword is to sell his cloak and buy one" (Luke 22:36). Of course, there is a substantial difference between protecting yourself or loved ones and wanton violence. For example, Jesus later rebuked Peter for striking Malchus, a servant of the high priest, while the soldiers sought to arrest Him: "Then Jesus said to him, 'Put your sword back into its place; for all those who take up the sword will perish by the sword'" (Matthew 26:52). Likewise, the Gospel of John reinforces this point. It wasn't about the sword itself; it was about Peter interfering with the divine plan. "So Jesus said to Peter, 'Put the sword into the sheath; the cup which the Father has given Me, am I not to drink it?'" (John 18:11).

This proves that Jesus wasn't inherently against the use of force. Rather it is the context which matters. HE could have said something like "throw your sword away" or "cast your blade into the sea!" Yet He didn't. Jesus told Peter to put his sword *back in its place*! Which meant sheathed, on his hip and ready for action. Jesus' rebuke came because Peter interfered with Christ's sacrifice. Jesus made His reasoning abundantly clear in the two verses that followed His rebuke. "Or do you think that I cannot appeal to My Father, and He will at once put at My disposal more than twelve legions of angels? How then would the Scriptures be fulfilled, which say that it must happen this way?" (Matthew 26:53–54). If you consider yourself a pacifist remember this:

even angels are capable of violence for righteous means! No mortal is holier or more enlightened than harbingers of the Lord himself.

Weapons are the only thing protecting us from the vicious teeth and claws of wolves. Joshua, David, and Nehemiah all understood this. However, a sword is useless if one lacks the courage to wield it! Nehemiah knew that fear manipulated his team with an icy grasp. He needed to boost morale fast, before all hope slipped away. He delivered an inspiring speech we should all remember. "When I saw their fear, I stood and said to the nobles, the officials, and the rest of the people: Do not be afraid of them; remember the Lord who is great and awesome, and fight for your brothers, your sons, your daughters, your wives, and your houses" (Nehemiah 4:14).

He understood the distinction between a warrior and a mercenary. By giving glory to God and redirecting attention to their loved ones he rekindled their fighting spirit! He tapped into their protective instincts. It's easy to embrace cowardice when you only think about yourself. Cowards are happy to sacrifice anyone and everyone for the chance to save their own skin. The Levite throwing his concubine to the wolves is a perfect example of this (Judges 19). Perhaps if the Levite's host had rebuked him, things might have gone very differently. When you are forced to consider the wellbeing of your loved ones, it's much harder to bow down to fear. This course correction led them back to their holy mission with resolve and clarity.

A little sunshine drives away the darkness as the creatures of the night lose their camouflage. When Nehemiah's enemies realized they lost the element of surprise, they scattered like roaches. "Now when our enemies heard that it was known to us, and that God had frustrated their plan, then all of us returned to the wall, each one to his work" (Nehemiah 4:15). The coast was clear for now, but wickedness never dies. It simply gets better at hiding. Slippery as an eel, it slithers to some dark hole and waits patiently. Nehemiah understood that vigilance and preparedness were the only way forward. "And from that day on, half of my servants carried on the work while half of them kept hold of the spears, the shields, the bows, and the coats of mail; and the captains were behind all the house of Judah. Those who were rebuilding the wall and those who carried burdens carried with one hand doing the work, and the other keeping hold of a weapon" (Nehemiah 4:16–17).

Every man stood armed and prepared to answer threats. Nehemiah and his team embodied the masculine ideal. He wasn't there to raid and pillage. He was there to protect and build. His team members put their lives on the line while building a better future for the generations to follow. They weren't in it to make a quick buck; they forged civilization for the better. They personified a good shepherd, providing security for the weak and weary while keeping the predators at bay.

None of this happened by accident. It was all the result of deliberate planning and hard work. Both are necessary to scale operations. Nehemiah reinforced the community spirit by ensuring communications amongst the spread-out teams. This was tactically very wise. Divide and conquer is one of the most fundamental strategies of warfare. Nehemiah's communication strategy got ahead of this thorny issue before it became a problem. He briefed his team on the simple and effective strategy. "And I said to the nobles, the officials, and the rest of the people, 'The work is great and extensive, and we are separated on the wall far from one another. At whatever place you hear the sound of the trumpet, assemble to us there. Our God will fight for us'" (Nehemiah 4:19–20).

Thus, the work continued. Invigorated by the Spirit, they labored day after day. From dawn until the stars painted the night's sky, they rebuilt with a sword at their side. Their vigilance remained constant and unwavering. "So neither I, my brothers, my servants, nor the men of the guard who followed me—none of us removed our clothes; each took his weapon even to the water" (Nehemiah 4:23). Sun Tzu would certainly approve of their defensive posture. According to the wise general, "To secure ourselves against defeat lies in our own hands, but the opportunity of defeating the enemy is provided by the enemy himself" (Tzu 7). Nehemiah refused to give the enemy an easy target!

The work on the fortifications progressed at a rapid rate, exceeding all expectations. In a paradoxical fashion, while the walls grew in strength, the social bonds within the walls deteriorated. Jerusalem looked like a pristine vase filled with rotting flowers. This was due to long-term financial exploitation and forgetting about the Lord's ways. The officials had strayed from the Mosaic Law. Some of the commoners elaborated on the abuse they faced from their own people: "We have borrowed money for the king's tax on our fields and our vineyards. And now our flesh is like the flesh of our brothers, our children like their children. Yet behold, we are forcing our sons and our daughters to be slaves, and some of our daughters are forced into bondage already, and we are helpless because our fields and vineyards belong to others (Nehemiah 5:4–5).

With neighbors like that, who needs enemies? Additionally, the officials charged the peasants interest on all of the above. This was blatantly forbidden under the Law. "If you lend money to My people, to the poor among you, you are not to act as a creditor to him; you shall not charge him interest" (Exodus 22:25). Moreover, Leviticus elaborates on how to treat people who fall on hard times. "Now in case a countryman of yours becomes poor and his means among you falter, then you are to sustain him, like a stranger or a resident, so that he may live with you. Do not take any kind of interest from him, but fear your God, so that your countryman may live with you. You shall not give him your silver at interest, nor your food for profit" (Leviticus 25:35–37).

Nehemiah didn't tolerate any of it! Imagine his outrage and frustration. He and his men put their lives on the line every day against mortal threats. Meanwhile, officials took advantage of their own people. Like opportunistic parasites, they exploited the most vulnerable, purely for profit. It's impossible to build the future when your own children are sold into slavery. Typically, an external threat brings people closer together. Vicious enemies surrounded Jerusalem on all sides. They wanted the Israelites wiped out entirely. Yet instead of worrying about the wolves at the door, the officials were more concerned with making a quick buck! The city couldn't flourish with such a parasitic relationship between the higher-ups and the common folk. Nehemiah understood this and wasted no time calling out their misdeeds. He assembled a large meeting of the officials and said the following: "And I said to them, 'We, according to our ability, have redeemed our Jewish brothers who were sold to the nations; now would you even sell your brothers that they may be sold to us?' Then they were silent and could not find a word to say" (Nehemiah 5:8).

Shame overwhelmed them as an awkward silence filled the room. Nehemiah demanded the property returned to the people and the end of usury. The officials didn't fight back or argue. They knew they had overstepped. Nehemiah then recruited priests to swear an oath upholding these laws. Nehemiah ended things in a dramatic fashion; a warning of consequences should they stray again. "I also shook out the front of my garment and said, 'So may God shake out every person from his house and from his possessions who does not keep this promise; just so may he be shaken out and emptied.' And all the assembly said, 'Amen!' And they praised the Lord. Then the people acted in accordance with this promise" (Nehemiah 5:13).

Now freed from the yoke of usury and slavery, Jerusalem had a fighting chance to prosper. While Nehemiah brokered these deals internally, the external threats grew more powerful. They continued plotting and scheming to trap Nehemiah. His team had completed the wall. It stood robust and powerful, yet the gates were not set yet. Nehemiah's enemies saw their window of opportunity slipping through their fingertips. Their attempts grew more desperate.

They lacked the forces to besiege Jerusalem, so they attempted to lure its leader outside the walls. This was a wise strategy, yet Nehemiah was no fool. They harassed Nehemiah with messengers to depart the city, which he promptly declined four times over. The fifth time, things escalated. Nehemiah's enemies attempted to provoke him into confrontation. They accused Nehemiah of mutiny and hiring false prophets. Using the Lord's name in vain was no small matter. Of course, this was nothing but slanderous lies. There is a logic to such a tactic. According to *The Art of War,* "If your opponent is of choleric temper, seek to irritate him. Pretend to be weak, that he may grow arrogant" (Tzu 4). Baiting your opponent into overstepping is a classic tactic in both warfare and martial arts. In fact, "a delicacy of honor which is sensitive to shame" is one of the most dangerous traps a general can fall into (Tzu 16).

Many people feel the need to defend their ego, which clouds their judgement and objectivity. Fiery emotions melt away ice cold logic. Nehemiah's humility protected him. Remember that pride is the mother of all sins. A humble heart protects us from reacting without thought or reason. When our ego is out of the way, we can *respond* tactically, rather than *react* impulsively. An unshakable opponent is much harder to defeat!

Fortunately for Israel, Nehemiah was no coward! He saw right through their dirty tricks. He delivered a curt reply confronting their slander: "Then I sent a message to him saying, 'Nothing like these things that you are saying has been done, but you are inventing them in your own mind'" (Nehemiah 6:8). Nehemiah balanced the need for honesty and justice without engaging in petty mudslinging. He elaborated on his reasoning and gave us a model to emulate. "For all of them were trying to frighten us, thinking, 'They will become discouraged with the work and it will not be done.' But now, God, strengthen my hands" (Nehemiah 6:9).

Did he pray for easier circumstances? Did he pray for his enemies to run away? Did he pray for God to do all the work for him? Not at all! In his own words, he prayed for strength. We must remember Nehemiah's prayer the next time adversity confronts us. Scripture clearly paints the path forward. Like Joshua and David before him, Nehemiah prayed for the strength and courage to endure epic feats and grueling hardships. He leveraged his faith to persevere through every trial and tribulation; *so can we.*

When his enemies failed to lure him outside the walls, they switched tactics. They opted for another PSYOP. Tobiah and Sanballat hired a false prophet to lead Nehemiah astray. Instead of luring Nehemiah outside, they opted to make him irrelevant and distracted. The false prophet warned Nehemiah of incoming enemies who wanted to kill him. The prophet advised Nehemiah to barricade himself inside the temple and wait. Clearly, they misjudged his warrior spirit. He responded resolutely, "Should a man like me flee? And who is there like me who would go into the temple to save his own life? I will not go in" (Nehemiah 6:11). Clausewitz would certainly approve of Nehemiah's refusal. According to the Prussian General, "in war, nothing is accomplished without risk, that the very nature of war does not give an unconditional opportunity to always predict in advance where you are headed" (Clausewitz 31). Nehemiah fit the bill perfectly.

Nehemiah was far from the only person who dealt with false prophets. In the Gospel of Matthew, Jesus Christ warned us of deceptive charlatans. "Beware of the false prophets, who come to you in sheep's clothing, but inwardly are ravenous wolves" (Matthew 7:15). So how are we to discern false prophets from divine messengers? Luckily, Christ also shows us the way. "You will know them by their fruits. Grapes are not gathered from thorn bushes, nor figs from thistles, are they? So every good tree bears good fruit, but the bad tree bears bad fruit. A good tree cannot bear bad fruit, nor can a bad tree bear good fruit" (Matthew 7:16–18). Nehemiah didn't have the benefit of Christ's teaching. Instead, he balanced his faith with reason. The prophet's advice directly conflicted with Nehemiah's holy mission and Biblical teachings. Therefore, the man proved himself nothing more than a fraud. Later, Nehemiah confirmed that his adversaries had indeed hired this fraudster for malicious means (Nehemiah 6:12–13).

Nehemiah remained unfazed by this deception. He was a man on a mission; nothing slowed him down. The newly renovated fortifications sent a clear and resounding message to the Lord's adversaries: *we shall prevail*. The overwhelming success of the wall took the wind right out of their sails! Rudderless, Israel's enemies stood disillusioned and downtrodden. "When all our enemies heard about it, and all the nations surrounding us saw it, they lost their confidence; for they realized that this work had been accomplished with the help of our God" (Nehemiah 6:16).

This echoes back to the time of Joshua, shortly before the taking of Jericho. Recall what Rahab said to Joshua's spies. " When we heard these reports, our hearts melted and no courage remained in anyone any longer because of you; for the Lord your God is God in heaven above and on earth below" (Joshua 2:11). No matter how powerful kings and emperors may perceive themselves to be, they pale in comparison to the glory of God! Reflect back to the humiliation of the Pharaoh during the time of Moses. He stood atop one of the greatest empires we have ever seen. The Egyptian empire created marvels of technology that we still don't understand, thousands of years later. Yet, the Lord effortlessly cast the Pharaoh's best soldiers and charioteers into the Red Sea. The waves swept them away like forgotten children's toys on a stormy shore. All their technology, ambition and fighting spirit was futile against our glorious God!

Despite his peaceful mission, Nehemiah still received nasty letters from his adversary, Tobiah. But harsh words mean little to a man accustomed to staring death in the face. The mission carried on.

Nehemiah appointed his brother as one of the chief leaders of Jerusalem—not out of nepotism, but due to his character. As Nehemiah stated, "then I put Hanani my brother, and Hananiah the commander of the citadel, in charge of Jerusalem, for he was a faithful man and feared God more than many" (Nehemiah 7:2). Wise leadership criteria indeed.

Nehemiah knew that no man lives on an island to himself. Delegation was essential for scaling operations, and integrity was non-negotiable. Most importantly, "The fear of the Lord is the beginning of knowledge; Fools despise wisdom and instruction" (Proverbs 1:7). He and his team had spent countless hours rebuilding Jerusalem's walls under constant threat. Appointing the wrong person would squander their sacrifice. That's why Nehemiah gave his brother and the leader of the citadel specific operating procedures focused on security. He instructed them, "The gates of Jerusalem are not to be opened until the sun is hot, and while they are standing guard, the gatekeepers are to keep the doors shut and bolted. Also appoint guards from the inhabitants of Jerusalem, each at his post, and each in front of his own house" (Nehemiah 7:3).

This two-pronged approach was brilliant in its simplicity. Delaying the gate's opening improved security without additional costs. Cities typically opened at dawn, but this protocol disrupted enemy plans and reduced the threat of surprise attacks. As Musashi wrote, "The important thing in strategy is to suppress the enemy's useful actions but allow his useless actions" (Musashi 23). Without surprise, Nehemiah's enemies became little more than spectators outside Jerusalem's walls. The city was a natural fortress; with the new fortifications, even more so.

Placing guards near their homes also had a profound psychological impact. Men fight harder when their loved ones are behind them. Nehemiah's policy created a culture of good shepherds, warriors with skin in the game and something sacred to protect.

The physical security now stood strong. Next came the harder task—rebuilding the city from within. More exiles returned to repopulate and restore the city. With the walls secured, the conditions were now fertile for a spiritual revival. Ezra the priest took point on this crucial mission. He understood that many of Israel's past hardships stemmed from abandoning God's ways. His challenge was to convey this truth to a people largely untrained in the faith.

The townspeople gathered in the public square to hear the Law. Ezra, faithful to his calling, stood atop a platform and read the

Scriptures from dawn until noon (Nehemiah 8:3–4). The people listened with reverence. "Then Ezra blessed the Lord, the great God. And all the people answered, 'Amen, Amen!' with the raising of their hands; then they kneeled down and worshiped the Lord with their faces to the ground" (Nehemiah 8:6).

Ezra also appointed Levitical priests to assist. "They read from the book, from the Law of God, translating to give the sense so that they understood the reading" (Nehemiah 8:8). As the truth sank in, sorrow overwhelmed the crowd. The people broke down and wept, convicted by how far Israel had strayed. Their grief echoed Nehemiah's own mourning before King Artaxerxes. But Nehemiah knew this sorrow could lead to joy. He told them, "Go, eat the festival foods, drink the sweet drinks, and send portions to him who has nothing prepared; for this day is holy to our Lord. Do not be grieved, for the joy of the Lord is your refuge" (Nehemiah 8:10). The Levites helped to calm the people and reinforced his message. It was a bittersweet moment, with tears of regret giving way to hope.

A wise pastor once said that the Bible is both a mirror and a map. Like a mirror, Scripture reflects back an honest picture. The standards are objective. If you don't like what you see, the mirror isn't to blame. Responsibility lies with the beholder. That's why sorrow swept over the crowd. They saw clearly. But that sorrow soon turned to joy. The priests showed them the way forward (Nehemiah 8:11–12). Simply telling people they're wrong isn't effective. That only provokes defensiveness. People need something to aim for, something beyond themselves. Scripture isn't a cudgel; it's a life raft. God's Word rescues us from this fallen kingdom, but only if we accept the invitation.

The revival was under way! The people of Jerusalem rediscovered long-forgotten customs. They confessed and renewed their covenant. Clothed in garments fit for mourning, they gathered together. The Levites led a powerful prayer of confession, beginning with the foundations of creation itself. "You alone are the Lord. You have made the heavens, The heaven of heavens with all their lights, The earth and everything that is on it, The seas and everything that is in them. You give life to all of them, And the heavenly lights bow down before You" (Nehemiah 9:6). That prayer still rings true today. It reminds us of the magnificence of the Earth and stars, and who is truly in charge.

The prayer was fundamentally a story of contrasts. The everlasting faithfulness of the Lord stood in sharp opposition to the myopic and rebellious nature of man. The priests recounted the entire history of Israel—from the call to Abram, the journey of Moses, the cycle of judges, and the fall of Jerusalem. They told the good, the bad, and the ugly. Despite Israel's wickedness and treachery, the Lord remained faithful to His covenant. The priests made one thing abundantly clear: God always honors His promises—it is people who falter. They took ownership of their failings and declared, "You are righteous in everything that has happened to us; For You have dealt faithfully, but we have acted wickedly" (Nehemiah 9:33).

Israel's history was nothing short of tumultuous. Marked by chaos and bloodshed, kingdoms rose and fell. Heroes occasionally emerged to lead the charge, only for the people to relapse into pride and disobedience. Yet amid the madness, one constant endured: the mercy and grace of God. His compassion always surpassed Israel's shortcomings.

Following their public confession, the people renewed their faith and vowed to obey God's commands. They understood that words alone were not enough—true repentance requires action. These renewed commitments included honoring the Sabbath, refraining from interfaith marriage, tithing, and supporting the temple and its priests. For a time, this spiritual revival held strong.

Yet some began to backslide during Nehemiah's absence, after he returned to serve King Artaxerxes. For example, certain officials allowed foreign merchants to set up shop during the Sabbath. Nehemiah swiftly intervened, stationing his own men at the gates to put an end to the offense. As Nehemiah recounted: "Once or twice the traders and merchants of every kind of merchandise spent the night outside Jerusalem. Then I warned them and said to them, 'Why do you spend the night in front of the wall? If you do so again, I will use force against you.' From that time on they did not come on the Sabbath" (Nehemiah 13:20–21).

Once again, Nehemiah displayed the critical difference between harmlessness and peacefulness. A harmless man can't lay down the law—Nehemiah had no trouble doing so. He also confronted Israelites who had married outside the faith, reminding them of the consequences. "Did Solomon the king of Israel not sin regarding these things? Yet among the many nations there was no king like him, and he was loved by his God, and God made him king over all Israel; nevertheless the foreign women caused even him to sin" (Nehemiah 13:26).

The wise learn from the mistakes of others. The average learn from their own mistakes, while fools learn nothing at all. Nehemiah clearly stood amongst the wise. His story reminds us that sanctification is a process, not a destination. We are always swimming upstream against the current. The world will never stop trying to pull us off course. Due to our fallen nature, we can never be perfect, but we can always strive to do better.

Nehemiah's accomplishments are a testament to hope and perseverance. Jerusalem rose from the ashes because one man had the boldness and courage to *try*. Despite all the mockery and threats, Nehemiah prayed for strength and got to work.

We should all strive to be more like Nehemiah.

CLOSING THOUGHTS

Nehemiah's mission was relatively straightforward; rebuild the wall. The Lord called him for a specific task at a specific time. It required great discipline, courage, and strategic thinking. The goal was action oriented and measurable, which always makes things a bit smoother. In contrast, the modern Christian mission is much broader. With the sacrifice and resurrection of Christ, everything changed. The Apostle Paul explained this in his letter to the Galatians: "Abraham believed God, and it was credited to him as righteousness. Therefore, recognize that it is those who are of faith who are sons of Abraham. The Scripture, foreseeing that God would justify the Gentiles by faith, preached the gospel beforehand to Abraham, saying, 'All the nations will be blessed in you.' So then, those who are of faith are blessed with Abraham, the believer" (Galatians 3:6–9). Moreover, "the blessing of Abraham would come to the Gentiles, so that we would receive the promise of the Spirit through faith" (Galatians 3:14). Jesus freed us from rigid legalism. While He freed us from legalism, we still have duties and responsibilities to one another.

Unlike Nehemiah's singular goal of rebuilding a city wall, Christians are called to rebuild, fortify, and maintain the spiritual lives of individuals and communities. This mission is not limited to a singular physical structure or territory. Rather, *we* are the walls protecting the flock from the wolves! We must use the light of Christ to drive away dark forces and evildoers. This means we must be willing to risk our own lives in service to the community. Christ remains the ultimate example of a noble sacrifice. Think back to the cowardly Levite sacrificing his concubine to evildoers. This is the exact opposite approach for us to follow. When the wolves come stalking in, we cannot pray for them to vanish. Like Nehemiah, we must pray for stronger hands; hands ready to act with resolute courage and faith. With our strength we must protect the flock and fight for what is right and just. Nehemiah was far from a weakling. He used his strength to protect his people from abuse and exploitation. In fact, strength and courage were required to protect his people. Contrast him with Gideon's son, Abimelech, who used violence for selfish means. The problem *isn't* fighting abilities, as pacifists naively believe. It's how those abilities are used. A hammer may be used to build homes for the needy, or to bash in a robbery victim's skull. It's how we use our gift of free will that matters.

As Christians, Christ calls us to follow His ways. Jesus was both the Lion of Judah and the Sacrificial Lamb. At first glance, this might appear confusing. However, these roles are *complementary,* not contradictory. In fact, this balance embodies the warrior ethos—the strength to fight evil tempered by compassion for others. Jesus Christ advocated for such a balance when He sent the disciples out on the Great Commission. In fact, the following verse inspired this book in the first place. "Behold, I am sending you out as sheep in the midst of wolves; so be as wary as serpents, and as innocent as doves" (Matthew 10:16). Jesus didn't advocate for one or the other. He advocated for a balance of *both* qualities. We are called to be wary as serpents, who only strike humans in self-defense. But when snakes defend themselves, they swiftly lash out with lethal venom or crushing strangulation. However, we must differentiate ourselves from mercenaries or opportunists. We must only resort to violence in the service of more righteous goals, hence the wariness Christ mentioned.

We *must* have this warrior resolve; to possess the lethality of the serpent tempered with the purity of the dove. Modern Christianity has a bevy of the latter and woeful deficit of the former. How can a warrior properly bear a shield when he remains ignorant in the ways of the sword? As Jesus emphasized, we need both!

The warrior fights for a cause grander than himself. For his family, his home, his country, for those who cannot stand up for themselves. David spent his youth fighting bears and lions to protect his flock. Little did he know he was training for Goliath! He needed to be dangerous to care lovingly for the most vulnerable. He needed to be lethal to keep the predators at bay. All the other Israelites trembled at the thought of Goliath, but not David! The seemingly small everyday decisions added up with compound interest. Training is a way of life, David understood this. He shocked the crowd when he effortlessly slayed Goliath. His opening shot crushed the bullseye on his armored foe. One shot, one kill, like a top marksman. The only reason this appeared effortless was because David had practiced for years beforehand!

In this same way, we must prepare ourselves for the giants in our own lives. David proved that with modest tools, a fighting spirit, and faith, we can accomplish what seems impossible. But only if we have the courage to try in the first place! Moreover, Christ was ferocious towards Satan and compassionate to the downtrodden. This balance is paramount for any society to last and flourish. If we forget our martial roots, we are helpless to the barbarians at the gates: mere sitting ducks. Should we forget our compassion for the vulnerable, *we become barbarians*! Become dangerous so you may lovingly serve and protect your loved ones and community. A good shepherd requires both qualities in balance. It all begins with a little bit of courage; the courage to step forward when others hesitate, to fight for what is right, and to protect the vulnerable.

Choose wisely, for your strength and compassion will shape the future of those you lead.

REFERENCE LIST

Burian, Michal, et al. *Operation Anthropoid 1941–1942*. Ministry of Defence of the Czech Republic – AVIS, 2007.

Carnegie, Dale. *How to Win Friends and Influence People*. Simon and Schuster, 1936.

Chesterton, G. K. *Illustrated London News*, January 14, 1911.

Clausewitz, Carl von. *Principles of War*, translated by Georg Hamm. e-book. Amazon Digital Services, 2010.

Donovan, Michael J. *Strategic Deception: Organization Fortitude*. US Army War College, 2022. https://apps.dtic.mil/sti/tr/pdf/ADA404434.pdf.

Eisenhower Presidential Library. *World War II: D-Day, the Invasion of Normandy*, n.d. https://www.eisenhowerlibrary.gov/research/online-documents/world-war-ii-d-day-invasion-normandy.

Eldredge, John. Wild at Heart: Discovering the Secret of a Man's Soul. Thomas Nelson, 2001.

Gorchow, Joe. *Man Picks Wrong Man to Attack; Ex-MMA Fighter Quickly Subdues Knife Wielder*. CBS Miami, November 3, 2023. https://www.cbsnews.com/miami/news/man-picks-wrong-man-to-attack-ex-mma-fighter-quickly-subdues-knife-wielder/.

Greene, R. *The Concise 48 Laws of Power*. Profile Books, 2002.

Hagler, Marvin. *It's Tough to Get Out of Bed to Do Roadwork at 5 am When You've Been Sleeping in Silk Pajamas*. https://www.azquotes.com/quote/868780.

Herbert, F. *Dune Messiah*. Putnam, 1969.

Hopf, G. Michael. *Those Who Remain*. CreateSpace, 2016

References

Lange, Katy. *5 Things You May Not Know About D-Day*. US Department of Defense, 2022. https://www.defense.gov/News/Feature-Stories/story/Article/3052217/5-things-you-may-not-know-about-d-day/.

Lewis, C. S. *Mere Christianity*, Revised & Enlarged edition. e-book. HarperOne/Amazon Digital Services, 2009.

Machiavelli, Niccolò. *The Prince*. e-book. Tech Tok, 2023.

Munez, Everett, and J. E. Luebering. *Operation Fortitude: World War II [1944]*. Britannia. https://www.britannica.com/event/Operation-Fortitude.

Musashi, Miyamoto. *The Book of Five Rings*. e-book. Amazon Digital Services, 2025.

Office of Strategic Services. *Simple Sabotage Field Manual*. Independently Published, 2022.

Roosevelt, Theodore. *Citizenship in a Republic*. Delivered 23 Apr. 1910, Sorbonne, Paris, France. Theodore Roosevelt Center, Dickinson State University, https://www.theodorerooseveltcenter.org/Research/Digital-Library/Record?libID=0280630.

Tzu, Sun. *The Art of War*. Amazon Classics, 2017.

Vallejo, Jessica. *Video: Retired MMA Fighter Javier Baez Fights Off Man with Knife in Cutler Bay*. NBC Miami, November 3, 2023. https://www.nbcmiami.com/news/local/video-retired-mma-fighter-javier-baez-fights-off-man-with-knife-in-cutler-bay/3150958/.

Vegetius, Flavius. *De Re Militari: Complete Official Edition*. e-book. Harper–McLaughlin–Adet, 2019.

Warren, Charles, Walter Morrison, and Arthur Stanley. *The Recovery of Jerusalem: A Narrative of Exploration and Discovery in the City and the Holy Land*. D. Appleton, 1871. https://archive.org/details/recoveryofjerusa00wilsuoft/page/191/mode/1up.

ABOUT THE AUTHOR

Jordan Hix is a United States Navy veteran and passionate martial artist with nearly a decade of competition and coaching experience. He served as a Surface Warfare Officer, including two and a half years stationed in Japan, where he deepened his study of Jiu Jitsu and Judo. In 2023, he was awarded the Navy and Marine Corps Achievement Medal, in large part for contributing over 500 hours of hand-to-hand combat training to military police and security personnel.

While in the Navy, Jordan wrote his first book, *Rapid Growth in Grappling: How to Use the Principles of Deliberate Practice to Get Better Faster!* He has since applied those same principles of focused improvement to his writing, culminating in the development of this book.